INTRODUCTION TO THEORETICAL METEOROLOGY

Introduction to THEORETICAL METEOROLOGY

SEYMOUR L. HESS

Florida State University

KRIEGER PUBLISHING COMPANY
MALABAR, FLORIDA

Original edition 1959
Reprint edition 1979

Printed and Published by
ROBERT E. KRIEGER PUBLISHING COMPANY, INC.
KRIEGER DRIVE
MALABAR, FLORIDA 32950

Printed in the United States of America

Library of Congress Cataloging in Publication Data

Hess, Seymour L
Introduction to theoretical meteorology.

Reprint of the ed. published by Holt, Rinehart, and Winston, New York.
Includes index.
1. Meteorology. I. Title. II. Title: Theoretical meteorology.
[QC861.2.H47 1979] 551.5 78-27897
ISBN 0-88275-857-8
10 9 8 7 6

To the memory of my parents

PREFACE

The present state of meteorology is so closely linked to fundamental theoretical developments that purely descriptive books are inadequate if not misleading introductions to the true nature of the science. However, many of the theoretical volumes in existence present the material in such a way that theoretical meteorology has gained the undeserved reputation of being "difficult." I have long believed it possible to present this subject in a reasonably rigorous fashion without the use of extremely refined mathematical tools. However esthetically satisfying such methods as vector analysis may be to the accomplished theoretician, my teaching experience indicates that the majority of beginners are confused by unfamiliar mathematical language.

Consequently this book makes no mathematical demand upon the reader other than a thorough knowledge of differential and integral calculus. Although on rare occasions differential equations are presented, the reader need not comprehend their solutions, merely verifying by substitution that the alleged solution satisfies the governing equation. In short, every effort has been made to present theoretical meteorology rather than advanced mathematics.

On the other hand, I have felt it essential to relate the subject closely to its basis in classical physics, and it is assumed that the reader has mastered the fundamentals of that science. The approach is largely deductive but examples from observational experience are given where appropriate. Because science is neither purely deductive nor purely inductive, many cases will be found in which the theoretical developments are suggested and strongly guided by observation.

This book was written primarily for those desiring an introduction to theoretical meteorology or a review of the subject. Consequently, references are given only to the extent necessary to document unproved assertions and to provide the reader with leads to material that will allow him to extend his knowledge. No attempt has been made to provide exhaustive bibliographies.

Almost all chapters have problem sets appended, on the principle that the student must learn to apply theory by solving challenging problems. A strenuous effort has been made to achieve a significant gradation of difficulty over a wide range of topics, including relatively simple problems in which a numerical result must be obtained from a previously derived expression, as well as some that should tax the student's ingenuity.

The text includes more material than can be covered conveniently in the usual one-year course, in order to afford each instructor an adequate choice of subjects. For example, Chapters 8, 9, and 10 on radiation may be omitted as a group. Any of Chapters 17 through 21 may be omitted without affecting the subsequent material.

I am indebted to a number of my professional colleagues who have offered suggestions or criticism on various parts of the book. Chief of these are Werner A. Baum, Warren A. Dryden, Thomas A. Gleeson, Noel LaSeur, Julius London, Stanley L. Rosenthal, and the late Leon Sherman. Above all, I am indebted to the successive classes of students who have struggled through the several editions of mimeographed class notes from which the present book was developed. Despite everyone's assistance, however, some errors and inadequacies no doubt exist, which of course are the author's sole responsibility.

S. L. H.

Tallahassee, Fla.
June, 1959

CONTENTS

INTRODUCTION TO THEORETICAL METEOROLOGY

CHAPTER 1

INTRODUCTION

1.1 The Physical Foundation

The theoretical branch of meteorology is based upon the fundamental postulate that the behavior of the atmosphere is capable of being analyzed and understood in terms of the basic laws and concepts of physics. The three fields of physics which are most applicable to the atmosphere are *thermodynamics*, *radiation*, and *hydrodynamics*.

Thermodynamics is the study of the initial and final equilibrium states of a system which has been subjected to a specified energy process or transformation. By "system" we mean a specific sample of matter. It has been found through experiment that the "equilibrium state" of a system can be completely specified for our purposes by a finite number of properties such as pressure, temperature, and volume. These properties are known as *variables of state*, or *thermodynamic variables*. An example of the kind of question that can be answered by reference to the laws of thermodynamics is the following: What is the final temperature of 1 gram of air saturated with water vapor, initially at a pressure of 1000 millibars and a temperature of 0° C, if it is allowed to expand without the addition or loss of energy from or to its surroundings until its pressure is 500 millibars? Note that a definite system is delineated, an initial state is specified, and an explicit process for the energy transformation is given (condensation of water vapor and no exchange of energy with the surroundings); with these data one must determine the final state of the system.

Radiation is the study of the emission by matter of energy in the form of electromagnetic waves, the propagation of that energy through space, and the absorption of the energy by matter. This kind of energy transfer is sharply distinguished from other forms (such as convection and conduction) by the fact that it travels at the speed of light and does not require an intervening material medium. The wave length of the radiation is its most important characteristic. As examples of the type of question that may be answered through knowledge of the laws of radiation together with certain experimental measurements, consider the following: A perfectly radiating object (black body) is at a temperature of 20° C. At what rate does each square cm of this body emit radiation between the wave lengths 8 microns and 12 microns? If this radiation is allowed to pass through 1 meter of air at 10° C, a pressure of 1000 millibars, and a relative humidity of 50 per cent, what proportion of the incident radiation between 8 and 12 microns will be transmitted? Because virtually all the energy available on our planet comes to us from a radiating body (the sun) and must pass through our atmosphere to reach the earth's surface, the significance of such problems should be obvious.

Hydrodynamics deals with the motion of fluids (liquid or gaseous) in relationship to the forces acting upon the fluids. Suppose air to be moving horizontally on the surface of the earth subject to no other forces than those arising from horizontal variations in pressure and the rotation of the earth. If these forces balance each other, what will be the direction and speed of the air? Such a question can be answered by a study of the principles of hydrodynamics.

It will be useful to set down explicitly, at the outset, those fundamental physical laws and concepts upon which the science of theoretical meteorology is based. Although many of these concepts can not be fully comprehended until a specific application to meteorology has been made, it is essential to know in advance precisely where the starting line is. The student should refer to this fundamental list as each concept is utilized in order to remain in close contact with the foundation while the superstructure is being raised.

Laws of Thermodynamics

From the field of thermodynamics we set down three important applicable laws:

1. *The equation of state of a perfect gas.*
2. *The first law of thermodynamics* (*conservation of energy*).
3. *The second law of thermodynamics* (*degradation of energy*).

The *equation of state* is a relationship among the thermodynamic variables that define the state of a system. The simple form appropriate to an idealized or perfect gas is applicable with sufficient accuracy to the real gas, air. The most common form of this useful equation expresses the interrelationship between the pressure, volume, and temperature of a given sample of gas.

The *first law of thermodynamics* is the law of conservation of energy for a thermodynamic system. As such it is a fundamental principle, incapable of theoretic proof but based upon experimental evidence.

The *second law of thermodynamics* specifies the direction in which heat may flow during a thermodynamic process. It too is a basic principle founded upon experience.

Laws of Radiation

The laws governing radiation are:

4. *Kirchhoff's law.* This law relates the intensity of emission of radiation by a body to its fractional absorption of incident radiation at the same wave length. The ratio of these two is a function only of the temperature of the body and the wave length, indicating that the better a substance absorbs, the better it emits. A perfect absorber (black body) is therefore also a perfect emitter.

5. *Planck's law.* The rate at which a black body emits radiational energy is related to the temperature of the body and the wave length of the radiation by Planck's law, from which one may derive Wien's displacement law relating the wave length of maximum energy emission to the reciprocal of the absolute temperature, and the Stefan-Boltzmann law relating the total rate of emission at all wave lengths to the fourth power of the absolute temperature.

6. *Beer's law.* When a beam of monochromatic radiation is transmitted a short distance through an absorbing medium, a certain fraction of the incident radiation is absorbed. Beer's law states that the fraction thus absorbed is directly proportional to the density of the medium and to the distance traversed. The constant of proportionality is the absorption coefficient of the medium.

7. *The equation of radiative transfer.* As radiation passes through an absorbing medium it is partially absorbed, but the energy is re-emitted by the medium, generally at another wave length, as required by Kirchhoff's and Planck's laws. Thus a complex transfer of energy results which is described by the equation of radiative transfer.

Laws of Hydrodynamics

In hydrodynamics we make use of:

8. *Newton's law of universal gravitation.* We customarily assume, however, that within the atmosphere the variations of attractive force with distance from the earth's center are negligibly small, so that this law reduces to the statement that the acceleration due to gravitation is nearly constant.

9. *The concept of equilibrium of forces.* This idea finds frequent application because of the common (though not invariable) tendency of systems subject to a number of stresses to move towards a state of equilibrium among these stresses. Much may be learned in a simple fashion by studying these balanced states.

NEWTON'S THREE LAWS OF MOTION:

10. *Bodies in motion remain in motion with the same velocity, and bodies at rest remain at rest, unless acted upon by unbalanced forces.* This law describes the observed property of inertia of bodies, and from it we derive the concept of mass.

11. *The rate of change of momentum of a body with time is equal to the vectorial sum of all forces acting upon the body and is in the same direction:* $d(m\mathbf{V})/dt = \Sigma \mathbf{F}$, where m is the mass of a body, $\mathbf{V}$ its velocity, t is time, $\mathbf{F}$ represents forces, and boldface type indicates a vector quantity. This relationship will be taken as the fundamental law of the dynamics of the atmosphere.

12. *To every action (force) there is an equal and opposite reaction (force).* Newton's third law of motion is important whenever we consider the mutual interaction of two media such as the influence of one portion of the atmosphere on another, or the interaction between atmosphere and earth.

13. *The law of conservation of mass.* This law is used to derive a basic restriction upon the behavior of fluids (the equation of continuity). It constitutes one of a short set of equations governing the atmosphere.

14. *Newton's law of viscosity.* This is an empirical statement relating the stress, exerted by a layer of moving viscous fluid on adjacent layers, to the shear of velocity at right angles to the stress. It is the basic fact of viscosity and the source of the definition of the coefficient of viscosity.

1.2 The Goal

The ultimate aim of theoretical meteorology is to express the above laws and ideas in those forms that are applicable to the atmosphere, and to apply the resultant statements to pertinent atmospheric situations. This application has a twofold purpose, first to increase our understanding of atmospheric processes and events, second to predict quantitatively the behavior of the atmosphere. The present state of theoretical knowledge is quite incomplete. We cannot claim to have approached these ends as closely as is demanded by the users of meteorological information; we can, however, block in the outlines of the field and exhibit a number of partial results which are of great theoretical and practical utility. We shall endeavor to point out the characteristics of the relationships that make the solutions so difficult, and to supply a background of deductive material upon which current practical procedure is based.

1.3 Units and Dimensions

Unless otherwise specified, all quantities are given in the centimeter-gram-second (c-g-s) system of units. The reader should be aware, however, that some meteorologists prefer to work in the meter-ton-second system (one metric ton $= 10^3$ kg), and that even the English unit of distance—the foot—is widely used in certain meteorological work. To these units must be added that of temperature, which are here expressed either in degrees Celsius* or in degrees absolute.

Careful attention to units and dimensions will generally produce a considerable increase in accuracy of computation and will facilitate verification of equations. Implicit in the choice of units expressed above is that all quantities to be dealt with (such as amounts of force, work, and specific heat capacity) may be expressed in terms of units of length, mass, time, and temperature. These fundamental *dimensions* will be represented, respectively, by the letters L, M, T, θ. All pure numbers, such as an angle, trigonometric function, or a logarithm, have no dimensions whatever and are mere numerical multipliers. All derived quanti-

* Degrees Celsius are identical with degrees Centigrade. By international agreement the designation "Celsius" has, in scientific work, replaced the more familiar "Centigrade," in honor of Anders Celsius, who first described the 0-100 temperature scale.

ties when first introduced are followed by their dimensions expressed as a combination of the four fundamental dimensions inside brackets, thus: force $[MLT^{-2}]$, work $[ML^2T^{-2}]$, specific heat capacity $[L^2T^{-2}\theta^{-1}]$.

Students sometimes experience difficulty in evaluating the dimensions of derivatives and integrals. The derivative of a function with respect to a variable is the limit of the ratio of an increment in the function to an increment in the variable. Thus the dimensions of a derivative are the dimensions of the function divided by the dimensions of the variable. For example, dW/dt, where W is work and t is time, has dimensions $ML^2T^{-2}/T = ML^2T^{-3}$ which are the dimensions of power.

An integral of a function with respect to a variable is a summation, in the limit, of the product of the function with an increment of the variable. Thus the dimensions of an integral are the dimensions of the function multiplied by the dimensions of the variable. For example, $\int c dT$, where c is specific heat capacity and T is temperature, has dimensions $(L^2T^{-2}\theta^{-1})\ \theta = L^2T^{-2}$, which are the dimensions of specific energy.

The greatest immediate advantage to be obtained from this concept of dimensions is its use in verifying the form of equations. Any physical equation represents not only a relationship between certain quantities but also the same relationship between their dimensions. For example, Newton's second law of motion requires that a quantity of force be equal to a quantity of rate of change of *momentum*. The corresponding dimensional equation is $MLT^{-2} = (MLT^{-1})/T$, which is obviously a correct equality. If it were erroneously supposed that Newton's law required the quantity of force to be equal to the quantity of rate of change of *velocity*, the corresponding dimensional equation would be $MLT^{-2} = (LT^{-1})/T$, which is clearly not true. Thus the process of checking the dimensions of both sides of an equation may reveal an error immediately and thus save the student considerable wasted effort. Even when the dimensions check, however, the equation is not necessarily correct, since a non-dimensional factor may have been omitted, or two errors may cancel dimensionally. Nevertheless the dimensional check will rule out a wide class of errors.

Although we shall not discuss here the broad topic of dimensional analysis, one should know that by the use of this powerful tool it is possible to derive partial relationships for otherwise unapproachable problems.[1]

---

[1] P. W. Bridgeman, *Dimensional Analysis*, New Haven, Yale Univ. Press, 1931.

There are a number of different units widely used to measure pressure. Pressure $[ML^{-1}T^{-2}]$ is the force normal to a surface per unit area of surface. In the c-g-s system the unit of pressure is the *dyne per cm²:* 1 dyne/cm² = 1 g cm s^{-2}/1 cm². Since sea-level atmospheric pressure is about 10^6 dynes/cm², this unit is too small for convenience. A quantity called the *bar* has been defined:

$$1 \text{ bar} = 10^6 \text{ dynes/cm}^2$$

However this unit is too large for convenience. Thus in practice the thousandth part of a bar, the *millibar* is used:

$$1 \text{ mb} = 10^3 \text{ dynes/cm}^2 = 10^{-3} \text{ bar}$$

In the meter-ton-second system the unit of pressure is the hundredth part of a bar, or *centibar.*

Because atmospheric pressure is so often measured with a mercurial barometer one frequently finds the pressure reported in units of length of the column of mercury. Inches and millimeters are the units most often used. Since the density of mercury is 13.60 g cm^{-3} and the standard acceleration of gravity is 980.6 cm s^{-2},

$$1 \text{ in. Hg} = 33.86 \text{ mb}$$
$$1 \text{ mm Hg} = 1.333 \text{ mb}$$

Pressure is sometimes reported in fractions of normal sea-level atmospheric pressure, or fractions of an atmosphere. One atmosphere is 1013.25 mb or 760.000 mm Hg or 29.9213 in. Hg.

1.4 The Earth

The highest mountains on earth are less than 10 km above sea level. This is less than 0.2 per cent of the mean radius of the earth, which means that our planet has a relatively smooth and regular surface. Aside from these minor irregularities, the earth is an oblate spheroid with a polar radius of 6356.91 km and an equatorial radius of 6378.39 km, the two radii differing by about 21 km. This difference is due to the centrifugal effect of the rotation of the planet. The earth rotates about its axis relative to the fixed stars with a *sidereal period* of 23 hr 56 min 4.091 s, where the time units are those used throughout this book—namely mean solar hours, minutes, and seconds. The sidereal period differs from the period of rotation relative to the sun (by definition the mean solar period is exactly 24 hr) because the planet is revolving about the sun as well as rotating on its axis. The angular velocity of rotation is 7.292×10^{-5} radian s^{-1}.

The planetary forces to which an object on the surface of the earth is subjected are Newtonian gravitation and the centrifugal force owing to the rotation. Gravitation is, of course, directed towards the center of the earth, but the centrifugal force is perpendicular to the axis of rotation and is directed outward. Thus the vector sum of the two is not directed towards the earth's center. Because of this the earth cannot be truly spherical, for on a sphere there would always be a tangential component of force urging particles equatorward. The equilibrium shape, in which gravitation plus centrifugal force is perpendicular to the surface, is the oblate spheroid.

It is impossible to measure purely gravitational force, as we always observe the combined effect of the two forces—gravitational and centrifugal—which we call gravity. Because of the varying distance from surface points to the center of the planet, and the varying direction of the centrifugal force with respect to the vertical, the observed acceleration of gravity varies with latitude. In the Meteorological Gravity System, the acceleration of gravity at sea level at latitude 45° is 980.616 cm s^{-2}. At other latitudes φ the variation in gravity is given by

$$g = 980.616\,(1 - 0.0026373 \cos 2\varphi + 0.0000059 \cos^2 2\varphi)$$

Thus gravity varies from 978.036 cm s^{-2} at the equator to 983.028 cm s^{-2} at the poles. This is a variation of only about 0.5 per cent but it is sufficiently important to be taken into account in many applications.

Gravity also varies vertically because of the inverse square property of Newton's law of gravitation.[2] So long as z, the elevation of a point above mean sea level, is small compared to a, the radius of the earth, the value of the acceleration of gravity at height z is

$$g_z \approx g_o[1 - 3.14 \times 10^{-7} z]$$

to a sufficient degree of accuracy. Here g_o is gravity at sea level and z is measured in meters.

Gravity also varies erratically from place to place, depending upon the type of material in the earth's crust nearby. These variations, though small, must nevertheless be taken into consideration for many calculations. For our purposes it will suffice to adopt a value of 980.6 cm s^{-2} and to neglect all variations.

---

[2] For a detailed discussion of gravity values and corrections, see R. J. LIST (ed.), *Smithsonian Meteorological Tables*, 6th rev. ed., Washington, D.C., Smithsonian Institution, 1951, p. 488.

1.5 The Atmosphere

Near the surface the earth's atmosphere consists of a mixture of permanent gases, certain variable gases, and solid or liquid particles. The chief variable constituent is water vapor, which may occupy as much as 4 per cent of the volume of a sample of air, or so little as to be virtually absent. One does not need to study meteorology formally, however, to realize that this minor constituent plays an important role in atmospheric processes despite its small concentration.

The composition of pure dry air in the lower atmosphere is given in Table 1.1. In addition, traces exist of such gases as carbon monoxide,

TABLE 1.1

COMPOSITION OF PURE DRY AIR UP TO 25 KM

Constituent	*Per cent by volume*	*Per cent by mass*	*Molecular weight*
Nitrogen	78.09	75.51	28.02
Oxygen	20.95	23.14	32.00
Argon	.93	1.3	39.94
Carbon dioxide (var)	~.03	~.05	44.01
Neon	180×10^{-5}	120×10^{-5}	20.18
Helium	52×10^{-5}	8×10^{-5}	4.00
Krypton	10×10^{-5}	29×10^{-5}	83.7
Hydrogen	5.0×10^{-5}	$.35 \times 10^{-5}$	2.02
Xenon	$.8 \times 10^{-5}$	3.6×10^{-5}	131.3
Ozone (var)	$\sim.1 \times 10^{-5}$	$\sim.17 \times 10^{-5}$	48.00

radon, nitrous oxide, and methane. Nitrogen and oxygen make up about 99 per cent of air but are quite passive in most meteorological processes. Water vapor plays a dominant role in thermodynamic processes (because it can change phase and become liquid or solid) and also in radiative processes. Carbon dioxide, present throughout the atmosphere, and ozone, found mostly at high levels (20-30 km), are important because of their radiative characteristics. All the other gases are of little meteorological significance. It is a curious fact that the most important meteorological gases are the minor constituents water, carbon dioxide, and ozone.

Except for these three gases the composition of air is remarkably constant up to great elevations. Above about 80 km oxygen begins to dissociate under the influence of ultraviolet radiation from the sun. At still higher levels nitrogen also dissociates. The amount of air remaining at these levels is so small that these phenomena probably have no direct meteorological importance. However, their influence on the propagation of radio waves is considerable.

The nongaseous constituents in the atmosphere are of varying kinds and varying importance. Dust, smoke, salt particles from the evaporation of sea spray, and condensed H_2O are the chief nongaseous materials present. All of these affect the transmission of radiation through the air, including the visibility. These solid or liquid particles may be present in a wide range of concentrations, varying from only a few to millions of particles per cubic meter of air. The importance of clouds and precipitation need not be elaborated. Sea-salt particles supply hygroscopic nuclei upon which condensation of water vapor is usually initiated. Smoke and dust are mainly annoyances and producers of restricted visibility, although industrial smokes may have important effects on health, and radioactive dust from nuclear explosions represents a potentially very serious hazard.

PROBLEMS

1. Determine the dimensions of:

(a) pressure
(b) volume
(c) kinetic energy
(d) angular velocity
(e) acceleration
(f) torque
(g) the natural base of logarithms
(h) angular momentum

2. Check the following equations dimensionally; determine which are dimensionally consistent and which are not.

(a) $F = m_1 m_2 / r^2$, where F is the force of gravitational attraction between two bodies of masses m_1 and m_2, separated by the distance r.

(b) $mgz = \frac{1}{2} mv^2$
(c) $mgz = mv^2$
where g is the acceleration of gravity, z is height, v is velocity of fall.

(d) $p_o = \int_o^\infty \rho g dz$

(e) $A = a^{1/2} b^{3/2}$ where A is the area enclosed by an ellipse, a is the semi-major axis, b is the semi-minor axis of the ellipse.

(f) $R = \dfrac{[1 + (dy/dx)^2]^{3/2}}{d^2y/dx^2}$ where R is the radius of curvature of a line, and dx and dy are increments of distance along the curve.

3. Calculate the number of millibars exerted at sea level at latitude 45° by a mass of:

(a) one pound per square inch;
(b) one metric ton per square meter;
(c) one grain per square micron.

4. Show from Newton's law of universal gravitation that at elevation z meters above sea level the acceleration of gravity is given approximately by

$$g_z = g_o \left[1 - \frac{2z}{a}\right] = g_o[1 - 3.14 \times 10^{-7} z]$$

At what elevation would gravity be 1 per cent less than its sea-level value?

CHAPTER 2

THE EQUATION OF STATE

2.1 The Variables of State

A small system, or infinitesimal sample of matter, can be described thermodynamically by its volume (ΔV), mass (ΔM), pressure (p), temperature (T), and composition (kind of matter included). As our systems go through various processes, we shall usually restrict ourselves to the case where they do not change composition or mass. That is, they always consist of the same particles. This leaves three variables—volume, pressure, and temperature. These are the basic variables of state and their values completely describe the state of a given system. Other variables of state which are functions of these three will be defined later.

We shall prefer to express the volume by the *specific volume;* $\alpha = \Delta V/\Delta M$ $[L^3M^{-1}]$, or *density,* $\rho = \Delta M/\Delta V$ $[ML^{-3}]$. It is obvious that $\alpha\rho = 1$ is the relationship between these two variables and that we may change from one to the other at our convenience.

To define pressure precisely, consider a point in a nonviscous gaseous system at which is located a small plane surface of area ΔA oriented in some fixed direction. The constant bombardment by gas molecules on one face of the area will produce a force F perpendicular to ΔA. For a wide range of sizes of the area involved, it is found experimentally that the force produced is directly proportional to the area. The chief restriction on the size of the area is that it must not be so small as to approach intermolecular dimensions. The proportionality constant is called the pressure, p, at the given point in the given direction, such that $p = \Delta F/\Delta A$. Thus pressure is a force per unit area $[ML^{-1}T^{-2}]$.

Theory indicates and experiment confirms that pressure is independent of the orientation of the surface area ΔA and is a scalar, not a vector quantity. Furthermore, we shall be dealing with infinitesimal systems in equilibrium so that the pressure will not vary significantly with position in the system. Thus for thermodynamic purposes the pressure of a system is equal to the pressures at all points within it and is independent of orientation.

Although an infinitely small variation of pressure across a system is unimportant in thermodynamics, we shall see that such a change of pressure with distance (pressure gradient) is of the utmost importance in hydrodynamics.

Temperature is a concept with which we are all familiar, in an imprecise fashion, from the evidence of our senses. We are aware of the difference between hot and cold bodies, and that this difference ultimately disappears when such bodies are placed in physical contact with each other, but it is impossible to reproduce results or detect small differences through our senses alone.

In order to define a more precise temperature scale we must find certain fixed and reproducible temperatures. A mixture of pure ice and water in thermal equilibrium under a pressure of one atmosphere is defined to have a temperature of 0° Celsius (0° C).[1] Steam in thermal equilibrium with pure water boiling at a pressure of one atmosphere is defined to have a temperature of 100° C. Next we can take advantage of what appears to be a nearly uniform rate of expansion with temperature of such a substance as mercury to construct a mercury thermometer. On such an instrument the level of the mercury can be marked when it is in equilibrium with the ice-water mixture and the water-steam mixture. The interval between can then be divided into 100 equal parts to give an instrument of far greater reproducibility and accuracy than our senses. As we shall see, such a thermometer is by no means the ultimate in accuracy and is not used as a primary temperature standard, but suffices to disclose some of the empirical laws to which gases are subject.

2.2 Charles' Law

The fundamental experiment in thermodynamics described here was first performed by Jacques Charles at about 1787 and Joseph Gay-Lussac in 1802. Take a sample of dry air at a fixed pressure and a

[1] See footnote, page 5.

temperature of 0° C. Measure its specific volume, α_o. Now measure its specific volume α_t at a number of other temperatures, keeping the pressure constant. It will be found that the temperature is directly proportional to the difference between α_t and α_o:

$$t = k(\alpha_t - \alpha_o) \tag{2.1}$$

where t is Celsius temperature and k is a constant. The constant can be evaluated by measuring α_{100}, the specific volume at 100° C. At that temperature

$$100^\circ = k(\alpha_{100} - \alpha_o) \tag{2.2}$$

Thus by eliminating k between these two equations we obtain

$$t = 100^\circ \frac{\alpha_t - \alpha_o}{\alpha_{100} - \alpha_o} \tag{2.3}$$

Thus, dry air also expands linearly with temperature. As a matter of fact, all gases, sufficiently remote from conditions that will produce liquifaction, obey Eq. (2.3) to a fair degree of approximation—an important fact because it indicates that all gases expand with increasing temperature at nearly the same rate, and t in Eq. (2.3) is essentially independent of the gas used. Indeed, certain gases are more accurate working substances for a thermometer than liquids like mercury. Liquid thermometers are widely used because of their convenience, but the international standard of thermometry is a gas thermometer using helium as the medium.

Eq. (2.3) can be rewritten in the form

$$t + \frac{100^\circ\,\alpha_o}{\alpha_{100} - \alpha_o} = \frac{100^\circ\,\alpha_t}{\alpha_{100} - \alpha_o} \tag{2.4}$$

which suggests that we can define a scale in which the temperature will be directly proportional to the specific volume itself rather than to $\alpha_t - \alpha_o$. We shall define this new temperature to be

$$T = t + \frac{100^\circ\,\alpha_o}{\alpha_{100} - \alpha_o} \tag{2.5}$$

The accepted current evaluation of the constant $100^\circ\alpha_o/(\alpha_{100} - \alpha_o)$ for the "permanent" gases is 273.16° C. Thus if we add this number to all Celsius temperatures, we get a scale in which zero is reached when the specific volume vanishes completely, which does not actually happen because all gases liquify before this point is reached and so no longer obey Charles' law. However, the new scale is a useful abstraction because its zero point is absolute—temperatures below this point are theoretically impossible. Hence the scale is called the "absolute temperature scale" or the "Kelvin temperature scale" after the physicist who introduced it. Temperatures in this system will be given in "degrees Kelvin" (°K). One °K is equal in size to one °C, but any given temperature is higher by 273.16° in the Kelvin system than in the Celsius system. For meteorological purposes it suffices to round off this difference to 273°.

From Eqs. (2.4) and (2.5) we find

$$\frac{\alpha_T}{T} = \frac{\alpha_0}{T_0} \tag{2.6}$$

where T_0 is 273° K. This form of Charles' law shows most clearly that at constant pressure the specific volume is directly proportional to the absolute temperature.

2.3 Boyle's Law

A second fundamental experiment in thermodynamics was first performed by Robert Boyle about 1660. Take a sample of dry air at a fixed temperature and measure its specific volume at several different pressures. The results indicate that pressure and volume are inversely proportional to each other:

$$p\,\alpha = C \tag{2.7}$$

where the constant C depends upon the fixed temperature of the gas. This inverse proportionality is known as Boyle's law and is closely satisfied by all the permanent gases.

The behavior of real gases follows Charles' law and Boyle's law only approximately. It is useful to introduce the concept of an *ideal* or *perfect* gas, which, for the present, can be defined as one which follows these two laws exactly. Later we shall add another requirement involving the internal energy of an ideal gas.

2.4 The Equation of State of an Ideal Gas

An equation of state is a relationship among the variables of state, p, α, and T, or quantities derived from them. Charles' and Boyle's laws can be combined to produce such a relationship.

Let the specific volume of an ideal gas at temperature T and pressure p be written $\alpha(T, p)$. Let T_o be 273° K and p_o be standard sea-level pressure, 1013.25 mb. From Charles' law (2.6):

$$\frac{\alpha(T, p_o)}{T} = \frac{\alpha(T_o, p_o)}{T_o}$$

From Boyle's law (2.7):

$$p\alpha(T, p) = p_o\,\alpha(T, p_o)$$

Eliminating $\alpha(T, p_o)$ between these two we get

$$p\alpha(T, p) = \frac{p_o\alpha(T_o, p_o)}{T_o}\,T$$

The quantity $p_o\alpha(T_o, p_o)/T_o$ is a constant for any given gas, since p_o and T_o are constants and the volume of one g of gas at p_o and T_o is also constant. Replacing this combination of constants by the symbol $R[L^2T^{-2}\theta^{-1}]$, which can be determined experimentally, we then have

$$p\alpha = RT \tag{2.8}$$

where α is the specific volume under pressure p and temperature T. Eq. (2.8) is the desired equation of state of an ideal gas, in which R is the *specific gas constant* for the gas being considered.

Another form of the equation of state can be obtained in the following way. Consider a sample of gas whose mass is equal to its molecular weight, m. This sample will have a volume $V = m\alpha$. Therefore the equation of state is

$$pV = mRT$$

Avogadro's law, which is also empirical, states that at equal pressures and temperatures a molecular weight (called a *mole*) of two different gases will occupy the same volume. For an ideal gas at 273.16° K and 1013.25 mb this volume is 22.415×10^3 cm^3. Therefore the quantity

$$mR = \frac{pV}{T}$$

is a constant for all gases. Let $R^* = mR$, where R^* is the *universal gas constant*, equal to 8.3144×10^7 erg mol^{-1} °K^{-1}. We may now write the equation of state in the form

$$p\alpha = \frac{R^*}{m}\,T \tag{2.9}$$

The two forms Eqs. (2.8) and (2.9) are completely equivalent to each other. Either is sometimes called the *ideal gas law.*

Other equations of state, different from the one appropriate to an ideal gas, have been proposed. These relationships describe the behavior of certain real gases better than does the ideal law. However, for dry air and for water vapor the ideal gas law is sufficiently accurate so we need not resort to more complex statements.

The relationship of Eq. (2.9) is often put into graphical form by plotting lines of constant temperature (isotherms) on a graph of specific volume versus pressure (α, p diagram). We shall prefer to make the

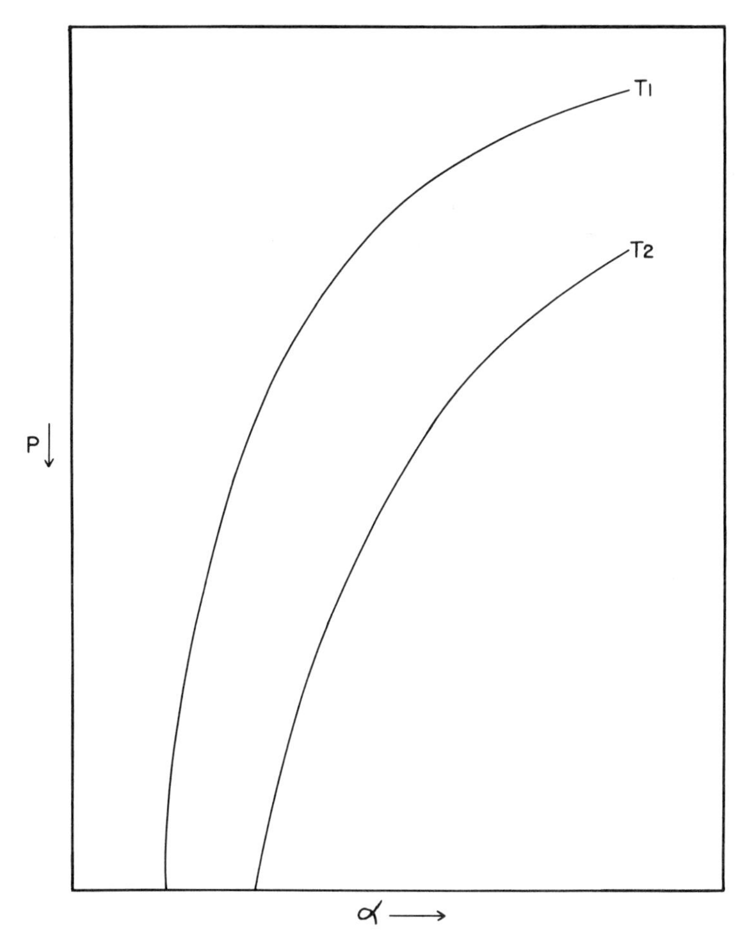

FIG. 2.1. Isotherms on an α, $-p$ diagram

ordinate $-p$ so that pressure will decrease upward as it does in the atmosphere. The resulting α, $-p$ diagram for an ideal gas is shown schematically in Fig. 2.1. The isotherms are rectangular hyperbolas. Each isotherm passes through those points (α, $-p$) which represent the states of the system with the given fixed temperature. On such a diagram it is possible to represent arbitrary changes in state. Of course the easiest processes to represent are those which are isobaric (constant pressure), isosteric (constant volume), or isothermal (constant temperature) because lines for each of these processes are present on the diagram. We shall have frequent occasion to use such a diagram and especially others developed from it.

2.5 Mixtures of Gases

Air is a mixture of gases which is observed to behave in a nearly ideal fashion, so long as condensation does not take place. In order to understand the behavior of mixtures we shall call upon another empirical statement called *Dalton's law*. This states that the total pressure exerted by a mixture of gases is equal to the sum of the partial pressures which would be exerted by each constituent if it alone filled the entire volume at the temperature of the mixture. That is, for a mixture of k components

$$p = p_1 + p_2 + \dots p_k = \sum_1^k p_n$$

where p is the total pressure and the p_n are the partial pressures. Let the volume of the mixture be V, let the mass of the nth constituent be M_n, and let its molecular weight be m_n. If each gas obeys separately the ideal gas law, then

$$p_n = \frac{R^*}{m_n} \frac{M_n}{V} T$$

From Dalton's law

$$p = \sum p_n = \frac{R^* T}{V} \sum \frac{M_n}{m_n}$$

or

$$p\alpha = R^* T \cdot \frac{\sum \frac{M_n}{m_n}}{\Sigma M_n} \tag{2.10}$$

where α is the specific volume of the mixture.

Equation (2.10) shows that a mixture of ideal gases obeys a gas law which is of the same form as the ideal gas law for a single constituent. It is of exactly the same form as the ideal gas law if we define a mean molecular weight

$$\frac{1}{\bar{m}} = \frac{\sum \frac{M_n}{m_n}}{\Sigma M_n} \tag{2.11}$$

for with such a definition Eq. (2.10) becomes

$$p\alpha = \frac{R^*}{\bar{m}} T$$

Eq. (2.11) states that the proper way to compute a mean molecular weight for a mixture is to take a mass-weighted harmonic mean. The mean molecular weight of dry air computed in this fashion is 28.966 g mol^{-1}. The specific gas constant for dry air is then $R = R^*/\bar{m} = 2.8704 \times 10^6$ erg g^{-1} $°K^{-1}$.

All of the approximate empirical laws governing real gases which were used in this chapter (Charles', Boyle's, Avogadro's and Dalton's laws) can be shown to be exact theoretical consequences of the kinetic-molecular theory of ideal gases. Thus the results we have obtained are well-founded both experimentally and theoretically.

PROBLEMS

1. The Fahrenheit temperature scale assigns the values 32° F to the melting point of ice and 212° F to the boiling point of water. Derive the relationship between temperature in °F and °C and compile a table giving the Fahrenheit temperature for these Celsius temperatures: −40°, −30°, −20°, −10°, 0°, +10°, +20°, +30°, +40°.

2. A sample of hydrogen is at a pressure of 1000 mb and a temperature of +10° C. Calculate its specific volume.

3. From the data given in Table 1.1 (p. 9) for the four most abundant constituents of dry air, calculate the mean molecular weight of air to four significant figures. Compare your result with the value given in this chapter.

4. Calculate and plot the isotherms of an α, $-p$ diagram for dry air for temperatures 200° K, 300° K, and 400° K. Let the pressure vary from 1000 mb to 200 mb and let α range from 1000 cm^3 g^{-1} to 2500 cm^3 g^{-1}.

CHAPTER 3

THE PRINCIPLES OF THERMODYNAMICS

3.1 Work

The physical definition of *work* is stated as follows: When a force of magnitude F is applied to a mass which then moves through a distance dS parallel to the force, the work done is $dW = FdS$. When the displacement dS makes an angle θ with the applied force, only the component of displacement along the force, $dS \cos \theta$, enters the computation of the work. In this case $dW = FdS \cos \theta$, and thus the dimensions of work are ML^2T^{-2}. In the c-g-s system the unit of work is the *erg*, which is the work done when a mass is moved 1 cm under the influence of 1 dyne of force applied parallel to the displacement.

The only force with which a nonviscous fluid can do work on its surroundings is that owing to the pressure along the surface of the fluid sample. For each element of surface area dA the fluid exerts a force on the surroundings equal to pdA; from Newton's third law of motion, the surroundings exert an oppositely directed force of magnitude pdA on the fluid. This force is normal to the area element. Suppose the fluid sample under consideration expands under the influence of the pressure force in such a fashion that the particles making up the surface element dA move at an angle θ to the direction of the pressure force through a distance dS. Then the increment of work done on the surroundings is $pdAdS \cos \theta$. But $dAdS \cos \theta$ is the volume swept out by the area element during its displacement. If we sum over the entire area of the sample of fluid, the work done becomes

$$dW = pdV \tag{3.1}$$

where dV is the infinitesimal change in volume of the sample because of infinitesimal surface displacements dS. We denote quantities that refer to the entire mass of a system by capital letters as in Eq. (3.1) and quantities which refer to a unit mass by small letters. Thus if we divide Eq. (3.1) by the mass of the sample of fluid we get

$$dw = p d\alpha \tag{3.2}$$

where dw, the specific work done, has dimensions L^2T^{-2}.

By the method of formulation adopted we have agreed to call dw positive if the system expands and does work on its environment. If the system is compressed by external pressure forces, then work has been done on the system and dw is negative. Obviously a system that neither expands nor contracts can do no work.

The amount of work done in an expansion from α_1 to α_2, different from each other by a finite amount, is

$$w = \int_{\alpha_1}^{\alpha_2} p d\alpha$$

where in general p may vary during the process. This may be shown graphically on an α, $-p$ diagram as in Fig. 3.1. The value of the integral

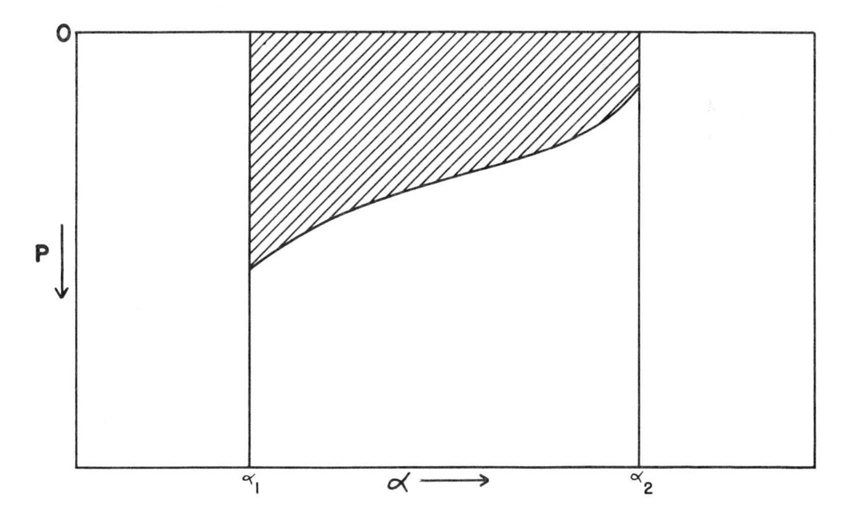

FIG. 3.1. Graphical representation of work done in an arbitrary process of expansion from α_1 to α_2

is given by the hatched area. Obviously the area will depend upon the specific path followed by the gas—that is, upon the process of expansion. There is an infinite number of curves connecting the initial state of the gas to its final state. If the gas is compressed back to α_1 by a different process, a different path will be followed on the α, $-p$ diagram. Such a *cyclic process* is shown in Fig. 3.2. Here the work done is clearly the hatched area between the curves.

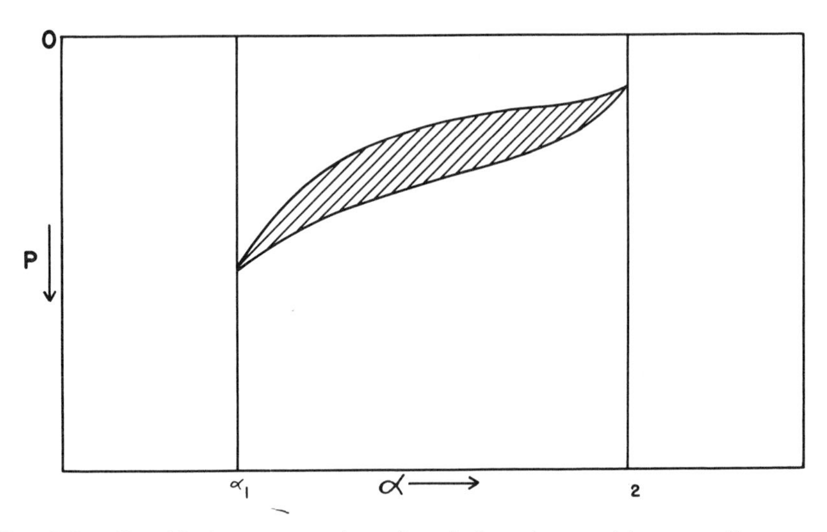

FIG. 3.2. Graphical representation of work done in an arbitrary cyclic process

3.2 Heat

As pointed out earlier, when two bodies of different temperature are brought into contact with each other the temperature difference disappears. The final temperature will be intermediate between the two initial temperatures. Thus if T_1 is the initial temperature of the warm body and T_2 the initial temperature of the cold body, the final temperature at thermal equilibrium, T_F, will be $T_1 > T_F > T_2$. Experiments with the same two bodies at various initial temperatures demonstrate that the final temperature is given always by the following formula to a good degree of precision:

$$C_1(T_F - T_1) + C_2(T_F - T_2) = 0 \tag{3.3}$$

where C_1 and C_2 are constants. If the mass of body 1 is varied in this experiment, the same formula holds with the same value of C_2, but the value of C_1 varies directly with the mass of body 1, $C_1 = c_1 M_1$, where c_1 is a constant for the material composing body 1.

These experiments suggest that something is being transferred unchanged between the two bodies because the two terms of Eq. (3.3) are of the same form but opposite in sign. This "something" that invariably flows from the warmer to the cooler body is called *heat.* The amount of heat, ΔH, lost by the warm body is

$$-\Delta H = C_1(T_F - T_1)$$

and the amount of heat gained by the cool body is equal in magnitude and given by

$$\Delta H = C_2(T_F - T_2)$$

The constants C_1 and C_2 are called the *heat capacities* of the two bodies.

Either of these two expressions for ΔH may be solved for the heat capacity in the form

$$C = \frac{\Delta H}{T_F - T} = \frac{\Delta H}{\Delta T}$$

If one passes to the limits as ΔT becomes infinitesimal, the expression for heat capacity becomes

$$C = \frac{dH}{dT} \tag{3.4}$$

It is advantageous to look upon this as the definition of C because heat capacities are not strictly constant with temperature. C must then be evaluated as a ratio of small increments of H and T in the neighborhood of a specific temperature.

In accordance with this, an internationally accepted unit of heat, the *15° gram calorie* or simply the *calorie,*[1] has been defined as that quantity of heat necessary to raise the temperature of one gram of pure water from 14.5° C to 15.5° C. In calculations the calorie may be looked upon as a new independent unit when convenient. In the next section we shall discuss its relationship to mechanical energy units.

With this definition of a heat unit we may return to Eq. (3.4) and state that heat capacity is measured in calories per °C. If we divide Eq. (3.4) through by the mass of the system considered we get

$$c = \frac{dh}{dT} \tag{3.5}$$

where c is the *specific heat capacity* of the material and dh is the heat added per unit mass. Clearly specific heat capacity is measured in calories per gram per °C, and, by definition, near 15° C the specific heat capacity of water is one cal g^{-1} $°C^{-1}$. For orientation purposes the specific heat capacities of various forms of water substance are given in Table 3.1.

[1] A slight change in the size of the calorie was made in 1929 but since it was increased by only .03 per cent, this will be completely unimportant to us.

TABLE 3.1

SPECIFIC HEAT CAPACITIES OF VARIOUS KINDS OF WATER SUBSTANCE

Substance	*Temperature* (°C)	*Salinity* (per cent)	*Specific heat capacity* (cal g^{-1} °C^{-1})
pure water	0.0	0	1.007
	5.0	0	1.004
	17.5	0	0.999
	30.0	0	0.998
sea water	17.5	2	0.951
	17.5	4	0.926
pure ice	0.0	0	0.503
	−20.0	0	0.468
sea ice	−10.0	2	0.57
	−20.0	2	0.52
	−10.0	4	0.64
	−20.0	4	0.55

3.3 The Law of Conservation of Energy

In physics the *energy* of a system is defined as the *capacity of the system to do work*. Energy is equal to the total work that can be done and is measured in the same units and dimensions $[ML^2T^{-2}]$. *Specific energy* is energy per unit mass and so has dimensions $[L^2T^{-2}]$.

The law of conservation of energy is an empirical statement which asserts that energy of all sorts can neither be created nor destroyed.[2] Thus when energy is added to (or extracted from) a system, the final energy is equal to the original energy plus the energy added (or minus the energy extracted). This proposition was originally advanced with respect to potential and kinetic forms of energy. Later it was shown that a similar principle seemed to hold for heat during thermal conduction, and this idea was used in the discussion of heat in Section 3.2.

Through the classical work of Joule it was shown experimentally that heat is a form of energy, that there is a fixed equivalence between heat measured in calories and mechanical energy measured in ergs, and that one generalized conservation principle governs both forms.

[2] Nowadays we must modify this statement to take into account the Einsteinian equivalence of mass and energy. Thus energy may be created by the destruction of matter, and vice versa. However, this fact is of no direct importance in meteorology.

The principle of Joule's experiment is to use a fixed amount of mechanical energy to agitate a known amount of water, thus warming it. The temperature rise is measured and thus the production of heat is known. Joule carefully repeated the experiment many times and found that the same amount of mechanical energy was required each time to add one calorie of heat to the water. The currently accepted value of the conversion factor is

$$1 \text{ calorie} = 4.186 \times 10^7 \text{ ergs}$$

This value represents the *mechanical equivalent of heat*. Because the erg is such an inconveniently small unit, requiring a great many to make a calorie, another unit, named in honor of Joule, is often used, 1 joule $= 10^7$ ergs. With this definition the mechanical equivalent of heat becomes 1 calorie $=$ 4.186 joules. We shall express energy in either mechanical or thermal units as dictated by convenience. It is extremely important, however, to convert all quantities in a problem into one or the other set of units, via the mechanical equivalent of heat, before proceding with numerical computation.

We shall formulate quantitatively the law of conservation of energy in differential form. That is, in the most general statement, any increment of energy added to a system is equal to the algebraic sum of the increments in organized kinetic energy, potential energy, internal energy (largely thermal), work done by the system on its surroundings, and whatever forms of electrical and magnetic energy may appear. We shall assume that only the first four forms of energy may validly require consideration, neglecting other possible forms of energy.

For meteorological purposes we may simplify even further. Problem 2 (p. 37) calls for a comparison of changes in kinetic energy with changes in internal energy and increments of work done in meteorological systems. The result, which the student should verify for himself, indicates that changes in kinetic energy may be neglected compared to the others. Furthermore, changes in potential energy of a parcel of air may be omitted since each parcel of air, for large-scale meteorological processes, is very nearly in buoyant equilibrium—the downward force of gravity is very nearly balanced by the upward force owing to the normal decrease of pressure with altitude. Thus, even though a rising parcel of air does increase its potential energy, this increment is almost exactly supplied by work done by the vertical gradient of pressure. We need not include either of these effects because they very nearly balance each other separately.

As a result we are left on the one hand, with an equality between energy added to the system from without (excluding work done on the system by vertical pressure gradient forces), and on the other, with the sum of internal energy and work done by the system on its surroundings. For all practical purposes the addition of energy to the system is now limited to addition of heat. An increment of heat added is called dH. This energy may be utilized partly in increasing the temperature or partly in overcoming the forces of attraction between the molecules, both of which constitute the *internal energy* of the system. An increment of internal energy is denoted by dU. Since we have denoted an increment of work done by dW, the conservation law becomes

$$dH = dU + dW \tag{3.6}$$

In classical thermodynamics, where no changes in organized kinetic energy and potential energy are permitted, Eq. (3.6) is the standard formulation. However, to reduce meteorological systems to this simple state requires the justification given above. Equation (3.6) is the formulation of the *first law of thermodynamics*.

If we divide Eq. (3.6) by the mass of the system we get another form of the first law:

$$dh = du + dw = du + p\,d\alpha \tag{3.7}$$

where dh is the heat added per unit mass and du is the specific internal energy.

In order to write Eq. (3.7) in a more useful meteorological form we shall have to examine more carefully the concept of specific heat capacity as applied to gases. The heat required to warm one gram of gas 1° C obviously will vary even for the same gas, depending upon whether we allow the gas to expand during the heating process and how much we allow it to expand. The reason for this is that some of the heat added will go to supply the energy needed to do work of expansion against the surroundings, and not to raise the temperature of the gas. Thus an infinite number of specific heat capacities exist for each gas, depending upon the limitations we may impose on expansion during the process. It turns out that two of this infinite array suffice for our purposes. One is measured when the heating is required to be a constant-volume process—that is, no expansion is permitted. The value so obtained is called the *specific heat capacity at constant volume*, c_v. The other is measured when the heating is a constant-pressure process—wherein only that amount of expansion is permitted which will keep the pressure unchanged. The value so obtained is called the *specific heat capacity at constant pressure*, c_p. From the general definition of specific heat capacity Eq. (3.5), we find

$$c_v = \left(\frac{dh}{dT}\right)_a; \quad c_p = \left(\frac{dh}{dT}\right)_p \tag{3.8}$$

where a subscript attached to a derivative means the property named in the subscript is held constant. It is clear that $c_p > c_v$, since in a constant-pressure process some of the added heat must be expended in work on the surroundings, while in a constant-volume process all the heat can be devoted to raising the temperature of the gas.

3.4 Internal Energy and Specific Heat Capacities of an Ideal Gas

The kinetic-molecular theory of gases assumes that an ideal gas consists of a large number of small masses called molecules which are in rapid random motion, which have negligibly small dimensions compared to the mean distances between them and negligibly small attractive forces for each other, and whose collisions with each other and the walls of the container are perfectly elastic. In such an idealized system the temperature is proportional to the mean kinetic energy of random motion of the molecules, and pressure is the force per unit area exerted by the bombardment of the molecules. It can be shown that such a system exactly obeys the ideal gas law. Furthermore, the internal energy of such a system is a function of the temperature only, since heat added at constant volume is applied only to increasing the random motion of the molecules. In real gases some of the energy may be expended in overcoming intermolecular forces, but by postulate these do not exist in an ideal gas. Thus in theory $du = c_v dT$. The two necessary conditions for a gas to be ideal are (1) that the perfect gas law be satisfied and (2) that the internal energy be a function of temperature only.

No real gas obeys these two conditions exactly, owing presumably to the finite sizes of actual molecules and to the existence of intermolecular forces in real gases. The degree to which these conditions are met can be determined experimentally. The kinds of experiments involved in verifying the ideal gas law have already been discussed. Broadly speaking, the lower the pressure or the higher the specific volume, the closer real gases come to obeying the ideal gas law. That is, gases come closer to ideal the closer the postulates of the kinetic-molecular theory are approached.

With respect to the dependence of internal energy on variables of state other than temperature, the following argument is offered prior to discussing the experimental results. For a real gas the internal energy might be, in general, a function of the state of the gas: $u = u\ (p,\ \alpha,\ T)$. But if the ideal gas law is applicable with sufficient accuracy, one of these variables may be eliminated and $u = u\ (\alpha,\ T)$, for example. From calculus we may write

$$du = \frac{\partial u}{\partial \alpha}\, d\alpha + \frac{\partial u}{\partial T} dT \tag{3.9}$$

where du is the change in internal energy during some process in which changes $d\alpha$ and dT take place. The partial derivative $\partial u/\partial \alpha$ means the rate of change of u with α when T is held constant, and the partial derivative $\partial u/\partial T$ means the rate of change of u with T when α is held constant. When we substitute Eq. (3.9) into the first law of thermodynamics, Eq. (3.7), we get

$$dh = \frac{\partial u}{\partial T}\, dT + \left(\frac{\partial u}{\partial \alpha} + p\right) d\alpha$$

If we require the process to be an isosteric one ($\alpha =$ const) then

$$c_v = \left(\frac{dh}{dT}\right)_a = \frac{\partial u}{\partial T}$$

Returning to Eq. (3.9) we then find

$$du = \frac{\partial u}{\partial \alpha}\, d\alpha + c_v dT \tag{3.10}$$

Now we are in a position to investigate the derivative $\partial u/\partial \alpha$ experimentally. The classical experiment is to set up two thermally insulated vessels connected to each other through a valve as in Fig. 3.3. Let one

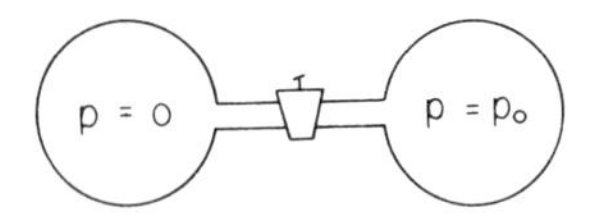

FIG. 3.3. Schematic representation of the apparatus for performing an expansion into a void

of the vessels be completely evacuated and the other filled with a gas at pressure p_o. When the valve is opened, the gas expands from one vessel into the other. Since the expansion is done against zero pressure, no work is done. Since the vessels are thermally isolated, no heat is added or taken away. Therefore the first law of thermodynamics requires that the change in internal energy be zero. Then Eq. (3.10) reduces to

$$\frac{\partial u}{\partial \alpha} = -\, c_v \frac{dT}{d\alpha}$$

Thus all we need do is measure the change in temperature during the known change in specific volume to calculate $\partial u/\partial \alpha$, since c_v is known from other experiments.

Early work of this type failed to yield any change in temperature, thus reducing Eq. (3.10) to what is expected of an ideal gas from the kinetic-molecular theory. Later and more precise measurements show that a small cooling does take place (Joule-Thomson effect), but $dT/d\alpha$ decreases with decreasing pressure or increasing specific volume of the gas. Thus once again we note that the closer conditions are to those assumed in the kinetic-molecular theory the more nearly is the gas ideal, in that $du = c_v dT$. In an ideal gas $\partial u/\partial \alpha = 0$ by extrapolation.

This result means that the internal energy of an ideal gas is a function of temperature only, provided c_v is at most a function of temperature only. Now c_v might be a function of two variables of state by the argument given earlier for u. That is, $c_v = c_v(\alpha, T)$ but since c_v was shown to be $\partial u/\partial T$

$$\frac{\partial c_v}{\partial \alpha} = \frac{\partial}{\partial \alpha}\left(\frac{\partial u}{\partial T}\right) = \frac{\partial}{\partial T}\left(\frac{\partial u}{\partial \alpha}\right)$$

But we have just seen that in an ideal gas $\partial u/\partial \alpha = 0$, therefore c_v is *not* a function of α and in an ideal gas c_v is a function of temperature only. Actually the variation of the specific heat capacities of dry air with temperature is very small, being about 0.01 per cent per °C. Experiments show that for an ideal gas c_v is constant. We shall consider dry air to be ideal in this respect also.

We may now write the first law of thermodynamics for an ideal gas in the form

$$dh = c_v dT + p d\alpha \tag{3.11}$$

where all the quantities can be measured or calculated. Another form of this energy equation can be written by substituting for $d\alpha$ from the equation of state. In differentiated form the equation of state is

$$p d\alpha + \alpha dp = R dT$$

When this is substituted into Eq. (3.11) we get after rearrangement

$$dh = (c_v + R)\, dT - \alpha dp \tag{3.12}$$

If we now consider an isobaric process ($dp = 0$) then

$$c_p = \left(\frac{dh}{dT}\right)_p = c_v + R$$

or

$$c_p - c_v = R \tag{3.13}$$

That is, for an ideal gas both c_p and c_v are constants whose difference is equal to the specific gas constant for the gas involved. Equation (3.13) demonstrates clearly that $c_p > c_v$, as indicated previously. It can be shown that for an ideal gas composed of diatomic molecules $c_p = 7R/2$ and $c_v = 5R/2$. Dry air is nearly such a gas for which the accepted values are

$$c_p = .240 \text{ cal g}^{-1}\,{}^\circ\text{C}^{-1}; \quad c_v = .171 \text{ cal g}^{-1}\,{}^\circ\text{C}^{-1}$$

Substitution of Eq. (3.13) into Eq. (3.12) gives still another form of the first law which will be used extensively:

$$dh = c_p dT - \alpha dp \tag{3.14}$$

This form is especially useful because the increments dT and dp are those most commonly measured in meteorology. Note that $c_p dT$ is *not* the change in internal energy, and αdp is *not* the work done.

3.5 Adiabatic Processes

An increment of heat, dh, may be added to a parcel of air by many processes. Among these are radiation, friction, condensation of water vapor, and turbulent transfer of heat. Most of these processes are difficult to deal with mathematically and physically. Furthermore, there is empirical evidence that for time periods up to a day or two these heating processes are often secondary in importance. Therefore it will be of value to investigate the implications of the first law of thermodynamics for *adiabatic processes*, in which no heat is added to or taken away from a sample of gas. The first law, Eq. (3.14), becomes

$$0 = c_p dT - \alpha dp$$

From the perfect gas law $\alpha = RT/p$, therefore

$$0 = \frac{dT}{T} - \frac{R}{c_p}\frac{dp}{p}$$

This is immediately integrable. Let us represent the nondimensional combination R/c_p by κ. Then

$$T = \text{const } p^\kappa \tag{3.15}$$

Equation (3.15) is one form of the adiabatic equation derived by Poisson over a hundred years ago, and is thus called *Poisson's equation.* Note the additional specification that the process be adiabatic permits a result that relates only two of the variables of state. Poisson's equation can easily be written as a relation between α and p, or α and T, in addition to the form in Eq. (3.15). The constant in Eq. (3.15) can assume different values, depending upon the initial pressure and temperature of the gas engaging in the adiabatic process. For example, if the initial pressure is 1000 mb and the initial temperature has a value θ, Eq. (3.15) becomes

$$\frac{T}{\theta} = \left(\frac{p}{1000}\right)^{\kappa} \tag{3.16}$$

The temperature θ is called the *potential temperature*. The potential temperature may be looked upon as the temperature a sample of gas would have if it were compressed (or expanded) adiabatically from a given state p and T to a pressure of 1000 mb. Obviously θ is a characteristic property of a parcel of air which is invariant during adiabatic processes. Such a quantity is referred to as a *conservative property*. Thus the potential temperature can be used to label a parcel of air in order to facilitate its identification as it moves about from day to day.

For dry air κ has the value 0.286.

Since θ is a function of two variables of state, it is itself a variable of state. As such it is possible to plot an α, $-p$ diagram in which lines of constant potential temperature are drawn in addition to the isotherms.

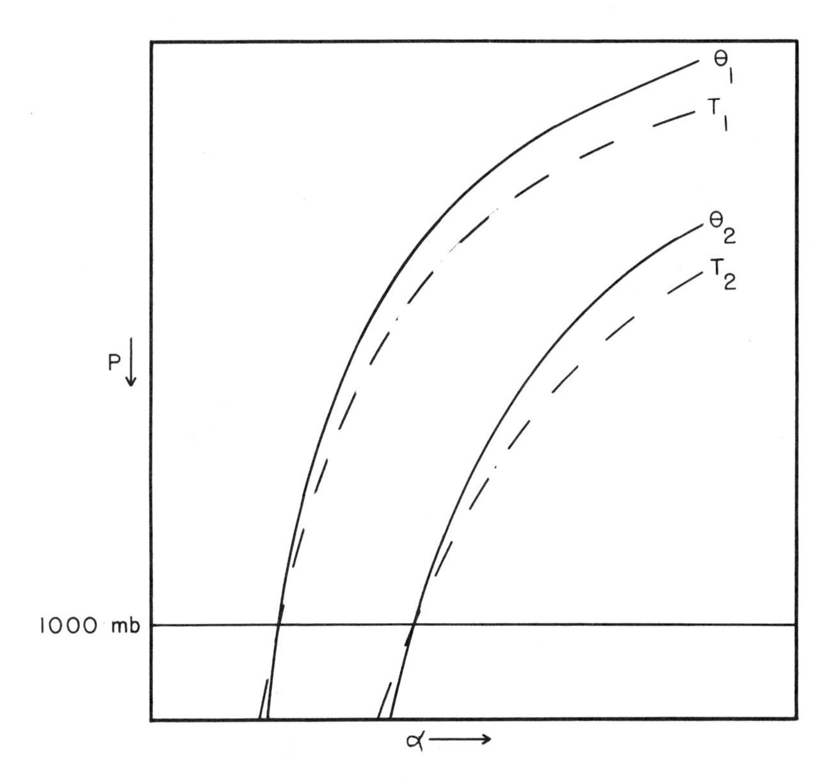

Fig. 3.4. Schematic representation of dry adiabats (solid lines) and isotherms (dashed lines) on an α, $-p$ diagram

A line of constant θ is called a *dry adiabat*. Figure 3.4 shows the schematic distribution of a few adiabats and isotherms on an a, $-p$ diagram. With the aid of dry adiabats one can portray graphically changes of state during adiabatic processes.

3.6 Entropy and the Second Law of Thermodynamics

We shall first make clear the difference between reversible and irreversible processes. In all thermodynamic events it is necessary to consider the changes that occur in the immediate environment of a system, as well as the alterations in the system under consideration. A *reversible process* is one in which each state of the system is an equilibrium state. Thus the system is thought to pass through the process via an infinite number of balanced states which are infinitesimally different from each other. In such a case one can reverse the process and cause the system to return to its original state; the environment will then be found to have returned to its original state also. *Irreversible processes* proceed in finite steps; if the system is restored to its original state by whatever means, the surroundings will have suffered a change from their original conditions. Thus the term "irreversible" does not mean that a system cannot be returned to its original state, but that the system *plus its environment* cannot be thus restored.

An example of an irreversible process is the "expansion into a void" discussed in Section 3.4. If after this isothermal expansion one wishes to restore the gas to its initial condition, it is necessary first to compress the gas to its original volume (thus doing work upon it), and then to cool the compressed gas to its original temperature by transferring heat to its environment. Thus in going back to the original state a certain amount of work is transferred to heat residing in the environment, and the surroundings are no longer in their original state. *All the other processes we have considered were assumed to be reversible.*

We may now explore those properties of the thermodynamic equations that lead to the important concept of entropy. Consider the first law of thermodynamics in the form

$$dh = c_p dT - \alpha dp$$

where we now assume the process is reversible. If we substitute for α from the equation of state and divide by the temperature we get

$$\frac{dh}{T} = c_p \frac{dT}{T} - R\frac{dp}{p}$$

which may also be written

$$\frac{dh}{T} = c_p d(\ln T) - Rd(\ln p) \tag{3.17}$$

The two terms on the right are differentials of functions of thermodynamic variables. In mathematics such differentials of functions are called *exact differentials.* Their importance can be seen easily. If one integrates an exact differential, such as d ($\ln T$), from an initial state where the temperature is T_1 to a final state where the temperature is T_2 the result, $\ln (T_2/T_1)$, depends only upon the initial and final conditions and not at all upon the process by which the change takes place. Such exact differentials are independent of the path followed in, for example, an α, $-p$ diagram. Furthermore, if one integrates an exact differential along some closed path (which necessarily returns to the same point) the result must be zero for any closed path chosen. These are the general considerations behind a theorem which is proved rigorously in advanced calculus:

A necessary and sufficient condition that a differential be exact is that its integral around a closed path be zero.

If we integrate the first law of thermodynamics around some closed path in an α, $-p$ diagram we get

$$\oint dh = \oint c_v dT + \oint p d\alpha$$

where the symbol $\oint$ denotes integration along the given closed curve. But for a perfect gas, $c_v dT$ is an exact differential and its closed integral is zero. Furthermore we know that $\oint p d\alpha$ is the work done during the process and is equal to the area contained within the closed curve on the α, $-p$ diagram. Hence it is not zero and

$$\oint dh = \oint p d\alpha = w \neq 0$$

Thus neither dh nor $pd\alpha$ are exact differentials, and their closed integrals depend upon the path chosen.

The mathematical properties of exact differentials are so valuable that we prefer to put our equations in such form that, if possible, all the differentials are exact. Consider Eq. (3.17); its integral around a closed path is

$$\oint \frac{dh}{T} = \oint c_p d(\ln T) - \oint Rd(\ln p)$$

or

$$\oint \frac{dh}{T} = 0$$

because the two right-hand terms are closed integrals of exact differentials. But this means that dh/T must also be an exact differential since its line integral is zero. Thus dividing by T converts the inexact differential dh into an exact differential.[3]

This is a remarkable result which could not have been anticipated. We may now define a quantity ϕ such that

$$d\phi = \frac{dh}{T} \tag{3.18}$$

From Eq. (3.17) we can conclude that $d\phi$ is a function of two variables of state and is therefore itself a variable of state. This new thermodynamic variable ϕ is called the specific *entropy* $[L^2T^{-2}\theta^{-1}]$ of the system. It arises from purely mathematical considerations, but can be interpreted physically by means of the second law of thermodynamics.

Entropy is related to potential temperature in a particularly simple fashion. If we logarithmically differentiate[4] Poisson's equation (3.16) we get, upon rearranging:

$$c_p \frac{d\theta}{\theta} = c_p \frac{dT}{T} - R\frac{dp}{p}$$

or

$$c_p d(\ln \theta) = c_p d(\ln T) - Rd(\ln p)$$

but the right side of the above is identical with the right side of Eq. (3.17). Therefore with Eq. (3.18) we have

$$d\phi = c_p d(\ln \theta)$$

That is, a change in entropy is a function only of the potential temperature and its change. Integrating,

$$\phi = c_p \ln \theta + \text{const}$$

Thus, apart from constants, the entropy of a system is given by the logarithm of the potential temperature. Since we shall never be concerned with the total entropy, but only with its changes, the additive constant is of no importance. These results show that a process in which the potential temperature remains constant is also one in which the entropy does not change. As a result, adiabatic processes are often referred to as *isentropic* processes. In meteorology we prefer to use potential temperature as a variable rather than entropy.

We shall discuss the physical meaning of entropy in a nonrigorous,

---

[3] The quantity $1/T$ is called an integrating factor.

[4] Logarithmic differentiation is a very useful process which is accomplished by first taking the natural logarithm of a quantity and then differentiating it. For example:

$$d \ln \left(\frac{XY^a}{Z}\right) = d(\ln X + a \ln Y - \ln Z) = \frac{dX}{X} + a\frac{dY}{Y} - \frac{dZ}{Z}$$

heuristic fashion. More detailed discussions are given in textbooks of thermodynamics.[5] This interpretation is derived from the second law of thermodynamics. Although the first law of thermodynamics asserts a quantitative equivalence among various forms of energy, it imposes no further restrictions. However, it is found through experience that whereas work can always be transformed completely into heat, heat cannot be transformed completely into work. Thus nature operates under restrictions which are not included in the first law but are expressed separately in the second law. One form of this empirical principle is the Kelvin-Planck version:

It is impossible to construct an engine which operates in a cycle and which produces no other effect than the extraction of heat from a heat reservoir and the performance of an equivalent amount of work.

This form of the second law can be illustrated by a fluid analogy. A body of water is incapable of doing any work unless it can flow downward through some sort of machine to a lower elevation. In much the same way a quantity of heat in a reservoir at temperature T_1 cannot yield any work unless it flows through some sort of engine and unless part of the heat is deposited in another reservoir at a lower temperature T_2. Thus the "other effect" of the second law is the deposition of some of the heat into a colder reservoir. Since the engine is extracting heat at T_1, converting part of it to work and depositing the remaining heat at T_2, the heat at T_2 is unavailable to the engine for conversion to work. Thus the second law implies a certain degree of unavailability of heat for the production of work. It can be shown that *when an irreversible process takes place some of the energy becomes unavailable for the production of work. This amount of unavailable energy is* $T\Delta\phi$, where $\Delta\phi$ is the difference in entropy between the final and initial states of the system and its surroundings. Thus *entropy is a measure of the unavailability of energy*. Since work is an orderly process of motion, we can also say that *entropy is a measure of disorder in systems*. Furthermore, since all natural processes are more or less irreversible, the second law implies that nature tends towards a state of greater disorder—as stated in the *principle of degradation of energy*. All of these statements are consequences of, and some of them are alternatives to, the second law of thermodynamics.

It is important to remember that the concept of entropy is defined for equilibrium systems, and the change in entropy from one state to another is associated with a reversible process connecting the two

[5] See, for example, M. W. Zemansky, *Heat and Thermodynamics*, New York, McGraw-Hill, 1943.

states. When a change between two given states occurs via an irreversible process, the change in entropy is exactly the same as for a reversible process, but $\int dh/T$ between the two states is not the same for irreversible and reversible processes. Indeed, it is an observed fact that $\Delta\phi < \int dh/T$, where dh is the heat increment in an irreversible process. This suggests that to accomplish a given change in entropy (or state) by an irreversible process requires more heat energy than when a reversible process is involved. This fact in turn suggests the proposition, which can be demonstrated, that reversible processes are the more efficient of the two.

3.7 Summary of Thermodynamic Variables

The fundamental variables of state used in the ideal gas law Eq. (2.8) are pressure, specific volume (or its reciprocal, density), and temperature. No more need be said about these basic quantities. In addition, a large number of other variables of state have been defined as functions of the basic set which are useful in one branch or another of thermodynamics. We shall not need all of these, but certain of them find sufficient place in meteorology to warrant discussion.

We have already defined potential temperature by means of Poisson's equation (3.16) and have interpreted it physically as the temperature a sample of dry air would attain if it were brought adiabatically to a pressure of 1000 mb. Potential temperature is widely used in practical and theoretical meteorology.

We have defined specific entropy in the differential form $d\phi = dh/T$, for a reversible process. Entropy is proportional to the logarithm of potential temperature and is to be interpreted physically as a measure of the unavailability of heat energy for conversion to work.

Internal energy in differential form is $du = c_v dT$. Thus for an ideal gas the internal energy is a function of the temperature only, being a measure of the energy of random motion of the molecules of a system. From the first law of thermodynamics,

$$dh = du + p d\alpha$$

we see that the internal energy is the heat transferred during isosteric processes.

Still another thermodynamic variable used in meteorology is the specific *enthalpy*, which we shall denote in differential form as $db = c_p dT$.

It is a measure of temperature changes also. From another form of the first law of thermodynamics,

$$dh = c_p dT - \alpha dp$$

we see that changes in enthalpy measure the heat transferred during isobaric processes. Since many meteorological processes are more nearly isobaric than isosteric, enthalpy is somewhat preferable to internal energy, although the difference is slight.

With respect to all the variables that are defined differentially—namely entropy, internal energy, and enthalpy—we must measure them from some convenient reference level. We shall not concern ourselves with the somewhat vexatious question of the location of the natural zero points of these functions. It suffices to call them zero at some arbitrary state. The reference state 200° K and 1000 mb that has been suggested is indeed convenient.

PROBLEMS

1. A sample of aluminum having a mass of 100 g, a specific heat capacity of 0.20 cal g^{-1} °C^{-1}, and a temperature of 100° C is put into 400 g of water at 20° C contained in a copper vessel with a mass of 100 g and a specific heat capacity of .095 cal g^{-1} °C^{-1}. What is the temperature of the system at thermal equilibrium?

2. Compute the specific energy changes in cal g^{-1} for the following processes: (a) Heating of dry air at constant volume from 0° C to +10° C. (b) Isothermal expansion of dry air from a volume of 900 cm^3 g^{-1} to 1000 cm^3 g^{-1} at a temperature of +7° C. (c) Increasing the horizontal speed of air from 10 ms^{-1} to 50 ms^{-1}. *All of these changes are possible meteorologically.* Compare the numerical results and indicate which process is relatively unimportant.

3. Show that the equation for an adiabatic process may be written

$$p\alpha^{\gamma} = \text{const} \quad \text{or} \quad \alpha T^{c_v/R} = \text{const}$$

where $\gamma = c_p/c_v$.

4. Calculate and construct the dry adiabats of an α, $-p$ diagram for air for potential temperatures 200° K, 300° K, and 400° K. Do this on the diagram you constructed for Problem 4 of Chapter 2.

5. Calculate the changes in specific internal energy, specific enthalpy, and specific entropy of dry air for the following processes:

(a) an isothermal expansion from α = 900 cm^3 g^{-1} to 950 cm^3 g^{-1}; (b) an isobaric heating from −10° C to +10° C; (c) an isosteric cooling from 0° C to —10° C; and (d) an adiabatic compression from $\alpha = 900$ cm^3 g^{-1} to 850 cm^3 g^{-1}.

6. A sample of 50 g of dry air is initially at a pressure of 1000 mb and a temperature of 280° K. Heat is added in an isobaric process during which the sample expands by 10 per cent of its original volume. Calculate the final temperature of the air, the work done against the surroundings (in joules), and the amount of heat added (in calories).

7. Consider an ideal gas confined in an insulated chamber of volume V_1 at a pressure p_1 and temperature T_1. Let this gas expand into an insulated evacuated vessel of volume V_2 until it fills both vessels. Show that the change in entropy which occurs is

$$\Delta\phi = R \ln \frac{V_1 + V_2}{V_1}$$

CHAPTER 4

THE THERMODYNAMICS OF WATER VAPOR AND MOIST AIR

4.1 Isotherms on an α, e Diagram

As we have seen, the isotherms of an ideal gas on a plot of specific volume versus pressure are rectangular hyperbolas. However, physical conditions on our planet are such that water may exist in gaseous, liquid, and solid phases. Whenever water vapor is sufficiently far from condensation or sublimation, it is nearly ideal, so that its isotherms are also close approximations of hyperbolas.

The marked deviations from the ideal gas law during condensation may be studied with the aid of Fig. 4.1. Here specific volume is plotted against water substance pressure (e). We allow e to increase upward in this diagram since there is no need to make the pressure scale roughly correspond with height in the atmosphere. Consider the point marked A. Here all the water is in the form of vapor. As we increase the pressure while keeping the temperature constant, the specific volume will diminish in approximate accord with the ideal gas law. Thus the state of the sample will be represented on the α, e diagram by an isotherm which is nearly hyperbolic. Eventually a state is reached, represented by point B, where a slight increase in pressure will cause some of the vapor to condense to liquid water. In this state only an infinitesimally small increment of pressure is needed to condense all the vapor into liquid at a constant temperature. Thus as the state of the sample moves from point B to C the pressure remains essentially constant, and the specific volume changes from that of the vapor at the given pressure and temperature to that of the liquid at the same pressure and temperature. This constant pressure is called the ***saturation vapor pressure*** at that temperature. At C all of the sample is liquid, which is only slightly compressible. Consequently, the isotherm thereafter rises almost vertically—that is, the specific volume remains nearly constant as the pressure increases.

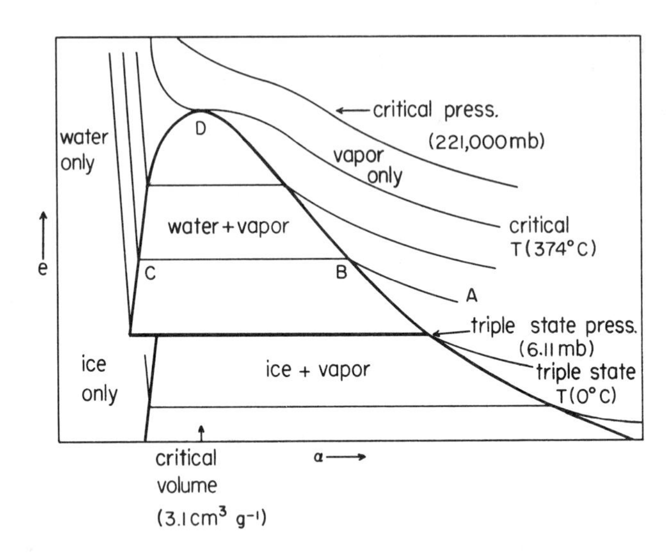

FIG. 4.1. Phase diagram of water substance with specific volume, α, and pressure, e, of water substance as coordinates. The heavy lines separate the various phases.

To the right of B the sample is completely gaseous. Between B and C vapor and liquid coexist, and to the left of C the sample is completely liquid. Because it delineates those areas in which various phases may exist, such a diagram is called a *phase diagram*. The heavy lines separate the various phase areas.

Similarly, if one follows the isotherm labelled "triple-state isotherm," a special situation is encountered. As the vapor is compressed, a point is again reached where a change of phase occurs, but this time the vapor condenses into water and ice. Equilibrium exists simultaneously among vapor, liquid, and solid phases along the horizontal portion of this isotherm. This condition occurs for a fixed temperature and pressure (that is, a specific state of the vapor), hence the name *triple state*. The pressure and temperature of the triple state for water are 6.11 mb and 0.01° C. For pressures and temperatures below the triple state, liquid does not occur and equilibrium exists between ice and vapor only. Supercooled water is an important exception which will be discussed later.

Another special case is the isotherm which passes through the point *D*. To the right of *D* along this isotherm only vapor can exist, while to the left of *D* only liquid can exist. Exactly at *D* the distinction between liquid and gas vanishes. Here the surface tension of the interface between liquid and vapor becomes zero; indeed the interface itself disappears. This anomalous state is called the *critical point*; for water it is found at a pressure of 221 bars and a temperature of 374° C. Thus water in the atmosphere is always far below the critical point.

The general significance of the critical point is this: Above the critical temperature it is impossible to liquify a gas by compressing it; above the critical pressure it is impossible to liquify a gas by cooling it. This explains the great difference in behavior of the permanent atmospheric gases on the one hand and water vapor on the other. The permanent gases do not condense because their critical temperatures are much below those of the atmosphere. Water vapor does condense because both its critical temperature and pressure are far above those found in the atmosphere. In addition there is often enough vapor (sufficiently small specific volume) to allow the state to lie in the multiple-phase portion of the α, e diagram.

Carbon dioxide also has a critical temperature (31° C) and critical pressure (74 bars) sufficiently high to permit condensation. However, the amount of CO_2 present is so small (specific volume is so high) that its atmospheric state always lies far to the right of the multiple-phase region. Thus CO_2 never condenses naturally in our atmosphere. The planet Venus, however, has so much more CO_2 in its atmosphere that it is not impossible for condensation to take place there.

4.2 Thermal Properties of Water Substance

Ice, the solid state of H_2O, is capable of existing in several different crystalline phases. Because only one of these, called ice I, can exist under normal atmospheric conditions, we may ignore the others. The specific volume of ice at 0° C is 1.091 cm^3 g^{-1}, which is larger than the specific volume of water at the same temperature, causing ice to float. It is most unusual for the solid phase of a substance to be less dense than the liquid phase, and this anomalous property of H_2O has far-reaching consequences for physical conditions on our planet. The specific volume of ice varies slowly with temperature, but for our purposes may be considered constant.

The specific heat capacity of ice at 0° C is 0.503 cal g^{-1} °C^{-1}. It, too, varies with temperature (see Table 3.1) but at so slow a rate that we may safely ignore it.

Liquid H_2O has a specific volume very close to 1.000 cm^3 g^{-1} from 0° C to 8° C, and even at 40° C is within 0.8 per cent of this value, so that its variations may likewise be ignored. The specific heat capacity of water is by definition exactly 1 cal g^{-1} °C^{-1} at 15° C. The variation with temperature (see Table 3.1) is again so small that we shall neglect it.

In discussing the thermal properties of water vapor we shall assume the vapor satisfies the equation of state of a perfect gas. In the range of conditions with which we shall deal this is a very good approximation. Since the saturation vapor pressure over water is the highest vapor pressure possible at any temperature, and since the departures from an ideal gas increase with pressure, the most severe test for conformity to the ideal gas law is for the saturation vapor pressure. Table 4.1 gives

TABLE 4.1

DEPARTURES OF WATER VAPOR FROM IDEAL CONDITIONS

Temperature (°C)	e_s (mb)	α_v, ideal/α_v
−10	2.86	1.0003
0	6.11	1.0005
10	12.27	1.0008
20	23.37	1.0012
30	42.43	1.0018
40	73.78	1.0027

the ratio of the specific volume from the ideal gas law (α_v, ideal) to the observed specific volume (α_v) for the indicated saturation vapor pressure (e_s) as a function of temperature. It is clear that water vapor satisfies the first condition for an ideal gas to within a fraction of 1 per cent throughout the range of meteorological conditions.

The specific heat capacities of water vapor at low vapor pressures are

$$c_{vv} = 3R_v = 0.331 \text{ cal g}^{-1}\,°\text{C}^{-1} \text{ and } c_{pv} = 4R_v = 0.441 \text{ cal g}^{-1}\,°\text{C}^{-1}$$

where the subscript v means the value for water vapor. Because these quantities may vary appreciably with temperature, water vapor does not satisfy this second condition for an ideal gas. However, we shall always be concerned with a mixture of air and a small amount of vapor, so the specific heat capacities of the mixture cannot be seriously affected by variations in the properties of the vapor.

4.3 The Equation of State of Moist Air

We have found that dry air and water vapor separately satisfy the equation of state of an ideal gas with sufficient accuracy for our purposes. We may now ask, what is an appropriate form of the equation of state for a mixture of the two? We found on page 18 that an appropriate mean molecular weight for a mixture is obtained by taking a weighted harmonic mean of the molecular weights of the individual constituents. We may look upon dry air as one constituent of the mixture and vapor as the other. The mean molecular weight of moist air is then given by

$$\frac{1}{\bar{m}} = \frac{1}{M_d + M_v}\left(\frac{M_d}{m_d} + \frac{M_v}{m_v}\right)$$

where a subscript d refers to dry air and a subscript v refers to water vapor. M denotes the mass of each constituent in a mixture; m denotes the molecular weight of that constituent. The absence of a subscript means that we are dealing with the characteristics of the entire mixture. Expressing $\bar{m}$ in terms m_d gives

$$\frac{1}{\bar{m}} = \frac{1}{m_d}\frac{M_d}{M_d + M_v}\left(1 + \frac{M_v/M_d}{m_v/m_d}\right) \tag{4.1}$$

The ratio M_v/M_d is the mass of vapor per unit mass of dry air. This nondimensional measure of the moisture content of air is called the *mixing ratio* because it indicates how much of the two constituents must be mixed in order to match a given sample of moist air. We represent the mixing ratio by

$$w \equiv \frac{M_v}{M_d} = \frac{\rho_v}{\rho_d} = \frac{m_v}{m_d}\frac{e}{p - e} \approx \frac{m_v}{m_d}\frac{e}{p} \tag{4.2}$$

The ratio m_v/m_d is a known constant which we shall replace with the symbol ϵ. If we make these substitutions and divide the numerator and denominator of the right side of Eq. (4.1) by M_d we get:

$$\frac{1}{\bar{m}} = \frac{1}{m_d}\frac{1 + w/\epsilon}{1 + w}$$

Thus the mean molecular weight of moist air has been expressed in terms of the molecular weight of dry air, the variable mixing ratio, and the constant ϵ. We may now write the equation of state of moist air as

$$p\alpha = \frac{R^*}{m_d}\left(\frac{1 + w/\epsilon}{1 + w}\right)T$$

The above expression is exactly the same as the equation of state for dry air except for the factor in parentheses, which is obviously a correction to be applied to the specific gas constant for dry air to obtain the value of the gas constant for moist air containing an amount of vapor given by w. However, this would mean a variable gas "constant," which is awkward. We prefer to apply the correction to the temperature itself and define a new temperature:

$$T^* \equiv \frac{1 + w/\epsilon}{1 + w} T \tag{4.3}$$

This new temperature is called the *virtual temperature*. It is a fictitious temperature that satisfies the equation of state for *dry* air:

$$p\alpha = \frac{R^*}{m_d} T^*$$

The virtual temperature is obviously the temperature that dry air would have if its pressure and specific volume were equal to those of a given sample of moist air. Accordingly we may continue to use the equation of state for dry air, provided that we do not use the observed temperature of moist air but a slightly different one—a very useful technique in meteorological practice.

Since water vapor is less dense than air, its presence always increases the specific volume. Hence the virtual temperature is always higher than the observed temperature. The known value of ϵ is 0.622, therefore Eq. (4.3) may be written

$$T^* = \frac{1 + 1.609w}{1 + w} T$$

Since w never exceeds 40 g of water vapor per kg of dry air (4×10^{-2}), the difference between the two temperatures is never more than 7° K and is usually less than 1° K.

4.4 Changes of Phase and Latent Heats

Whenever a substance changes phase (melts, freezes, evaporates, condenses, or sublimes), a quantity of heat must be supplied to or taken away from the substance even though the temperature remains constant. This quantity is called the *latent heat of the phase change*. It is measured in cal g^{-1} and has dimensions $[L^2T^{-2}]$. Let the two phases in a transition be denoted by subscripts 1 and 2. The latent heat of a change from phase 1 to 2 will be represented by L_{12} and the reverse change by L_{21}. From the first law of thermodynamics, during a phase change

$$dh = du + e_s d\alpha$$

where by definition

$$L_{12} = \int_1^2 dh$$

and e_s and T remain constant during the integration. Thus

$$L_{12} = u_2 - u_1 + e_s(\alpha_2 - \alpha_1) \tag{4.4}$$

We have shown that u is a function of temperature only for the ideal vapor phase, but for liquids and solids the internal energy depends upon the specific volume as well. If phases 1 and 2 are water and vapor respectively, L is the latent heat of evaporation (positive) and e_s is the saturation vapor pressure. When phases 1 and 2 are ice and vapor respectively, L is the latent heat of sublimation (positive) and e_s is again the saturation vapor pressure. In the remaining case when phases 1 and 2 are ice and liquid respectively, L is the latent heat of melting (positive) and e_s is the "saturation" or equilibrium pressure of the mixture of ice and water. At the triple state these saturation pressures are equal, therefore when Eq. (4.4) is written for each of these changes and the three equations are combined we get:

$$L_{\text{sublimation}} = L_{\text{melting}} + L_{\text{evaporation}}$$

At other temperatures this is only approximately true since the latent heats vary with temperature at different rates.

We now calculate from theory the dependence of $L_{\text{evaporation}}$ on temperature. From Eq. (4.4)

$$L_{\text{evaporation}} = u_v - u_w + e_s(\alpha_v - \alpha_w)$$

The specific volume of water is very small compared to that of vapor, so we may neglect α_w. Substituting from the equation of state for α_v,

$$L_{\text{evaporation}} = u_v - u_w + R^*T/m_v$$

If we now differentiate with respect to temperature we obtain

$$\frac{dL_{\text{evaporation}}}{dT} = \frac{du_v}{dT} - \frac{du_w}{dT} + \frac{R^*}{m_v}$$

By Eq. (3.3)

$$c_{vv} = \left(\frac{dh}{dT}\right)_\alpha = \left(\frac{du_v}{dT}\right)$$

and the specific heat of the liquid is

$$c_w = \frac{dh}{dT} = \frac{du_w}{dT} + e\,\frac{d\alpha_w}{dT} \approx \frac{du_w}{dT}$$

since $d\alpha_w/dT$ is negligible.

Thus

$$\frac{dL_{\text{evaporation}}}{dT} \approx c_{vv} - c_w + \frac{R^*}{m_v}$$

or, for an ideal gas,

$$\frac{dL_{\text{evaporation}}}{dT} = c_{pv} - c_w \tag{4.5}$$

Thus *the rate of change of the latent heat of evaporation with absolute temperature is equal to the difference between the specific heat at constant pressure of the vapor and the specific heat of the liquid.* When numerical values are inserted in Eq. (4.5) we obtain

$$\frac{dL_{\text{evaporation}}}{dT} = -0.566 \text{ cal g}^{-1}\,{}^{\circ}\text{C}^{-1}$$

The above value is correct within less than 1 per cent over the meteorological range of temperature. Since at 0° C the latent heat of evaporation is 597.3 cal g^{-1}, this amounts to a rate of change of less than 0.1 per cent per °C and often may be neglected.

A similar argument may be given for the latent heat of sublimation with the result that an even smaller variation with temperature is found. Furthermore, melting occurs in the atmosphere over such a narrow range of temperatures near 0° C that we need not concern ourselves with the variation with temperature of the latent heat of melting. Thus our final result is: *for most meteorological purposes, the latent heats of water substance may be assumed to be constant.* At 0° C the values are

$$L_{\text{evaporation}} = 597.3 \text{ cal g}^{-1}$$

$$L_{\text{sublimation}} = 677.0 \text{ cal g}^{-1}$$

$$L_{\text{melting}} = 79.7 \text{ cal g}^{-1}$$

4.5 The Clausius-Clapeyron Equation

We now derive an important equation, based upon the concept of entropy, relating the saturation vapor pressure to the latent heat of a phase change. If a phase change occurs reversibly, no alteration in the entropy of the system can occur except through the addition or subtraction of heat. Thus the heat added is

$$L_{12} = \int_1^2 dh = \int_1^2 T d\phi$$

Since a reversible phase change is observed to be isothermal this may be written

$$L_{12} = T(\phi_2 - \phi_1) \tag{4.6}$$

The problem now is to evaluate $\phi_2 - \phi_1$. We shall do this by considering a cycle of reversible changes that carries a substance through two phase changes back to the original state. The cycle (Fig. 4.2) is as follows:

State A, where the substance is in Phase 1 at equilibrium pressure, temperature, and volume e_s, T, α_1.

First transition—a slight increase in temperature, ΔT, without any change of phase leading to:

State B, where the substance is in Phase 1 at $(e_s + \Delta e_s)$, $(T + \Delta T)$, $(\alpha_1 + \Delta \alpha_1)$.

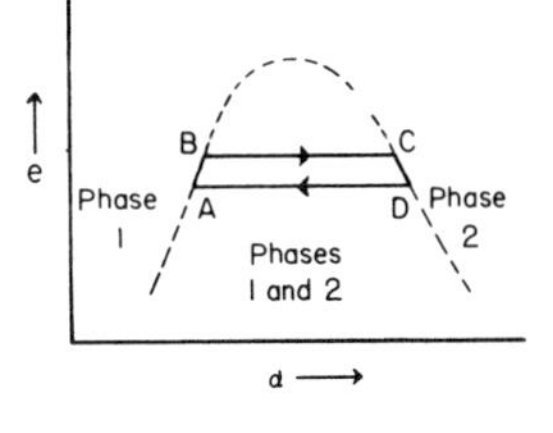

FIG. 4.2. A cycle of phase changes on an α, e diagram

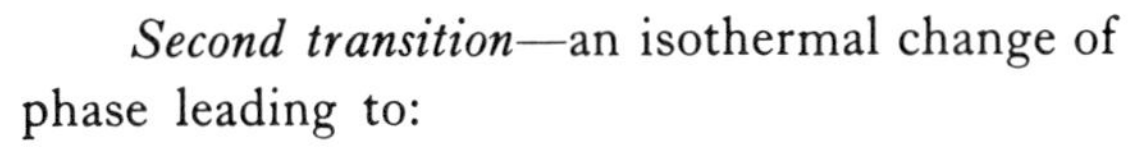

Second transition—an isothermal change of phase leading to:

State C, where the substance is in Phase 2 at $(e_s + \Delta e_s)$, $(T + \Delta T)$, $(\alpha_2 + \Delta \alpha_2)$.

Third transition—a slight decrease of temperature, $-\Delta T$, without any change of phase leading to:

State D, where the substance is in Phase 2 at e_s, T, α_2.

Fourth transition—an isothermal change of phase back to State A.

Since at all stages in this cycle the substance is at the equilibrium pressure for a transition between Phases 1 and 2, we may write the first law of thermodynamics in the form

$$dh = Td\phi = du + e_s d\alpha$$

If we integrate this around the closed cycle of Fig. 4.2 we get

$$\oint Td\phi = \oint e_s d\alpha \tag{4.7}$$

since the closed line integral of du is zero. The right side of Eq. (4.7) is the area enclosed by the cycle and is approximately $(\alpha_2 - \alpha_1)\,\Delta e_s$. Thus

$$\oint Td\phi = (\alpha_2 - \alpha_1)\,\Delta e_s \tag{4.8}$$

To evaluate the left side consider the fact that

$$\oint d(T\phi) = \oint Td\phi + \oint \phi dT = 0$$

since $d(T\phi)$ is an exact differential. Therefore,

$$\oint Td\phi = -\oint \phi dT$$

We now evaluate this integral along each of the four parts of the cycle. From State A to State B we get

$$-\int_A^B \phi dT = -\phi_1 \Delta T$$

where ϕ_1 is the entropy of Phase 1 at a state between A and B. When we let ΔT approach zero, this will become identical with the entropy of Phase 1 at e_s, T, α_1. For the second transition

$$-\int_B^C \phi dT = 0$$

since the process is isothermal. For the third transition

$$-\int_C^D \phi dT = \phi_2 \Delta T$$

where in the limit as ΔT approaches zero, ϕ_2 will be the entropy of Phase 2 in the state e_s, T, α_2. Finally for the fourth transition

$$-\int_D^A \phi dT = 0$$

since the process is isothermal. Thus we find that

$$\oint T d\phi = -\oint \phi dT = (\phi_2 - \phi_1)\,\Delta T$$

and from Eq. (4.8) we get

$$(\phi_2 - \phi_1)\,\Delta T = (\alpha_2 - \alpha_1)\,\Delta e_s$$

We now divide by ΔT and let ΔT approach zero. This gives

$$\phi_2 - \phi_1 = (\alpha_2 - \alpha_1)\frac{de_s}{dT}$$

When this is substituted in Eq. (4.6) we get

$$L_{12} = (\alpha_2 - \alpha_1)\,T\frac{de_s}{dT}$$

and solving for de_s/dT we find

$$\frac{de_s}{dT} = \frac{L_{12}}{T(\alpha_2 - \alpha_1)} \tag{4.9}$$

Equation (4.9) is called the *Clausius-Clapeyron equation*, or sometimes simply the Clapeyron equation, after B. P. Clapeyron, who first obtained it in 1832, and Rudolf Clausius who later rederived it from a more modern viewpoint. This expression gives the slope of the curve of saturation vapor pressure versus temperature as a function of the latent heat of the phase change, the temperature, and the difference in specific volume of the two phases. If L_{12}, α_1, and α_2 are known as functions of temperature, then Eq. (4.9) can be integrated to give e_s as a function of temperature.

We shall integrate Eq. (4.9) for the cases of evaporation and sublimation, for which we have already shown that the latent heat is nearly constant. Thus Phase 2 will be the vapor phase and $\alpha_2 \gg \alpha_1$ so that Eq. (4.9) may be written

$$\frac{de_s}{dT} = \frac{L_{12}}{T\alpha_2} = \frac{m_v L_{12} e_s}{R^* T^2}$$

or

$$\frac{de_s}{e_s} = \frac{m_v L_{12}}{R^*} \frac{dT}{T^2}$$

Therefore

$$\ln e_s = -\frac{m_v L_{12}}{R^* T} + \text{const}$$

where the constant of integration may be evaluated by measuring the saturation vapor pressure at some standard temperature such as 273° K. Since $e_s = 6.11$ mb at $T = 273°$ K for both evaporation and sublimation we may write

$$\ln \frac{e_s}{6.11} = \frac{m_v L_{\text{evaporation}}}{R^*} \left(\frac{1}{273} - \frac{1}{T}\right)$$

for the evaporation curve and

$$\ln \frac{e_s}{6.11} = \frac{m_v L_{\text{sublimation}}}{R^*} \left(\frac{1}{273} - \frac{1}{T}\right)$$

for the sublimation curve.

These two results, plus the corresponding one for melting which we have not derived, permit us to plot another kind of phase diagram with T and e as coordinates, as in Fig. 4.3. The evaporation curve begins at the triple point and curves upward to the right. Along it water and vapor are in equilibrium, while for states to its left only water can exist and to its right only vapor can exist. The evaporation curve ends at the critical point where the distinction between liquid and vapor ceases. The boiling point occurs where the vapor pressure becomes equal to the total atmospheric pressure.

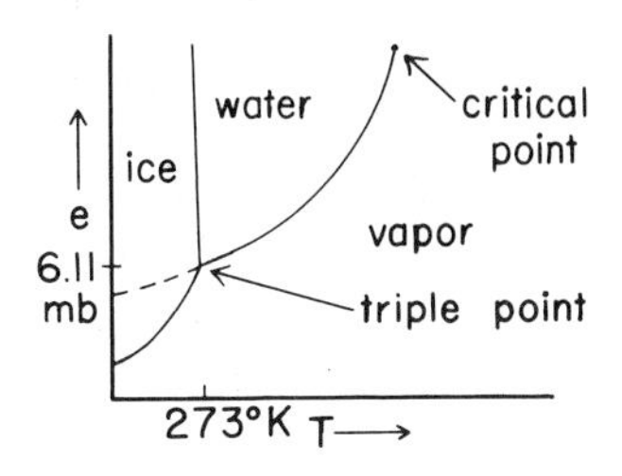

FIG. 4.3. T, e phase diagram for water substance

The sublimation curve also begins at the triple point and curves downward to the left. From Eq. (4.9) we can see that near the triple point it must have a steeper slope than the evaporation curve, because the latent heat of sublimation is larger than that of evaporation. For states along the sublimation curve, ice and vapor are in equilibrium. To its right only vapor can exist and to its left only ice can exist (except for supercooled water which will be discussed later).

Although we have not derived the melting curve from the Clausius-Clapeyron equation, it is easy to see qualitatively what it must be like. Since the specific volume of ice (α_1) is only slightly larger than that of water (α_2), the right side of Eq. (4.9) is large in magnitude. If the two specific volumes were equal, the melting curve would be a straight vertical line in Fig. 4.3. Since α_1 is slightly larger than α_2, the slope is slightly negative and the melting line is only slightly curved. Along the melting curve, water and ice are in equilibrium, while to its right only water can exist and to its left only ice can exist.

It is a well-verified fact that when water is cooled below 273° K it often does *not* freeze, for reasons that more properly concern the field of physical meteorology. It suffices to point out the observation and to recognize that in the free atmosphere this failure to freeze is the general

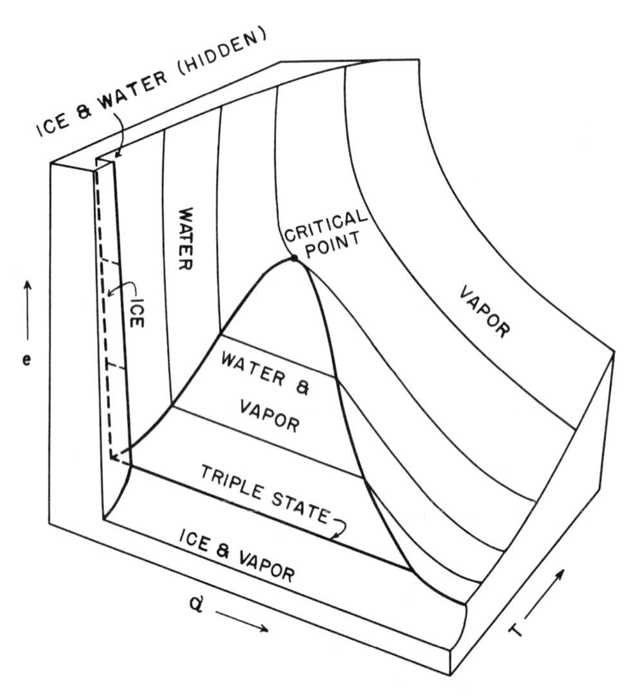

FIG. 4.4. Three-dimensional thermodynamic surface of water substance

rule rather than the exception.[1] Liquid water below 0° C is called *supercooled water*: The equilibrium between supercooled water and its saturated vapor is not given by the sublimation curve but the dashed smooth extension to the evaporation curve in Fig. 4.3. Note that the saturation vapor pressure over supercooled water is higher than that over ice at the same temperature. As we have seen, this is because

[1] Laboratory experiments indicate that H_2O may remain liquid until about —39° C is reached. At that temperature spontaneous freezing always occurs. Between —39° C and 0° C liquid water may exist. The closer to 0° C the more likely it is that water rather than ice will be present.

$L_{\text{sublimation}} > L_{\text{evaporation}}$. This means that if we introduce a crystal of ice into a cloud of supercooled water droplets, the vapor will begin rapidly to condense on the ice. Thus the crystal will grow in size and fall out as a small snowflake. This process is the physical basis of the Bergeron-Findeisen mechanism for the initiation of precipitation.

We are now in a position to represent completely the various states which water substance may adopt. To do this we must specify the relationship among e, α, and T for all states and phases. This may be done by constructing a three-dimensional plot of e, α, and T. Each state of water substance will be a point in this three-dimensional space, and all the possible states of water substance will define a surface in this space. The resultant *thermodynamic surface* of H_2O is shown as a perspective drawing in Fig. 4.4. Each area on the surface where the isotherms are horizontal is an area of equilibrium between two or more phases. Note that if the thermodynamic surface is viewed parallel to the T axis, it reduces to the α, e diagram of Fig. 4.1, and if viewed parallel to the α axis it reduces to the T, e diagram of Fig. 4.3. Thus we have already studied front and side elevations of the three-dimensional thermodynamic surface.

It is difficult to gain an appreciation of the thermodynamic surface from a drawing on two-dimensional paper. The construction of a three-dimensional clay model is very helpful.

4.6 Adiabatic Processes of Saturated Air

We have already discussed adiabatic processes in dry air and have found that the temperature is governed by Poisson's equation (3.16)—that is, the potential temperature remains constant during dry adiabatic processes. The results of Problem 5 of this chapter indicate that even for moist unsaturated air the completely dry adiabatic result is satisfactory. As soon as one allows condensation of the vapor to occur, however, large differences appear because the latent heat of condensation is released. The rate at which saturated air cools as it expands adiabatically is smaller than the rate at which unsaturated air cools adiabatically, because part of the cooling is canceled by the latent heat released. We may think of two extreme cases in the atmosphere. In the first case we shall suppose that all the condensation products (water droplets or ice crystals) remain suspended in the air. Thus any atmospheric sample under consideration consists always of the same material and the processes through which the sample may go can be truly reversible. The latent heat released by the phase change will warm the dry air, the water vapor, and the condensation products. Such a process is adiabatic in the sense that no heat is added from outside the air parcel, even though latent heat appears as sensible heat within the parcel. This is a *moist-adiabatic* or *saturated-adiabatic process.*

In the second case we shall suppose that all the condensation products which may form fall out of the air parcel immediately. Thus the sample changes in mass and composition, and the process is not reversible. For example, consider a parcel of saturated air which initially is rising and expanding as the surrounding pressure decreases. Latent heat will be released at a certain rate as condensation products form and drop out. If the parcel later changes direction and begins to sink, the compression will cause a rise in temperature. This warming cannot be put back into latent heat because the water (or ice) is no longer present to be evaporated. Thus the parcel will increase in temperature at the dry-adiabatic rate, which is different from the rate at which it cooled during the expansion. In this case the parcel cannot be returned to its original state without alterations in its environment, a condition that represents the essence of irreversibility. Since the condensed water that falls out carries some heat with it, this process is not truly adiabatic, and accordingly is known as a *pseudoadiabatic process.* The latent heat released during such a process heats the dry air and the vapor it contains, but not the condensation products.

The real atmospheric situation often lies between these two extremes, since some of the condensed water may drop out and some of it remains suspended as cloud particles in the air. Fortunately the mixing ratio of air is small so the amount of condensed water that can fall out cannot carry much heat with it and the subsequent effects of its absence will be small. For example, suppose 5 g of water condense out of 1 kg of air during a moist-adiabatic expansion. This is a fairly large amount of condensation. In any further cooling, ΔT, the ratio of the heat which could be supplied by these 5 g of water to that supplied by the remaining 1 kg of air is

$$\frac{5\text{ g} \times c_w \times \Delta T}{10^3\text{ g} \times c_v \times \Delta T} = \frac{5 \times 1.0}{10^3 \times 0.17} = \frac{1}{34}$$

Thus even if such a large amount of condensed water were to drop out, its absence would have only a 3 per cent effect on subsequent processes. This is negligible in itself, but because we have neglected from the outset nonadiabatic effects such as radiation and turbulent mixing, we are especially justified in neglecting this small difference between the moist-adiabatic and pseudoadiabatic processes. We conclude that *the rate at which rising saturated air cools in a pseudoadiabatic process is essentially equal to the cooling rate in a truly moist-adiabatic process.*

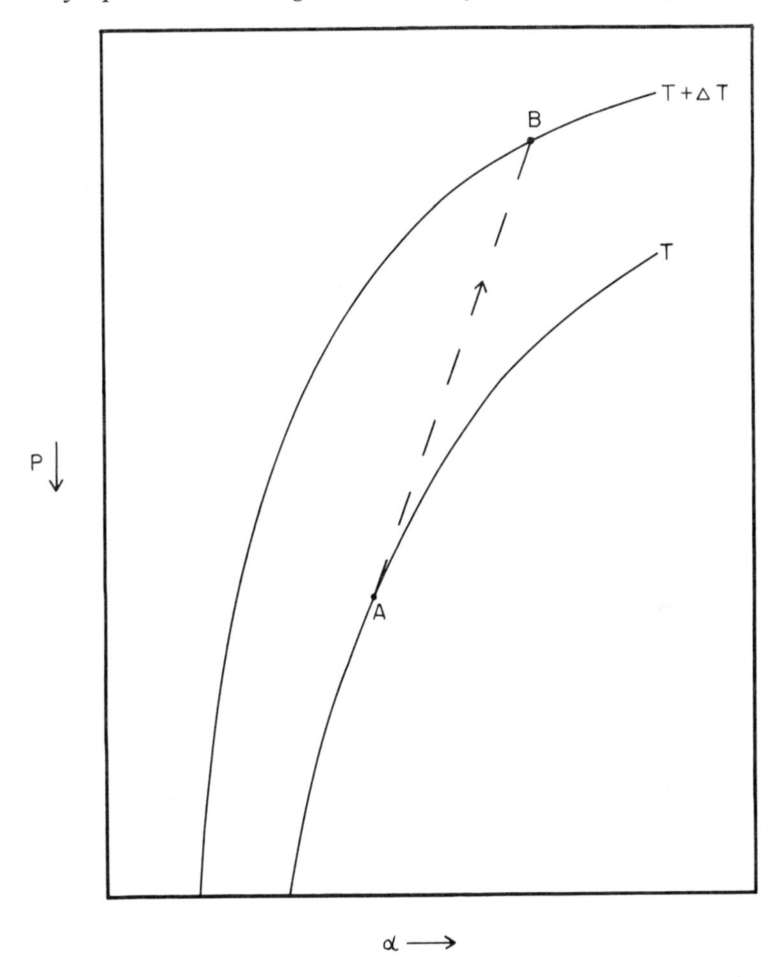

FIG. 4.5. A schematic pseudoadiabatic expansion from A to B

We shall now derive an equation for the pseudoadiabatic process of saturated air. Consider a parcel consisting of 1 g of dry air and w_s grams of saturated vapor. The initial state of this parcel is given by its pressure and temperature (p and T) and the amount of vapor, w_s. Let it undergo a small adiabatic expansion to the state $p + dp$, $T + dT$, $w_s + dw_s$, where dp, dT, and dw_s are all negative. This is represented graphically in Fig. 4.5. The condensation of $- dw_s$ grams of water vapor releases a quantity of heat given by $-L\,dw_s$. We shall think of this amount as heating the moist air. Since $w_s \ll 1$ the amount of heat taken up by the remaining vapor is small compared to that taken up by the dry air. Therefore we shall assume all the heat goes into the 1 g of dry air. From the first law of thermodynamics we find, approximately:

$$-L dw_s = c_p dT - \frac{R^* T}{m_d} \frac{d(p - e_s)}{p - e_s} \tag{4.10}$$

Since $w_s = \epsilon e_s/(p - e_s)$ and e_s is a known function of temperature only, Eq. (4.10) is a differential relationship between p and T during a pseudo-adiabatic process. It may be integrated numerically or graphically and the results tabulated. A sample of results which have been obtained are given in Table 4.2.

TABLE 4.2

PRESSURE (mb) AND TEMPERATURE (°C) ALONG SELECTED PSEUDOADIABATIC LINES

t°C \ θ_e°K	309.6	299.6	290.8	283.4
+12	1000	1093		
+ 8	912	1000	1087	
+ 4	834	917	1000	1082
0	766	844	923	1000
− 4	706	781	855	928
− 8	652	723	793	862
−12	604	671	738	804

Equation (4.10) may be simplified further by neglecting e_s compared to p wherever the combination $p - e_s$ appears. No great error results. If we divide by T, Eq. (4.10) becomes

$$-L \frac{dw_s}{T} = c_p \frac{dT}{T} - \frac{R^*}{m_d} \frac{dp}{p}$$

But logarithmic differentiation of Poisson's equation (3.16) yields

$$c_p \frac{d\theta}{\theta} = c_p \frac{dT}{T} - \frac{R^*}{m_d} \frac{dp}{p}$$

Therefore the equation for a pseudoadiabatic process may be written

$$-L \frac{dw_s}{T} = c_p \frac{d\theta}{\theta} \tag{4.11}$$

Since dw_s is negative (condensation) during an expansion, Eq. (4.11) shows that during a pseudoadiabatic expansion the potential temperature is not conserved, but increases. Obviously this is because latent heat is being released. The temperature continues to decrease during the expansion (otherwise no condensation would take place), but less rapidly than for unsaturated air.

Equation (4.11) may be integrated formally if one makes a further simplification.

Consider

$$d\left(\frac{w_s}{T}\right) = \frac{w_s}{T}\left(\frac{dw_s}{w_s} - \frac{dT}{T}\right) \tag{4.12}$$

We shall show that the first term in parentheses is large compared to the second term in parentheses. The ratio of these two terms is

$$\frac{T}{w_s}\frac{dw_s}{dT}$$

But

$$w_s \approx \frac{\epsilon e_s}{p}$$

and logarithmic differentiation followed by multiplication by T gives

$$\frac{T}{w_s}\frac{dw_s}{dT} = \frac{T}{e_s}\frac{de_s}{dT} - \frac{T}{p}\frac{dp}{dT}$$

Substitution of the Clausius-Clapeyron equation into the first term on the right yields

$$\frac{T}{w_s}\frac{dw_s}{dT} = \frac{m_v L}{R^* T} - \frac{T}{p}\frac{dp}{dT}$$

where dp/dT is the rate at which pressure changes with temperature during a pseudoadiabatic process. This rate of change may be estimated from Table 4.2. In the vicinity of 0° C we find, from the first column of Table 4.2,

$$\frac{T}{w_s}\frac{dw_s}{dT} \approx 20 - 6 = 14$$

This numerical result varies somewhat with p and T, and increases as p and T decrease. Thus the ratio is large throughout the atmosphere. We may therefore neglect the second term in parentheses of Eq. (4.12) compared to the first term. Thus

$$d\left(\frac{w_s}{T}\right) \approx \frac{dw_s}{T}$$

and Eq. (4.11) may be written

$$-Ld\left(\frac{w_s}{T}\right) = c_p\frac{d\theta}{\theta}$$

This expression is now integrable and we obtain

$$-\frac{Lw_s}{T} = c_p \ln\theta + K$$

where K is a constant of integration. We may evaluate K at some very low temperature where $w_s \approx 0$. At that point let the potential temperature be denoted by θ_e. Then

$$\theta = \theta_e e^{\frac{-Lw_s}{c_pT}} \tag{4.13}$$

The quantity θ_e is the potential temperature of a parcel of air when its mixing ratio has been reduced to zero and all the latent heat turned into sensible heat. A parcel may attain the temperature θ_e by the following process. From an initial saturated state T and p let the air expand pseudoadiabatically until all the vapor has condensed, released its latent heat, and fallen out. Then let the air be compressed dry adiabatically until a pressure of 1000 mb is reached. It will then have temperature θ_e. This is a kind of potential temperature since it is reached by a dry adiabatic compression to 1000 mb. All the latent heat has then been converted to its equivalent in sensible heat. Thus θ_e is called the *equivalent potential temperature.*[2] Clearly the equivalent potential temperature of an air parcel is conserved during both dry and saturated adiabatic processes. Therefore it is a valuable supplement to the potential temperature as a means of identifying air samples.

It is possible to express the equivalent potential temperature of a saturated air parcel as a function of any two thermodynamic variables, such as p and α. This can be done by combining Eq. (4.13) with Poisson's equation for θ, the definition of w, Eq. (4.2), the Clausius-Clapeyron equation, and the equation of state. Thus lines of constant value of θ_e can be plotted on an $\alpha, -p$ diagram. We shall not exhibit these lines here, but later we shall consider their appearance on a number of more practical diagrams developed from the $\alpha, -p$ diagram. It suffices at the present time to state that the θ_e lines are asymptotic to the θ lines at low pressure and temperature, and that θ increases along lines of constant θ_e.

Since the equivalent potential temperature is constant during a pseudoadiabatic process, each θ_e line on a diagram represents the successive states of an air parcel which is engaged in a pseudoadiabatic process. Conversely, each pseudoadiabatic process is characterized by

---

[2] There are other processes than the pseudoadiabatic one by which the latent heat may be realized. For example, one could think of the moisture as being condensed by an isobaric process. The resultant potential temperature might be called an *isobaric equivalent potential temperature* while the present θ_e could be called a *pseudoadiabatic equivalent potential temperature.* In practice there is no significant difference between these two. However, the latter is preferable, since there is a physically possible process by which it can be attained. No such physical process exists for the former temperature.

one and only one value of θ_e. Consequently each column of Table 4.2 has been labeled with its value of θ_e.

It is possible, and often convenient, to define also an *equivalent temperature*, T_e. The physical process which will bring saturated air to its equivalent temperature is (1) a pseudoadiabatic expansion to zero mixing ratio, and (2) a dry adiabatic compression back to the original pressure rather than to 1000 mb. Mathematically we may derive an expression for the equivalent temperature as follows. Solving Eq. (4.11) for θ_e gives

$$\theta_e = \theta e^{\frac{Lw_s}{c_p T}} \tag{4.14}$$

Now T_e is related to θ_e by a dry adiabatic process. Therefore, from Poisson's equation

$$\frac{T_e}{\theta_e} = \left(\frac{p}{1000}\right)^{\kappa}$$

But T is related to θ by a dry adiabatic process also, therefore

$$\frac{T}{\theta} = \left(\frac{p}{1000}\right)^{\kappa}$$

Thus

$$\frac{T}{\theta} = \frac{T_e}{\theta_e}$$

and Eq. (4.14) becomes

$$T_e = T e^{\frac{Lw_s}{c_p T}} \tag{4.15}$$

Thus equivalent temperature bears the same relationship to actual temperature as equivalent potential temperature bears to potential temperature. Clearly $T_e > T$, as it should be, since air at its equivalent temperature has realized all its latent heat. Equivalent temperature is not a conservative property during adiabatic processes.

In applying the above theory it should be kept in mind that T represents the temperature in any state of a saturated adiabatic process and w_s is the saturation mixing ratio in that state. When we have a parcel of air which is initially unsaturated but becomes saturated through a dry adiabatic expansion, then the w_s and T in these expressions may be looked upon as the saturation mixing ratio and temperature at the point of initial saturation. This w_s is equal to the mixing ratio of the unsaturated air since w is conserved during unsaturated adiabatic processes.

It is possible in this development to distinguish between condensation above 0° C (called the rain stage) and sublimation below 0° C (called the snow stage). The reason for making the distinction is that the latent heats of condensation and sublimation differ by about 10 per cent. However, the process of supercooling prevents the atmosphere from making a sharp transition at 0° C. Instead, a gradual blending is achieved in actuality. The theory is not developed in two separate stages here because in meteorological practice the difference is usually ignored.

4.7 Moisture Variables

Meteorologists have defined a large number of quantities that directly express or depend upon the amount of water vapor in air. Some of these we have already described and used in this book, while others remain to be described. In this section we shall list and describe the most important such variables.

Since some of these variables can best be understood by means of graphical operations we introduce a more convenient diagram than the α, $-p$ plot. One reason for not using an α, $-p$ diagram in practice is that α ordinarily is not measured directly. Instead, pressure and temperature are the commonly measured thermodynamic variables. Other more important reasons are developed in Chapter 5. The diagram used is a plot of temperature on a linear scale as the abscissa and p^{κ} as the ordinate with pressure increasing downward. Such a plot is called a Stüve diagram. Its immediate advantage is that the family of lines representing dry adiabatic processes are straight lines of varying slope passing through the point $T = 0°$ K, $p = 0$. The elements of a Stüve diagram are shown in Fig. 4.6. In order to simplify the diagram only one line for a pseudoadiabatic process, and only one saturation mixing ratio line are shown. For daily meteorological operations such diagrams have a full complement of isobars, isotherms, dry adiabats, pseudo-adiabats, and saturation mixing ratio lines which may be printed in several colors to minimize confusion.

The more important moisture variables are the following:

(a) *Vapor pressure*, e: When H_2O is the only gas present, the vapor pressure is simply the pressure of this gas. As we have seen, the vapor pressure obeys the ideal gas law quite well in the meteorological range of conditions. When water vapor is mixed with air, the pressure exerted by the H_2O is also called the *vapor pressure*. The amount of vapor that can remain gaseous is limited, the upper limit being represented by the saturation vapor pressure, e_s, which is a function of temperature only.

This dependence upon temperature may be calculated from the Clausius-Clapeyron equation.

(b) *Absolute humidity*, ρ_v, is simply the density of water vapor.

(c) *Mixing ratio*, w, is the ratio of the mass of water vapor present to the mass of dry air containing the vapor. Thus

$$w = \frac{\rho_v}{\rho_d} = \epsilon \frac{e}{p - e}$$

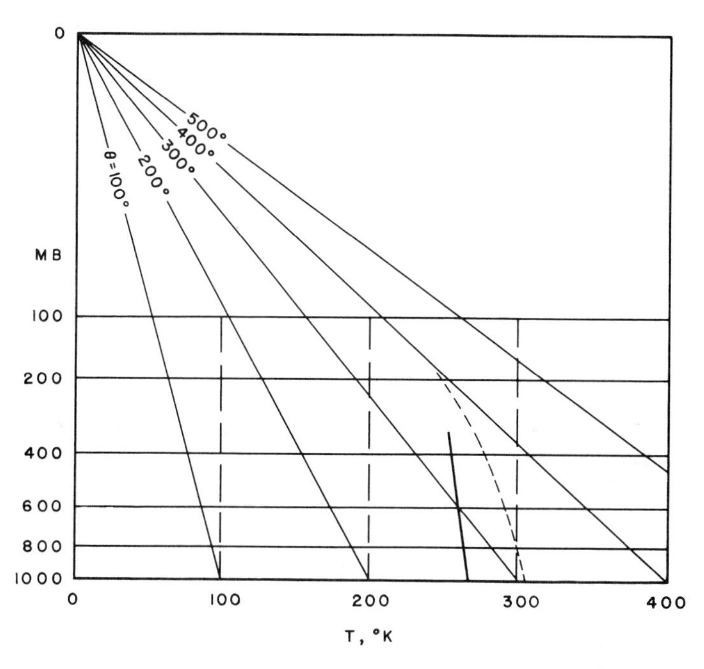

FIG. 4.6. A skeleton Stüve diagram. The only saturation mixing ratio line entered (nearly vertical heavy solid line) is for $w_s = 1.0$ g kg^{-1} and the only pseudoadiabat entered (dashed line) is for $\theta_e = 400°$ K. Note that this pseudoadiabat is asymptotic to the dry adiabat $\theta = 400°$ K.

Since $e \ll p$ we may also write

$$w \approx \epsilon \frac{e}{p}$$

The saturation mixing ratio is given by

$$w_s \approx \epsilon \frac{e_s}{p}$$

and is therefore a function of temperature and pressure. Because w_s is a function of two thermodynamic variables, lines of constant value of w_s can be plotted on an $\alpha, -p$ diagram or any diagram derived from it (see Fig. 4.6). Mixing ratio is expressed in practice as the number of grams of water vapor per kilogram of dry air.

(d) *Specific humidity*, q, is defined as the ratio of the mass of water vapor to the mass of *moist* air containing the vapor. Thus

$$q = \frac{M_v}{M_v + M_d} = \frac{\rho_v}{\rho} \approx \epsilon \frac{e}{p}$$

As can be seen, specific humidity and mixing ratio are nearly equal and have similar properties.

(e) *Relative humidity*, r, is the ratio of the actual mixing ratio of a sample of air at a given temperature and pressure to the saturation mixing ratio of the air at that temperature and pressure:

$$r = \frac{w}{w_s}$$

Thus relative humidity simply gives the fraction of the saturation limit of vapor content which is actually possessed by a given air sample. It is usually multiplied by 100 and expressed in per cent, and is a function of w, p, and T. In addition since the saturation vapor pressures over ice and water differ, it is necessary to specify which is being used at temperatures below 0° C.

(f) *Virtual temperature*, T^*, is the temperature at which dry air would have to be in order to have the same density as a sample of moist air, assuming both have the same pressure. Thus when the virtual temperature is used instead of the actual temperature the equation of state for dry air may be used for moist air also. The equation for virtual temperature is

$$T^* = T \frac{1 + w/\epsilon}{1 + w}$$

Thus T^* is a function of T and w.

(g) *Dew-point temperature*, T_d, is defined as the temperature to which moist air must be cooled during a process in which p and w remain constant, in order that it shall become just saturated with respect to water. Obviously at T_d the mixing ratio of the air becomes its saturation mixing ratio. Thus the mixing ratio lines on any thermodynamic diagram can be used to determine dew points graphically. A similar definition can be given for the *frost-point temperature*, the only difference being that saturation with respect to ice must be achieved.

A meteorological process by which air can be brought to its dew point is radiational cooling of a layer of air near the ground during the night. Indeed, this is the primary process by which dew forms.

(h) *The lifting condensation level, LCL*, is the level to which unsaturated air would have to be raised in a dry adiabatic expansion to produce condensation. At the *LCL* the mixing ratio becomes equal to the saturation mixing ratio. This level is found most easily by graphical means on a diagram. It is the level at which the dry adiabat through the initial pressure and temperature of the parcel intersects the saturation mixing ratio line whose value of w_s is equal to the actual initial mixing ratio of the parcel. For example, in Fig. 4.6, air initially at $T = 300°$ K, $p = 1000$ mb, $w = 1$ g kg^{-1} would have an *LCL* at about 600 mb because at this pressure the w_s line for 1 g kg^{-1} intersects the dry adiabat for $\theta = 300°$ K. In practical applications the *LCL* is referred to frequently.

(i) *The wet-bulb temperature*, T_w, may be defined in two ways which differ but slightly numerically. The first is in terms of an isobaric process. The wet-bulb temperature is defined to be the temperature to which air may be cooled by evaporating water into it at constant pressure until it is saturated. The latent heat is to be thought of as coming from the air. Note that w is not kept constant in this definition, so the wet-bulb temperature is different, in general, from the dew-point temperature.

When a *psychrometer* is used to measure moisture, the air is caused to move rapidly past two thermometer bulbs. One of these is dry and indicates the air temperature. The other is covered with a moist cloth and comes to thermal equilibrium at a temperature below that of the air, because the wet bulb is cooled by evaporation of some of its moisture. When thermal equilibrium has been reached at T_w, the loss of heat by the air flowing past the wet bulb must be equal to the sensible heat which is transformed to latent heat. That is,

$$(T - T_w)(c_p + w\, c_{pv}) = (w' - w)\, L \tag{4.16}$$

where T is the temperature of the air approaching the wet bulb, T_w is the temperature of the saturated air leaving the wet bulb (wet-bulb temperature), c_p is the specific heat at constant pressure of dry air, c_{pv} is the specific heat of water vapor, w is the mixing ratio of the approaching unsaturated air, w' is the mixing ratio of the leaving saturated air (w_s at T_w), and L is the latent heat of evaporation. Since w' is a tabulated function of T_w and the known pressure, we need only to know the temperature T (from the dry-bulb thermometer), the mixing ratio w, and some constants in order to calculate T_w from Eq. (4.16). The calculation may be performed by numerical approximation or graphically. The reverse problem is more easily solved. If T and T_w are measured with a psychrometer, w can be found directly.

It is immediately apparent that T_w lies between T and T_d. First, $T_w < T$ (unless the air is already saturated). Second, T_w is the saturation temperature for w', while T_d is the saturation temperature for w. Since $w \leq w'$ we conclude that $T_d \leq T_w \leq T$. The equalities apply when the air is initially saturated.

The second way of defining the wet-bulb temperature is via an adiabatic process and is best described in terms of operations on a diagram such as the Stüve diagram as in Fig. 4.7. Consider an unsaturated

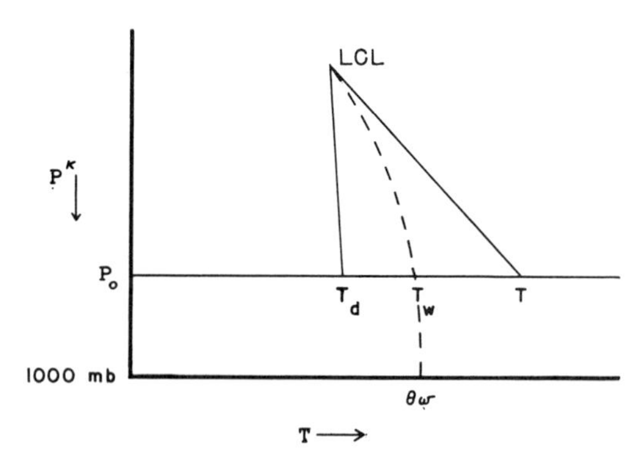

FIG. 4.7. The relationship of T, T_w, T_d, θ_w, and *LCL* on a Stüve diagram

parcel at temperature T, pressure p_o, and a certain mixing ratio w. Let the parcel expand dry adiabatically until the *LCL* is reached. If the w_s line through this point is followed down to the original pressure, the temperature found there will be the dew-point temperature, T_d. If the pseudoadiabat passing through the condensation point at the *LCL* is followed down to the original pressure, the temperature found there will be an *adiabatic wet-bulb temperature.* The relationship between this and the isobaric wet-bulb temperature is quite obscure. Comparisons indicate, however, that the two temperatures are rarely different by more than a few tenths of a degree Celsius, and the adiabatic version is always the smaller of the two for unsaturated air. Since the difference is so small, it is usually neglected in practice.

A meteorological process by which air may be brought to its wet-bulb temperature is the evaporation of falling rain into an initially unsaturated air layer.

(j) *The wet-bulb potential temperature,* θ_w, is most easily defined in terms of graphical operations. Consider the pseudoadiabat which passes through the wet-bulb temperature. Follow this adiabat until it intersects the 1000 mb isobar. The temperature at this intersection is θ_w

(see Fig. 4.7). Clearly each pseudoadiabat is characterized by one and only one value of θ_w, just as each such line is characterized by a unique value of θ_e. Thus a pseudoadiabat may be labeled with its value of either wet-bulb potential temperature or equivalent potential temperature. Furthermore, the wet-bulb potential temperature is conserved during both moist and dry adiabatic processes just as the equivalent potential temperature is conserved during these processes. Consequently which of the two is used is usually a matter of personal preference.

(k) *Equivalent temperature*, T_e, is the temperature a sample of air would have if all its latent heat were converted to sensible heat by means of a pseudoadiabatic expansion to low pressure and temperature followed by a dry adiabatic compression to the original pressure. Mathematically,

$$T_e = Te^{\frac{Lw_s}{c_pT}}$$

Graphically, T_e can be determined on a Stüve diagram as shown in Fig. 4.8. Here the original state of the parcel is T and p_o.

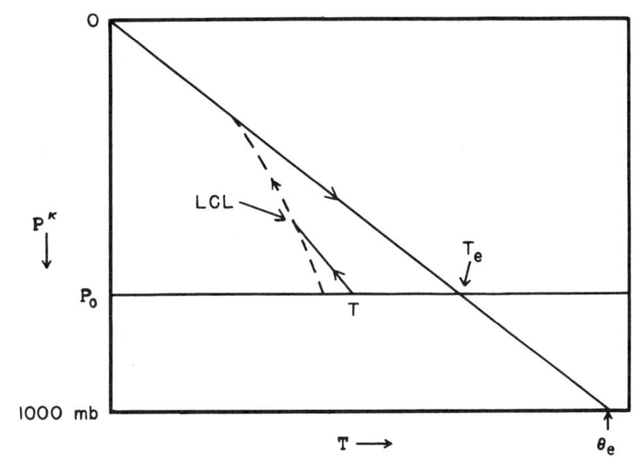

FIG. 4.8. Graphical determination of T_e and θ_e on a Stüve diagram

(l) *Equivalent potential temperature*, θ_e, is the temperature a parcel of air would have if it were taken from its equivalent temperature to a pressure of 1000 mb via a dry adiabatic process (see Fig. 4.8). Mathematically,

$$\theta_e = \theta e^{\frac{Lw_s}{c_pT}}$$

The corresponding properties of θ_e and θ_w have been pointed out already.

PROBLEMS

1. Compute the change in internal energy and the work done when 1 g of liquid water is vaporized reversibly at constant pressure at a fixed temperature of 0° C.

2. Show, for ordinary meteorological conditions, that $T^* \approx (1 + .61\, w)\, T$ to a sufficient degree of accuracy.

3. A sample of air is saturated at a pressure of 1000 mb and temperature 0° C. Calculate its virtual temperature.

4. Show that the specific heat at constant pressure, c_p, of moist air with mixing ratio w is $c_p = c_{pd}\,(1 + .84\, w)$, where c_{pd} is the specific heat at constant pressure of dry air. Calculate the percentage difference between c_p and c_{pd} for the extremely high value $w = 40$ g kg^{-1}.

5. Show that for unsaturated adiabatic processes of moist air the exponent which should be used in Poisson's equation (3.16) is $\kappa \approx \kappa_d\,(1 - .23\, w)$, where κ_d is the exponent for dry air.

Calculate the potential temperature of a parcel of air with a temperature of 0° C and a pressure of 500 mb, assuming (a) the air is perfectly dry; (b) the air has the maximum possible mixing ratio (at saturation) of 7.7 g kg^{-1}. From your results decide whether there is a significant difference between dry adiabatic processes and moist, unsaturated adiabatic processes in practical meteorological calculations.

6. Show for water near 0° C whose vapor is ideal that

$$L_{\text{evaporation}} = 597.3 - 0.566\,(T - 273) \text{ cal g}^{-1}$$

and

$$L_{\text{sublimation}} = 677.0 - 0.062\,(T - 273) \text{ cal g}^{-1}$$

7. Consider a parcel of air which is saturated at temperature and pressure −3° C and 500 mb. *Calculate* the values of e_s, w, q, T_e, and θ_e for this parcel.

8. The temperature of human blood is about 99° F. There is an elevation in the atmosphere where the pressure falls so low that a man in an unpressurized airplane will die because his blood will boil. Calculate that pressure in mb.

CHAPTER 5

THERMODYNAMIC DIAGRAMS

5.1 General Considerations

The primary function of a thermodynamic diagram is to provide a graphical display of the lines representing the major kinds of processes to which air may be subject, namely isobaric, isothermal, dry adiabatic, and pseudoadiabatic processes. Lines of constant value of saturation mixing ratio are also needed to permit the various kinds of graphical operations discussed in Chapter 4. We may plot on such a diagram the observed state of any set of air parcels and then be in a position to evaluate graphically the effect of any of these processes. Since energy changes are of primary importance, the first desirable characteristic of such a diagram is that the area enclosed by the lines representing any cyclic process be proportional to the change in energy or the work done during the process. This is such an important property that the designation *thermodynamic diagram* is often reserved for those in which area is proportional to work or energy.

The second desired characteristic of a diagram is that as many as possible of the fundamental lines be straight. The more a diagram satisfies this criterion the easier it will be to use.

The third desideratum is that the angle between the isotherms and the dry adiabats shall be as large as possible. As we shall see later, when soundings of the upper atmosphere are plotted on these diagrams the slope of the sounding is often compared to the slopes of the lines on the diagram. Thus the greater the difference in slope between an isotherm and an adiabat the easier it is to detect variations in slope. An isotherm-to-adiabat angle near 90° is considered very good.

We have already seen that the fundamental expression for an

element of specific work, $dw = pd\alpha$, suggests that we use p and α as the coordinates in order to satisfy the first criterion. However, the angle between the isotherms and adiabats of an $\alpha, -p$ diagram is quite small so such a diagram does not satisfy the third criterion. We must seek a means of setting up other suitable diagrams in which the coordinates are two functions of thermodynamic variables, subject to the restriction that the area enclosed by any cycle in the new diagram shall be equal to the area enclosed by the same cycle on a $\alpha, -p$ diagram. Such a diagram is called an *equal-area* transformation of the $\alpha, -p$ diagram. We may then examine these new diagrams to see how well they satisfy the other two criteria.

Consider two variables A and B. Let each be a function of one or more thermodynamic variables. Since a thermodynamic variable is determined by the state of a system it suffices to know α and p for a

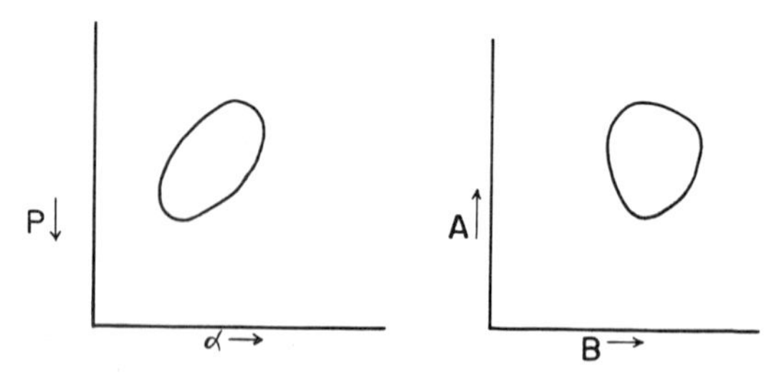

FIG. 5.1. Representation of a cycle on an $\alpha, -p$ diagram and its equal-area transformation to an A, B diagram

parcel in order to determine A and B. Thus each point on an $\alpha, -p$ diagram corresponds to a point on an A, B diagram and any closed cycle in one is a closed cycle (perhaps of different shape) on the other, as in Fig. 5.1. We shall require that the area enclosed on one diagram be equal to the area enclosed on the other. This insures that an A, B plot will be a thermodynamic diagram. Thus

$$-\oint p d\alpha = \oint A dB$$

for any given cyclic process. Thus

$$\oint (p d\alpha + A dB) = 0$$

But for this closed lines integral to be zero the integrand must be an exact differential—for example, ds:

$$p d\alpha + A dB = ds$$

where we shall look upon s as a function of α and B. But from calculus

$$ds(\alpha, B) = \left(\frac{\partial s}{\partial \alpha}\right)_B d\alpha + \left(\frac{\partial s}{\partial B}\right)_\alpha dB$$

Therefore, sufficient conditions for an equal-area transformation are

$$p = \left(\frac{\partial s}{\partial \alpha}\right)_B \quad \text{and} \quad A = \left(\frac{\partial s}{\partial B}\right)_\alpha$$

If we now differentiate the first term partially with respect to B and the second with respect to α we get

$$\left(\frac{\partial p}{\partial B}\right)_\alpha = \frac{\partial^2 s}{\partial \alpha \partial B} \quad \text{and} \quad \left(\frac{\partial A}{\partial \alpha}\right)_B = \frac{\partial^2 s}{\partial \alpha \partial B}$$

Therefore, if

$$\left(\frac{\partial A}{\partial \alpha}\right)_B = \left(\frac{\partial p}{\partial B}\right)_\alpha \tag{5.1}$$

then areas will be equal on the two diagrams. If we now specify the nature of the thermodynamic variable B, it is possible to determine what A must be in order to have an equal-area transformation from α, $-p$ to A, B.

5.2 The Emagram

Consider the case where $B = T$. This is a logical choice for one of the coordinates of a diagram, since one of the atmospheric properties we measure is the temperature. From Eq. (5.1),

$$\left(\frac{\partial A}{\partial \alpha}\right)_T = \left(\frac{\partial p}{\partial T}\right)_\alpha$$

The right side can be evaluated from the equation of state for air, $p\alpha = RT$. We find

$$\left(\frac{\partial A}{\partial \alpha}\right)_T = \frac{R}{\alpha}$$

or

$$\left(\frac{\partial A}{\partial \alpha}\right)_T d\alpha = R\frac{d\alpha}{\alpha}$$

Upon integration we obtain

$$A = R \ln \alpha + F(T) \tag{5.2}$$

where instead of a constant of integration we obtain an unspecified function of T because the partial derivative we integrated required T to be held constant. This $F(T)$ is completely at our disposal, just as a constant of integration may be specified at will. We now take the logarithm of the equation of state:

$$\ln \alpha = -\ln p + \ln R + \ln T$$

and substitute into Eq. (5.2) to get

$$A = -R \ln p + [R \ln R + R \ln T + F(T)]$$

We shall choose $F(T)$ such that the terms in brackets cancel completely. We are free to do this since the terms in brackets consist of constants and functions of temperature only. Therefore there is an $F(T)$ which will reduce the brackets to zero. Finally we obtain

$$A = -R \ln p$$
$$B = T$$

as the coordinates of a thermodynamic diagram. This plot was called the *emagram* by Refsdal as an abbreviation for "energy-per-unit-mass diagram." It consists of a linear scale of temperature along the abscissa and a logarithmic scale of pressure along the ordinate decreasing upward, as in the atmosphere. Consequently the isobars and isotherms are straight and perpendicular to each other. This is desirable since pressure and temperature are the two most commonly measured atmospheric variables.

Since pressure is on a logarithmic scale, the line $p = 0$ is at infinity and the diagram must be terminated at some conveniently low pressure such as 400 mb. This is not a great obstacle since the spacings of a logarithmic scale are repetitive. For example, the distance between the isobars 800 mb and 400 mb on this diagram is proportional to $\ln 800 - \ln 400 = \ln 2$, while the separation between isobars 400 mb and 200 mb is also proportional to $\ln 2$. Thus each isobar can be relabeled with any fraction of its original value. Of course the values of θ and θ_e attached to the adiabats and pseudoadiabats would have to be changed appropriately. (See Problem 1.)

The shape of the dry adiabats on an emagram can be deduced by taking the logarithm of Poisson's equation considering θ constant:

$$-\ln p = -\frac{1}{\kappa} \ln T + \text{constant}$$

Since $-\ln p$ is one of the coordinates of the diagram but $\ln T$ is not, the dry adiabats are logarithmic curves. They become steeper with decreasing temperature but do not depart markedly from straight lines in the usual meteorological range. A similar repetitive spacing exists for these logarithmic lines as for the isobars. The pseudoadiabats are markedly curved but the saturation mixing ratio lines are gently curved. Thus, considering the second criterion of a diagram, the isobars and isotherms are exactly straight, the adiabats and w_s lines are only slightly curved in the usual range of values, and the θ_e lines are more definitely curved.

The adiabat-isotherm angle can be changed by varying one or the other of the coordinate scales. However, convenience of use and economy of paper limits one's freedom here. In practice the angle between adiabats and isotherms on an emagram is near 45°. This is appreciably better than the corresponding angle on an α, $-p$ diagram.

In summary, the emagram has (1) area proportional to energy; (2) four sets of lines which are exactly or nearly straight and one set which is curved; (3) an adequately good angle between adiabats and isotherms. Therefore it is a convenient diagram which is in wide use.

5.3 The Tephigram

This diagram may be developed by letting $B = T$ as for the emagram. Thus

$$A = R \ln \alpha + F(T)$$

as before. But this time instead of substituting from the equation of state let us introduce potential temperature from Poisson's equation

$$\frac{T}{\theta} = \left(\frac{p}{1000}\right)^{\kappa} = \left(\frac{RT}{1000\alpha}\right)^{\kappa}$$

By taking logarithms and solving for $\ln \alpha$, we get

$$\ln \alpha = \frac{1}{\kappa} [\ln \theta - \ln T] + \ln T + \ln R - \ln 1000$$

or

$$R \ln \alpha = c_p \ln \theta + G(T)$$

where the function $G(T)$ includes the constants 1000 mb and R. Therefore

$$A = c_p \ln \theta + F(T) + G(T)$$

This time we shall choose the arbitrary function $F(T) = -G(T)$ so the coordinates become

$$A = c_p \ln \theta$$
$$B = T$$

Since $c_p \ln \theta$ is equal to entropy, apart from an additive constant, Sir Napier Shaw, who introduced this diagram, called it the T-ϕ diagram or *tephigram* for short.

The equation of the isobars on a tephigram may be obtained by taking the logarithm of Poisson's equation. For a constant value of p

$$\ln \theta = \ln T + \text{const}$$

Since one coordinate of this diagram is $\ln \theta$ but the other is a linear scale of T, the isobars are logarithmic curves which slope upward to the right and decrease in slope with increasing temperature. In the rather restricted range of meteorological conditions the isobars have only gentle curvature and are nearly straight. It is possible to rotate the diagram clockwise so the isobars are essentially horizontal with pressure decreasing upward as it does in the atmosphere. However, this is not absolutely necessary.

The pseudoadiabats are appreciably curved, but the saturation mixing ratio lines are nearly straight on a tephigram.

By the very nature of the diagram, the angle between isotherms and adiabats is exactly 90°. Thus this diagram is one in which changes of slope of a sounding are easily detected and comparison of slopes is readily accomplished. This large angle, which is roughly double that of the emagram, is the greatest advantage of the tephigram.

In summary, the tephigram has (1) area proportional to energy; (2) four sets of lines which are exactly or nearly straight and only one set which is quite curved; (3) an isotherm-to-adiabat angle which is large. This diagram comes very close to satisfying all three criteria perfectly and consequently it is used widely.

5.4 The Skew T-Log p Diagram

This diagram represents an attempt to modify the emagram so as to make the isotherm-adiabat angle more nearly 90°. It was first suggested by Herlofson.[1] We let $B = -R \ln p$, so this coordinate is identical with one coordinate of the emagram. Then Eq. (5.1) becomes

$$\left(\frac{\partial A}{\partial \alpha}\right)_{\ln p} = -\frac{1}{R}\left(\frac{\partial p}{\partial \ln p}\right)_{\alpha}$$

or

$$\left(\frac{\partial A}{\partial \alpha}\right)_{\ln p} = -\frac{p}{R}$$

When we multiply by $d\alpha$ and integrate, holding $\ln p$ (and therefore p) constant, we obtain

$$A = -\frac{p\alpha}{R} + F(\ln p)$$

or

$$A = -T + F(\ln p)$$

[1] N. Herlofson, The T, log p-diagram with Skew Coordinate Axes. *Meteor. Ann.*, 2, pp. 311-342, 1947.

We shall choose the arbitrary function to be

$$F(\ln p) = -K \ln p,$$

where K is a constant we may choose at will. We are not concerned with the sign of an area since this only involves the direction in which a cycle is carried out. Thus the coordinates may be written

$$A = T + K \ln p,$$
$$B = -R \ln p$$

The diagram is constructed with B as the ordinate and A as the abscissa. Consequently the ordinate is identical with that of the emagram.

On this diagram an isotherm has the equation

$$A = \text{const} + K \ln p$$

or

$$A = \text{const} - \frac{K}{R} B$$

or

$$B = -\frac{R}{K} A + \text{const}$$

That is, the isotherms are straight parallel lines whose slope depends upon the value of K selected. When K is chosen to make the isotherm-adiabat angle close to 90° then the isotherms slope upward to the right of the diagram at an angle of about 45° with the isobars.

The equation of the dry adiabats is obtained, as before, by taking the logarithm of Poisson's equation while holding θ constant:

$$\ln T = \frac{R}{c_p} \ln p + \text{const}$$

The quantity $R \ln p$ is one of the coordinates of the diagram, but $\ln T$ is not. Therefore the adiabats are not straight. In the meteorological range of conditions the adiabats are visibly but gently curved lines running from the lower right to the upper left of the diagram. They are concave upwards. As was pointed out earlier, the constant, K, is so chosen that the isotherm-adiabat angle is near 90° everywhere in the meteorological range.

As on all diagrams discussed here, the pseudoadiabats are distinctly curved. This is a characteristic which can be avoided only be sacrificing the energy-area proportionality. The saturation mixing ratio lines are essentially straight on this diagram.

In summary, the skew T-log p diagram has (1) area proportional to energy; (2) three sets of exactly or closely straight lines, one set of gently curved lines, and one set of markedly curved lines; (3) an adiabat-isotherm angle which varies with position on the diagram but is about

90°. This diagram satisfies the three basic criteria almost as well as the tephigram and is therefore a very good thermodynamic diagram. It has been adopted for official use by the Air Weather Service of the U.S. Air Force.

Figure 5.2 gives skeleton versions of the emagram, tephigram, and skew T-log p diagram.

5.5 The Stüve Diagram

This diagram, which was outlined briefly in Chapter 4, consists of p^{κ} on the ordinate with p increasing downward, and T on a linear scale as the abscissa. This choice of coordinates insures that the dry adiabats will be straight lines. As usual, the pseudoadiabats are curved but the saturation mixing ratio lines are essentially straight. The adiabat-isotherm angle is usually near 45°. As may be demonstrated (see

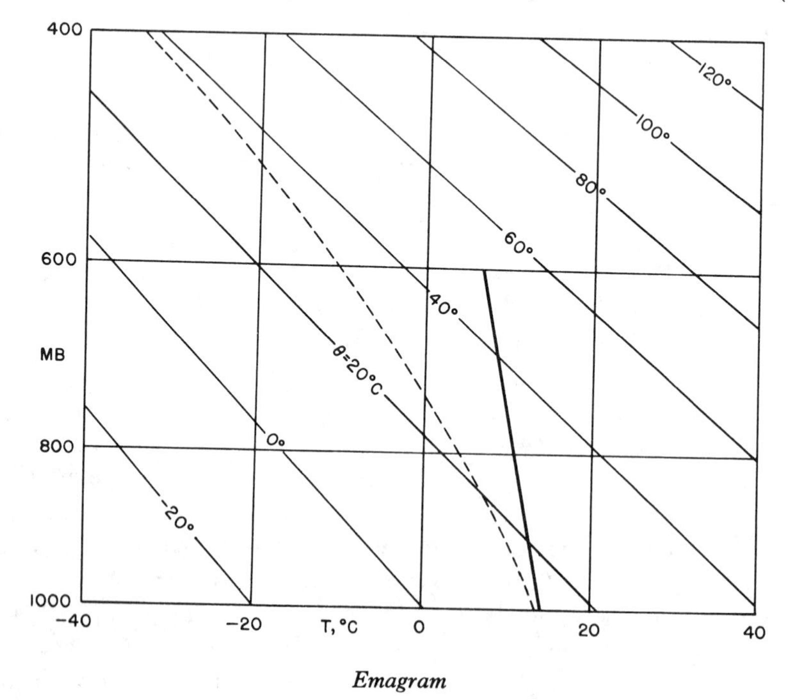

Emagram

FIG. 5.2. Isobars, isotherms, dry adiabats, saturation mixing ratio line, and pseudoadiabat on an emagram, a tephigram, a skew T-log p diagram. In all cases only the lines for $w_s = 10$ g kg^{-1} and $\theta_e = 40°$ C are shown.

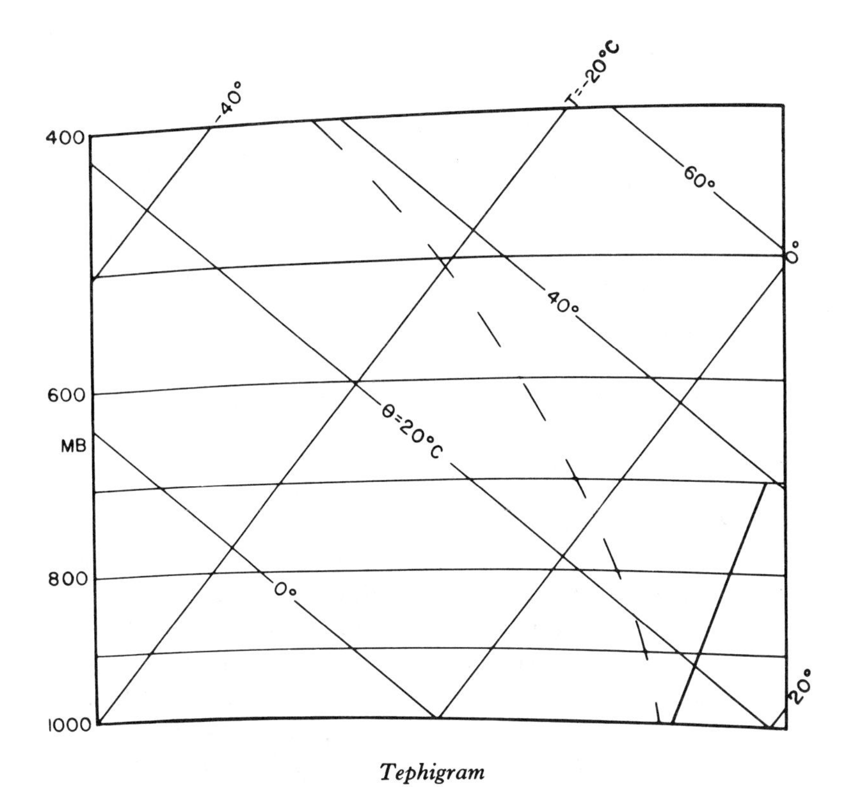

Tephigram

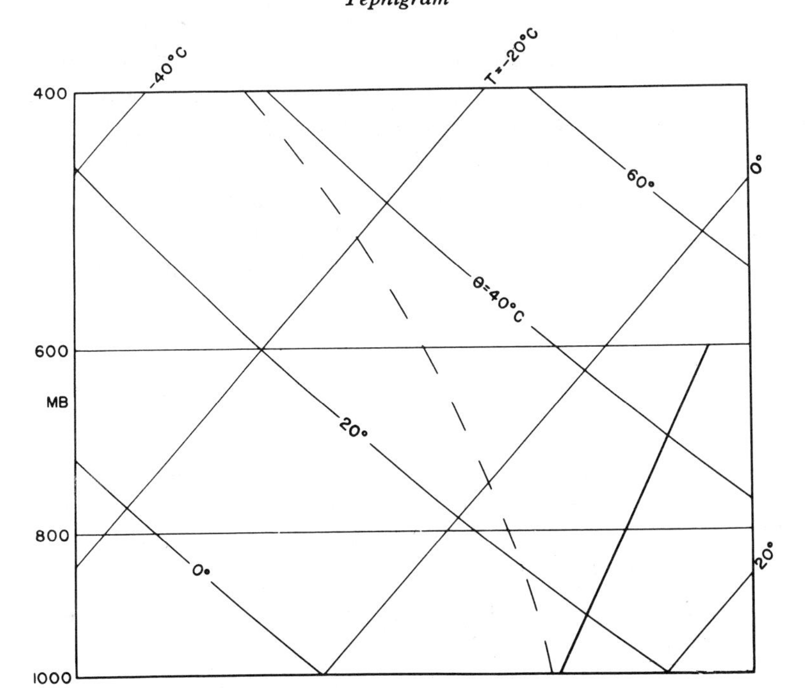

Skew T-*log* p *diagram*

Problem 2), the Stüve diagram is *not* an equal-area transformation of the α, $-p$ diagram. That is, area is not strictly proportional to energy.

In summary, the Stüve diagram (1) does not have area proportional to energy; (2) has four sets of lines which are exactly or nearly straight and only one set which is perceptibly curved; (3) has an adiabat-to-isotherm angle of 45°. This diagram is clearly not as good as some of the others, although it was introduced very early in the history of modern meteorology. It is used enough so that one should know its properties, but it is gradually being replaced.

5.6 Choice of a Diagram

A number of diagrams other than those described here have been devised. Despite the objective criteria which have been set up to determine the desirability of a diagram, it seems true that most meteorologists have an aversion to diagrams other than the one with which they are most familiar. This is understandable and even defensible, since there really is not a great deal of difference in practice among the various diagrams. Sometimes one diagram is preferred over another because of the excellence of the printing, the skillful use of color, and minimization of eye strain. In other cases the deciding factor may be the presence or absence of some auxiliary nomogram, such as one for rapid computation of the distance between pressure levels.

From an over-all point of view, the tephigram and the skew T-log p diagram seem to be superior to all the others by a small margin. However it is likely that all the major diagrams will continue in use for many years.

PROBLEMS

1. Suppose all the isobars on an emagram were relabeled with one-half their original pressures. Show that a dry adiabat which was originally labeled with potential temperature θ_1 should be relabeled with potential temperature $\theta = 2^{\kappa}\theta_1$.

2. Show that the Stüve diagram does not satisfy the sufficient condition for a thermodynamic diagram.

3. The "Refsdal aerogram" has coordinates $-RT \ln p$ and $\ln RT$. Show that this is an equal-area transformation of the α, $-p$ diagram.

CHAPTER 6

HYDROSTATIC EQUILIBRIUM

6.1 The Hydrostatic Equation

Hitherto we have been concerned with the properties and physical processes of individual air parcels. We shall now consider some aspects of the distribution in space of these properties.

It is well verified that the thermodynamic variables describing the state of air usually change more rapidly in the vertical than in the horizontal. It is therefore reasonable to devote our attention initially to these vertical changes. Consider the forces per unit mass which act along the vertical on air. The first is, of course, the force of gravity. This is a large force amounting to 980.6 dynes g^{-1}. Yet we do not observe that the atmosphere is accelerating downward under the influence of gravity as would be required by Newton's second law of motion if gravity were the only force acting. Clearly another force or forces must be acting upward to cancel gravity. The source of this force can be found by considering the vertical distribution of atmospheric pressure.

The pressure at sea level is for all practical purposes equal to the weight per unit horizontal area of all the air from sea level to outer space. If one moves upward from sea level the amount of air remaining above decreases and the pressure must also decrease. This is why we experience pressure sensations in the ears when ascending or descending in an elevator or airplane. Thus if we consider a small segment of length dz in a vertical air column there must be a small decrease of pressure, dp (negative), from the bottom to the top of the segment (see Fig. 6.1). Because

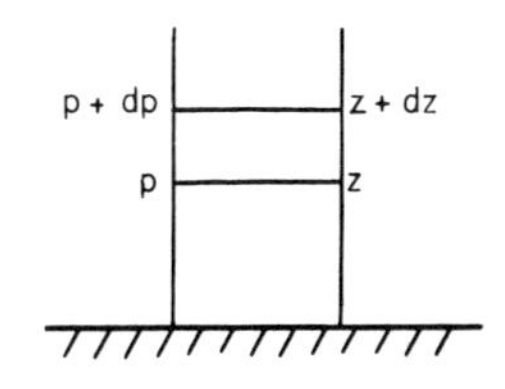

FIG. 6.1. Change of pressure with elevation

of this difference in pressure there is a force operating upward on the segment. The rate of change of pressure with height, or *vertical pressure gradient*, is dp/dz. Since pressure is a force per unit horizontal area, dp/dz is a force per unit volume. To convert to force per unit mass we multiply by the specific volume of the air at the given elevation. Thus $\alpha dp/dz$ is the magnitude of the upward directed force owing to the normal decrease of pressure with elevation.

Measurements indicate that this pressure gradient force almost exactly balances the gravity force in virtually all cases. Thus we need consider only the simple case of equilibrium between these two forces and we find that

$$-\alpha\frac{dp}{dz} = g \quad \text{or} \quad \frac{dp}{dz} = -\rho g \tag{6.1}$$

The minus sign is necessary because dp/dz is intrinsically negative and g is a positive number. Either form of Eq. (6.1) is called the *hydrostatic equation*, which is one of the best approximations in theoretical meteorology. It may be applied safely to all meteorological phenomena except those, like the tornado, in which appreciable vertical accelerations may exist.

We may write Eq. (6.1) in the form

$$dp = -\rho g dz$$

Since g varies somewhat from place to place it is convenient to define a new variable:

$$d\psi = g dz \tag{6.2}$$

where ψ is called the *geopotential* $[L^2T^{-2}]$. If we integrate Eq. (6.2) from $z = 0$, where ψ is taken to be zero, to some elevation z, where the geopotential has some value ψ, we get

$$\psi = \int_0^z g dz$$

The quantity ψ is to be interpreted as the potential energy imparted to a unit mass if it is lifted from sea level to some height z. Consider an imaginary surface in the atmosphere along which ψ is constant. A particle will have the same potential energy at any point on this surface. Since particles tend to move towards minimum potential energy, an object constrained to remain on such a surface will have no tendency to slide along the surface. This is the same as saying that the force of gravity (a combination of true Newtonian gravitation and the centrifugal effect of the earth's rotation) is perpendicular to surfaces of constant ψ. This convenient property is not shared by surfaces of constant elevation because the value of g varies, primarily with latitude.

This is equivalent to saying that the force of gravity is not normal to surfaces of constant z. Since we wish to have the simplicity of having reference levels to which observed gravity is perpendicular we refer all measurements to surfaces of constant geopotential.

Since $g \approx 9.8$ m s^{-2} the geopotential of a unit mass 1 m above sea level is about 9.8 m^2 s^{-2}, that is, ψ is closely 9.8 times the geometric height. Therefore a unit of geopotential has been defined, called the *geopotential* meter, such that height in geopotential meters is equal to $\frac{1}{9.8}\int_0^z g\,dz$. It should be clearly understood that a geopotential meter is a unit of specific energy, not a unit of geometric height. However, it has been defined so that specific energy in these units is numerically very close to geometric height in meters. The differences between the surface of constant potential energy equal to X geopotential meters and the surface of constant geometric height equal to X meters are primarily those which are due to the latitudinal variation of g. In order to convert a geopotential "height" from meters to any other unit, such as feet, we apply the same conversion factor used to convert from geometric meters to the desired geometric units.

Example: A pressure of 700 mb is found on a certain day to be at an "elevation" of 3,000 geopotential meters. What is the "elevation" in geopotential feet? Since 1 ft = .3048 m, 3000 geopotential meters = $\frac{3000}{.3048}$ = 9,843 geopotential feet. If this measurement were made at a location on latitude 45° where g = 980.6 cm s^{-2} the geometric height of the point where the pressure is 700 mb would be

$$3000 \text{ gpt m} = \frac{1}{9.8}\int_o^z 9.806\,dz$$

or

$$3000 = \frac{9.806}{9.8}z$$

Therefore

$$z = 2998 \text{ m}$$

Thus, at this point, the numerical difference between geopotential height and geometric height is only 2 meters.

6.2 Height Computations for Upper-air Soundings

Forecasting by modern techniques requires extensive data on the distribution of pressure, temperature, and humidity in the upper atmosphere. At present, such upper-air soundings are taken by releasing a balloon carrying a *radiosonde*. This is an inexpensive instrument which measures pressure, temperature, and humidity and transmits radio signals to a ground station. Such soundings are taken at least twice a day at several hundred locations over the world.

A basic problem in making these data useful is to determine the elevation at which the various data are observed. To do this one makes use of the hydrostatic equation in the form

$$d\psi = -\alpha dp$$

where $d\psi$ means the small difference in geopotential between two points in a vertical column whose pressures differ by the increment dp. If we substitute for α from the equation of state for moist air we get

$$d\psi = -RT^* \frac{dp}{p}$$

where R is the gas constant for dry air and T^* is the virtual temperature. We now integrate between geopotential levels ψ_1 and ψ_2 where the pressures are p_1 and p_2 and obtain

$$\psi_2 - \psi_1 = -R \int_{p_1}^{p_2} T^* \frac{dp}{p}$$

or

$$\psi_2 - \psi_1 = -R \int_{\ln p_1}^{\ln p_2} T^* \, d(\ln p) \tag{6.3}$$

Since the radiosonde instrument measures temperature, pressure, and moisture content, one can determine a curve of T^* as a function of pressure along the vertical. Thus it is possible to evaluate the integral graphically. If one plots the results of a sounding on an emagram as in Fig. 6.2, the integral to be evaluated is proportional to the area between

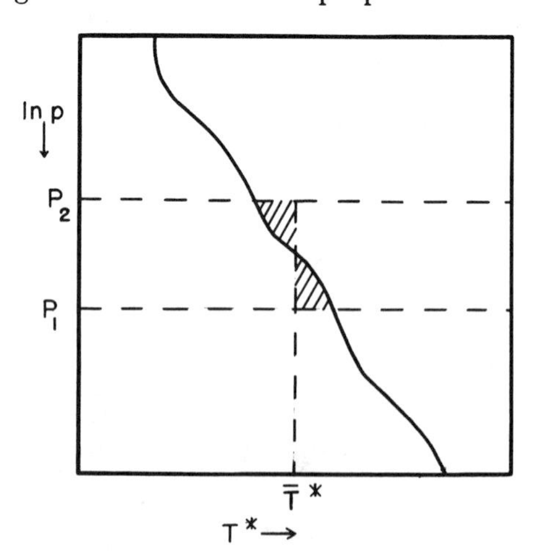

FIG. 6.2. Determination of mean vertical temperature of a layer on an emagram

p_1 and p_2 to the left of the virtual temperature curve. But one can always find some mean virtual temperature, $\bar{T}^*$, which will have the same area to its left as the observed temperature curve. On the emagram this $\bar{T}^*$ can be found quickly by selecting a vertical isotherm which is cooler than the sounding temperature for part of the pressure increment and warmer for the remainder. If this isotherm is selected so that the shaded areas in Fig. 6.2 are equal, then Eq. (6.3) reduces to

$$\psi_2 - \psi_1 = -R\bar{T}^* \ln \frac{p_2}{p_1} \tag{6.4}$$

The choice of a sufficiently accurate value of $\bar{T}^*$ can be made very rapidly by eye estimate using this method of equal areas. It then is possible to calculate the difference in geopotential between the two known pressures.

In practice the sounding curve is measured from the surface of the earth up to the point where the balloon bursts. The geopotential of the ground can be measured quite accurately for each radiosonde station. Therefore all one need do is plot the measured curve of T^* versus pressure on an emagram, divide the sounding into a number of suitable finite layers, graphically determine $\bar{T}^*$ for each layer, and calculate the geopotential of the top of each layer starting from the known geopotential at the ground.

Equation (6.4) gives the geopotential difference in units of specific energy. To obtain the result in geopotential meters express R in units of meters, seconds, and degrees Kelvin and divide by 9.8. Then Eq. (6.4) becomes

$$\Psi_2 - \Psi_1 = -\frac{R\bar{T}^*}{9.8} \ln \frac{p_2}{p_1}$$

or

$$\Psi_2 - \Psi_1 = -\frac{287.04}{9.8} \bar{T}^* \ln \frac{p_2}{p_1} \tag{6.5}$$

where Ψ represents the potential measured in geopotential meters. Equation (6.5) can be solved numerically or by means of special nomograms. Such nomograms are often printed directly on thermodynamic diagrams. For the important case where p_1 and p_2 are standard isobaric levels (for example, 1000 and 900 mb) the results have been tabulated and may be looked up without calculation.

This formulation of the problem of height determination from soundings has an important characteristic which is not immediately apparent. The only use we have made of the direct pressure measurements from the radiosonde is to plot the sounding on an emagram so that a mean virtual temperature for each selected layer may be estimated. Even if the pressure measurements of the radiosonde are very rough the estimate of $\bar{T}^*$ will still be correct to within a fraction of one per cent of the absolute temperature. Thus it is not necessary that the barometric element of a radiosonde be a precision instrument.

A typical requirement in practice is that the barometric element be capable of measuring pressure within $\pm$ 10 mb. Nevertheless, by means of Eq. (6.5) the pressure at some fixed geopotential height may be determined within $\pm$ 1 mb. It is only essential that at the ground the geopotential and the pressure be measured accurately. The geopotential is determined but once, when a station is established; the pressure is measured before each radiosonde ascent with a standard mercurial barometer. This procedure makes it possible to manufacture sounding instruments rather cheaply, for a crude aneroid barometer costs far less than a precise one. If it were not for this fact our present extensive aerological network might be too expensive to maintain.

It should be noted that the basic property of the emagram which permits graphical evaluation of the mean virtual temperature of a layer by the equal-area method is that a given area any place on the diagram represents the same amount of energy ($R\bar{T}^* \ln p_2/p_1$ has the dimensions of specific energy). Thus the same procedure may be followed on any truly thermodynamic diagram without further approximation. The technique cannot be applied exactly on a Stüve diagram, since it is not an equal-area transformation of the $\alpha, -p$ diagram. However, the factor of proportionality between area and energy varies sufficiently slowly over a Stüve diagram that no intolerable error results provided layers no thicker than 200-300 mb are used.

6.3 The Hydrostatics of Special Atmospheres

It is profitable to explore the consequences of the hydrostatic equation for certain special atmospheric cases. Five such special cases are considered in this section.

A. *The homogeneous atmosphere.*—The first is the simple case in which the density of the atmosphere is assumed to be a constant independent of height. If we divide such an atmosphere into layers of equal geometric thickness each layer will have the same weight per unit area, so that each layer will contribute equally to the total pressure at the bottom. Since the surface pressure cannot be infinite there must be only a finite number of such layers—that is, the homogeneous atmosphere has only a finite total height, which we call H, the *height of the homogeneous atmosphere* or *scale height*. If we integrate the hydrostatic equation from sea level, where the pressure is p_o, to H where the pressure is zero, we get

$$\int_{p_o}^{o} dp = -\rho g \int_{o}^{H} dz$$

or

$$p_0 = \rho g H$$

Thus the height of the homogeneous atmosphere is

$$H = \frac{p_o}{\rho g}$$

But the density is constant throughout this atmosphere so we may substitute for ρ from the equation of state applied to sea level:

$$\frac{p_0}{\rho} = RT_0$$

Thus

$$H = \frac{RT_0}{g} \tag{6.6}$$

We see, therefore, that the height of the homogeneous atmosphere is a function only of the bottom temperature and known constants. For dry air on earth with $T_0 = 273°$ K, $H = 8$ km.

Since pressure decreases with height in this case but density remains constant, the equation of state requires that temperature also decrease with elevation. To calculate the rate of decrease we write the ideal gas law for dry air, $p = \rho RT$, and differentiate with respect to elevation, holding ρ constant. We obtain

$$\frac{dp}{dz} = \rho R \frac{dT}{dz}$$

Substitution into the hydrostatic equation leads to the result:

$$\frac{dT}{dz} = -\frac{g}{R} = -34.1° \text{ C km}^{-1}$$

Thus the rate of change of temperature with height in a homogeneous atmosphere is constant and very large. This result is about six times larger than the temperature decrease normally observed in the lower atmosphere. Clearly the reason is that density is not constant in the real atmosphere but decreases with height. The quantity, $\gamma = -dT/dz$, for any atmospheric case is called the *temperature lapse rate*. We shall see later that the lapse rate in a homogeneous atmosphere is unstable, that is, the atmosphere will overturn as convection sets in.

Obviously the case of the homogeneous atmosphere is not a realistic one. Nevertheless, the concept is useful theoretically. We shall see in what follows how its characteristics appear as parameters in other more realistic cases.

B. *The isothermal atmosphere.*—When we assume that temperature, rather than density, is independent of elevation we find that the hydrostatic equation in the form

$$dp = -\frac{pg}{RT}\,dz$$

is again easily integrated. This may be rewritten

$$\frac{dp}{p} = -\frac{g}{RT}\,dz$$

We shall integrate from sea level ($z = 0$, $p = p_0$) to some arbitrary level z where the pressure is p. Since T is a constant,

$$\int_{p_0}^{p} \frac{dp}{p} = -\frac{g}{RT}\int_{o}^{z} dz$$

or

$$\ln\frac{p}{p_0} = -\frac{gz}{RT}$$

Since $RT/g = H$, the constant scale height in this case, our final result is

$$p = p_0 e^{-\frac{z}{H}} \tag{6.7}$$

Thus in an isothermal atmosphere, pressure decreases exponentially with height. As a result, there is no definite upper boundary to this atmosphere; it thins out gradually with elevation and the pressure goes to zero only at $z = \infty$. The height of the corresponding homogeneous atmosphere appears as a parameter. Its significance is that when $z = H$ the pressure will be $1/e$ of its surface value.

By definition the lapse rate of temperature in an isothermal atmosphere is zero.

C. *The constant-lapse-rate atmosphere.*—In this case we shall assume that T varies linearly with height:

$$T = T_0 - \gamma z$$

where T_0 is the sea level temperature and γ is the constant lapse rate of temperature. With this specification the hydrostatic equation in the form

$$dp = -\frac{pg}{RT}\,dz$$

becomes

$$\frac{dp}{p} = -\frac{g}{R}\,\frac{dz}{T_0 - \gamma z}$$

Once again this is easily integrated. We shall integrate between the limits $z = 0$, where $p = p_0$, and an arbitrary height z, where the pressure is p. The result is

$$\int_{p_0}^{p} \frac{dp}{p} = -\frac{g}{R}\int_{0}^{z} \frac{dz}{T_0 - \gamma z}$$

or

$$\ln\frac{p}{p_0} = \frac{g}{R\gamma}\ln\frac{T_0 - \gamma z}{T_0}$$

When we take antilogarithms we obtain

$$p = p_0\left(\frac{T_0 - \gamma z}{T_0}\right)^{\frac{g}{R\gamma}}$$

or

$$p = p_0\left(\frac{T}{T_0}\right)^{\frac{g}{R\gamma}} \tag{6.8}$$

In the usual tropospheric case where T decreases with z (γ positive) Eq. (6.8) requires that pressure decrease with elevation, in agreement with the hydrostatic equation. In the less common case (which nevertheless does occur) that T increases with elevation (γ negative), the ratio T/T_0 is greater than unity above the surface but the exponent of this ratio is negative, therefore p must decrease with height again, as required by the hydrostatic equation. Formula (6.8) cannot be applied in the special case $\gamma = 0$ (isothermal atmosphere) because during the present derivation we have divided by γ, and division by zero is not a permissible operation.

One should note that the exponent in Eq. (6.8) is the ratio of the constant lapse rate in the hypothetical homogeneous atmosphere to the actual lapse rate. Thus a property of the homogeneous atmosphere again appears as a parameter.

An atmosphere with a constant positive lapse rate (decrease of temperature with height) has only a finite extent. When the temperature falls to the absolute zero the pressure also vanishes. This top to the atmosphere occurs at a height equal to T_0/γ. In the very unrealistic case of a constant negative lapse rate (increase of temperature with height) there is no upper limit to the atmosphere.

D. *The dry adiabatic atmosphere.*—We shall now investigate the temperature lapse rate when the potential temperature is independent of height. Poisson's equation is

$$\frac{T}{\theta} = \left(\frac{p}{1000}\right)^{\kappa}$$

Logarithmic differentiation holding θ constant yields

$$\frac{1}{T}\frac{dT}{dz} = \frac{\kappa}{p}\frac{dp}{dz}$$

When we substitute from the hydrostatic equation and rearrange we obtain

$$\frac{dT}{dz} = -\frac{g}{c_p}\left(\frac{R\rho T}{p}\right)$$

But from the equation of state the term in parentheses is unity, therefore

$$\frac{dT}{dz} = -\frac{g}{c_p}$$

If we assign the symbol Γ_d to the lapse rate in this case we may write

$$\Gamma_d = \frac{g}{c_p} \tag{6.9}$$

An atmosphere with such a lapse rate is called a dry adiabatic atmosphere because a parcel engaging in a dry adiabatic process also has constant potential temperature. Since g and c_p are constants Γ_d must also be a constant. Thus this is a special case of the atmosphere with constant lapse rate. The value of Γ_d for dry air is 9.76° C km^{-1}. In practice this is usually rounded to 10° C km^{-1}.

These properties of the dry adiabatic atmosphere will be important when we study the hydrostatic stability of the atmosphere, for it will turn out that this case distinguishes stable from unstable conditions.

E. *The U. S. Standard Atmosphere.*—None of the special cases discussed so far can be made to fit the real atmosphere over a sufficiently great depth. There is, however, a real need for a complete description of the average characteristics of the atmosphere up to considerable heights. Although the atmosphere varies from day to day and with location, such a mean atmosphere is very useful in aeronautics. Consequently the U. S. Weather Bureau at the request of the National Advisory Committee for Aeronautics has computed the properties of such an average atmosphere. It is meant to represent normal conditions over the United States at 40° N and is called the *U. S. Standard Atmosphere* or the *NACA Standard Atmosphere*.

The following are the basic specifications of the U. S. Standard Atmosphere up to an altitude of 32 km:

(1) The surface temperature is +15.0° C and the surface pressure is 1013.250 mb.

(2) The air is assumed to be dry and to obey the ideal gas law.

(3) The acceleration of gravity is assumed to be constant and equal to 980.665 cm s^{-2}.

(4) From sea level to 10.769 km the temperature decreases at the constant rate 6.50° C km^{-1}. This region is the troposphere.

(5) From 10.769 km to 32 km the temperature is constant at —55.0° C. This region is the stratosphere.

From these specifications it can be seen that the standard atmosphere is a combination of two cases we have already discussed. From 0 to 10.769 km it has a constant lapse rate and from 10.769 km to 32 km it is isothermal. Therefore the pressure at any level can be computed from Eq. (6.7) and (6.8). The calculated characteristics of this atmosphere are given in Table 6.1.

TABLE 6.1

PROPERTIES OF THE U. S. STANDARD ATMOSPHERE UP TO 20 KM

Altitude (km)	*Pressure* (mb)	*Temperature* (°C)	*Density* (units of 10^{-3} g cm^{-3})
0	1013.25	15.0	1.226
1	898.71	8.5	1.112
2	794.90	2.0	1.007
3	700.99	− 4.5	.909
4	616.29	−11.0	.819
5	540.07	−17.5	.736
6	471.65	−24.0	.660
7	410.46	−30.5	.590
8	355.82	−37.0	.525
9	307.24	−43.5	.466
10	264.19	−50.0	.413
(10.769)	(234.53)	(−55.0)	(.375)
11	226.19	−55.0	.361
12	193.38	−55.0	.309
13	165.33	−55.0	.264
14	141.35	−55.0	.226
15	120.86	−55.0	.193
16	103.30	−55.0	.165
17	88.34	−55.0	.141
18	75.53	−55.0	.121
19	64.57	−55.0	.103
20	55.21	−55.0	.088

There is in existence an alternative formulation called the ICAN (International Commission for Air Navigation) Standard Atmosphere, which is used by many countries. It differs from the U. S. version by using a value of gravity of 980.62 cm s^{-2} and by having an isothermal layer which begins at exactly 11 km with a temperature of —56.5° C. Tentative extensions of the U. S. Standard Atmosphere to much higher levels have been prepared for use in rocketry engineering.

6.4 Altimetry

The most accurate method of determining the elevation of some point above sea level (altimetry) is by means of triangulation using surveyor's instruments. Most of the elevations given in geographical atlases have been determined in this manner. This painstaking method, however, is not feasible whenever conditions prevent the necessary investment of effort and time. One such case is the determination of the elevation of a very high mountain top; another is the measurement of the elevation of an airplane.

In these situations one resorts to pressure altimetry. That is, the atmospheric pressure at the point in question is measured with an aneroid barometer (for convenience) and the elevation of that pressure in a standard atmosphere is determined. Clearly there are two important sources of error in such a procedure. First, sea-level pressure at the moment of observation may not be 1013.25 mb as assumed in the standard atmosphere; second, the mean virtual temperature of the column from sea level to the point of observation may be different from that in the standard atmosphere.

The application of pressure altimetry to aircraft is described here, since it is the most important use for meteorologists. The instrument employed is an *altimeter* and consists of an accurate aneroid barometer whose scale is graduated linearly in feet rather than pressure. Of course, the equivalence between height and pressure is that prescribed in the standard atmosphere. In order to adjust for the first kind of error described above, the instrument is so constructed that the height scale and the barometer with its pointers may be rotated relative to each other. For example, suppose the aircraft is at rest at sea level at some airfield. A precise measurement at the weather station indicates that the pressure is 30.00 in. (All altimetry in the U. S. uses inches of mercury to measure pressure.) If the aircraft altimeter is set for standard conditions it will indicate that the aircraft is below sea level—which is manifestly false. Therefore the pilot adjusts the pointers so that the altimeter reads zero elevation. This is done by means of a knob which sets an auxiliary dial to 30.00 in. The value to which this dial is adjusted is called the *altimeter setting*. Clearly the altimeter setting is true sea level pressure in this case.

When an airfield is above sea level the procedure is to measure the pressure at the field and then, from the known elevation of the field,[1] to compute what sea level pressure would be if the atmosphere extended down to sea level and had the standard vertical temperature distribution. This may be done by means of the pressure-height relationship in an atmosphere of linear lapse rate. In practice the results of the calculation are tabulated in advance for each airfield and are merely looked up as needed.

In this way a pilot may correct his altimeter before take-off so that deviations of station pressure from the standard pressure for the station altitude are corrected. Sea level pressure, however, varies with time and location so that in the course of a flight the altimeter may once again give incorrect readings. A variation of .01 inch in altimeter setting means a variation of nearly 10 feet in altimeter reading near sea level. For example, if a pilot sets his altimeter to the latest value of 29.90 in. at the point of take-off, flies to his destination where the altimeter setting is 29.80 in., and lands without resetting the instrument, the plane will touch ground when the altimeter indicates it is still 100 ft above the runway. Hence it is good practice for a pilot to radio for new altimeter settings during flight and especially before landing.

The second kind of error described above results from differences between the true mean temperature of the column of air below the plane and the mean temperature assumed for the column in the standard atmosphere. Such errors are very difficult to correct because there is no routine way of knowing the true mean temperature below. Estimates may be made from the temperature at flight level or from radio reports of surface temperatures, but these estimates will be very crude. The effect may be important since in the lowest 15,000 ft of the atmosphere a deviation of the true mean temperature from the assumed standard, of 2.8° C will cause about 1 per cent error in altimeter reading. If the column below is colder than standard the altimeter will read too high. If it is warmer than standard the altimeter will read too low. For example,

[1] In practice one determines the pressure at a point 10 feet above the field runway, since this is more closely where the altimeter is located, and then reduces to sea level.

if an airplane with a correct altimeter setting is flying at a true altitude of 10,000 ft but the air below the plane is 11° C warmer than standard, the altimeter will read about 4 per cent too low, or 9,600 ft.

A third kind of error in pressure altimetry is sometimes encountered. All pressure altimetry assumes the validity of the hydrostatic equation, but there are occasions when significant departures from hydrostatic equilibrium may occur. One of these, which is especially dangerous for aircraft, is when strong winds blow over a mountain top. In such a situation the pressure may drop below the hydrostatic value by a measurable amount. Since a pilot flying at a constant altimeter reading is actually flying along a constant-pressure surface, such a decrease in pressure may cause the airplane to crash into the mountain even through the altimeter indicates sufficient altitude to clear the obstacle.

Many aircraft are now equipped with a radio altimeter as well as a pressure altimeter. This device measures the elevation of the aircraft above the terrain by emitting radio waves and measuring the time the signal takes to be reflected from the ground to the aircraft. Over the ocean the radio altimeter indicates true altitude above mean sea level (MSL). When both instruments are present a pilot can fly at a constant pressure-altitude over the sea while noting the variations in true altitude. In this way information can be obtained about the variations in height of a pressure surface. Since this is one of the most important things measured by the radiosonde instrument we can see how the aerological soundings can be supplemented by data from weather reconnaissance aircraft.

6.5 Reduction of Pressure to Sea Level

One of the basic weather maps constructed routinely by all weather services is the sea-level or surface map. On this chart are entered the various weather data observed at each weather station and it is the job of the meteorologist to analyze or interpret these observations. One of the basic analyses he performs is the construction of lines of equal pressure, the isobars. This is because the *horizontal* variations of pressure are of great importance in determining the future progress of the weather. However, the various land observing stations are located at widely varying altitudes above sea level. Since the vertical variation of pressure is large compared to the horizontal variation, it is absolutely necessary to correct all the observed station pressures to a common fixed elevation. It is internationally agreed that this reference level shall be mean sea level.

In principle, therefore, at a station some known geopotential distance above sea level one assumes the existence of a hypothetical column of air extending from the station barometer down to mean sea level and calculates the pressure difference between the top and bottom of this fictitious column from the hydrostatic equation. This pressure increment is then added to the observed station pressure to yield the sea-level pressure needed.

We have already developed the hydrostatic equation in a form, Eq. (6.4), suitable for such a reduction to sea level. Equation (6.4) can be solved for p_1 the pressure at the bottom of a layer:

$$p_1 = p_2 e^{\frac{\Psi_2 - \Psi_1}{R\bar{T}^*}}$$

Thus the sea level pressure at the bottom of the column depends only upon the measured station pressure p_2, the known geopotential difference between sea level and the station barometer, and the mean virtual temperature of the fictitious column of air. Unfortunately there is no definitive way of determining $\bar{T}^*$ for a column of air that does not exist. Because of this there are several methods in existence in the various national weather services and the results are not strictly comparable.

Actually, we have already discussed one possible way to achieve the reduction to sea level. The process for obtaining an altimeter setting is such a means. The assumption in this method is that the mean virtual temperature is that given by the U. S. Standard Atmosphere between sea level and the geopotential of the barometer. The trouble with this technique is that it does not take into account the actual temperature of the air at the surface. Consequently there may be appreciable differences between the altimeter setting and the sea-level pressure which would exist if the air column down to sea level were really there. Other systems attempt to introduce the observed temperature at the surface and a reasonable estimate of the lapse rate in the fictitious column. Since the surface temperature varies diurnally it is considered inappropriate to use merely the current temperature. Instead, it is common to take for the temperature at the top of the column the average of the current temperature and the temperature 12 hours previous. This practices is followed in the United States and many other places but is not world-wide.

Any one of a number of reasonable choices may be made for the assumed lapse rate. One may assume the lapse rate is one-half the dry adiabatic value, namely 5° C km^{-1}; this is about the normal value of lapse rate in the free atmosphere. Or one may assume $\gamma = 6.5°$ C km^{-1}, which is the value adopted in the U. S. Standard Atmosphere, and which also is near the average. In the U. S. an even more complex system is employed in which different representative lapse rates are used for each station. These were determined empirically by Bigelow.[2] In some cases it is assumed also that a normal rate of change of water vapor with height exists in the fictitious column and a virtual temperature correction is applied based upon the surface value of the dew point.

One further empirical adjustment is used in the U. S., Canada, and Alaska called the "plateau correction." This correction is proportional to the difference between the mean virtual temperature for the time of observation and the annual normal of this $\overline{T}^*$. Its purpose if to minimize differences in sea-level pressure between mountain stations and nearby low-level stations.

It should be clear that the methods for reduction to sea level are not uniform over the world and are especially complex in the United States. It would be desirable to make the procedure uniform and simple, but because many years of climatological records are based on the current unwieldy system it is unlikely that any revision will be made.

All methods for reduction to sea level give unsatisfactory results in certain situations. For example, the widespread system of using a surface temperature which is an average of the current value and the one 12 hours ago gives spurious results whenever the air is exceptionally cold or exceptionally warm for periods of 12 hours or more. When the air is very cold, as it may be in high valleys in winter, the method yields a very low $\overline{T}^*$ and therefore a very high sea-level pressure. This is partly responsible for the very rapid rate at which sea-level pressure so often seems to rise as one passes from the Great Plains to the Rocky Mountains in winter time. Conversely, on a high plain in summer the sea-level pressures may be spuriously low because of the high $\overline{T}^*$. This contributes to certain "heat lows" found in summer.

PROBLEMS

1. Calculate the thickness in geopotential feet of the layer from 800 to 700 mb if the mean temperature of the layer is —3° C and the average mixing ratio of the layer is 3 g kg^{-1}.

[2] For a brief survey and bibliography see R. J. List, (ed.), *Smithsonian Meteorological Tables*, 6th rev. ed., Washington, D.C., Smithsonian Institution, pp. 203-5, 1951.

2. Suppose the absolute temperature decreases exponentially with height according to the formula

$$T = T_0 e^{-\frac{z}{H}}$$

where $T_0 = 273°$ K is the temperature at $z = 0$ and H is the height of the homogeneous atmosphere for T_0. Show that

$$p = p_0 e^{\left(1 - e^{\frac{z}{H}}\right)}$$

where p_0 is the pressure at $z = 0$. Find the height in km at which the lapse rate of temperature becomes equal to the dry adiabatic lapse rate.

3. Derive a formula for the dependence of density upon height in an atmosphere of constant lapse rate of temperature.

4. Consider a layer of air with pressures p_1 and p_2 at the bottom and top respectively. If the mean virtual temperature of the layer remains constant and the upper pressure changes by an increment, dp_2, show that the change in the lower pressure, dp_1, is given by

$$\frac{dp_1}{p_1} = \frac{dp_2}{p_2}$$

5. Consider an atmosphere of constant lapse rate with surface temperature T_0 which varies with time while the surface pressure remains constant. Show that the rate of change of pressure with time at fixed elevations is a maximum at the height of the homogeneous atmosphere corresponding to T_0.

6. A station barometer is 3280.8 geopotential ft above mean sea level. The average temperature over the last 12 hours was 0° F. The fully corrected station pressure is 890.0 mb. Calculate the sea-level pressure if the air is assumed to be dry with a lapse rate of 6.5° C per geopotential km.

CHAPTER 7

HYDROSTATIC STABILITY AND CONVECTION

7.1 General Considerations

In this chapter we shall be concerned with an atmospheric environment which obeys the hydrostatic equation—that is, an atmosphere which is in hydrostatic equilibrium. We shall inquire whether a sample of air displaced by some impulse from its initial position will be forced back, will be accelerated away, or will be subject to no further forces by the environment. In the event that such a parcel becomes subject to restoring forces, we shall consider the atmosphere to be *stable* at that level; if the sample is subject to forces which push it farther away, we shall consider the atmosphere to be *unstable* at that level; and if the displaced parcel is subject to no net force, we shall consider the atmosphere to be in a state of *neutral equilibrium* at that level.

Clearly, criteria for determining the stability of vertically moving air will be important in understanding and predicting significant atmospheric phenomena such as convection and turbulence. We shall consider several models of vertical convection, starting with the oldest and simplest and proceeding to the newest and most complex theories. We shall see that atmospheric convection as exemplified by cumulus clouds is a complicated matter, but that certain concepts from the most elementary model have permanent value in understanding the newest ideas.

7.2 The Dry and Moist Adiabatic Lapse Rates

The criteria for stability at which we shall arrive involve the adiabatic lapse rates for a parcel moving dry adiabatically and moist adiabatically. Therefore we shall first derive expressions for these lapse rates.

When unsaturated air moves vertically it changes state in a very nearly dry adiabatic fashion and conserves its potential temperature. Therefore, by logarithmic differentiation of Poisson's equation, holding θ constant, we get:

$$\frac{1}{T}\frac{dT}{dz} = \frac{\kappa}{p}\frac{dp}{dz}$$

We assume the moving parcel is in dynamic equilibrium with its environment. Therefore the vertical pressure variation, dp/dz, depends upon the density of the environment and not upon the density of the parcel. We denote properties of the environment by a prime and properties of the parcel by the absence of a prime. Thus, from the hydrostatic equation

$$\frac{dp'}{dz} = -\rho' g = -\frac{p'g}{RT'} = -\frac{pg}{RT'}$$

Therefore

$$\frac{dT}{dz} = -\frac{g}{c_p}\frac{T}{T'}$$

Thus unsaturated rising air does not cool at exactly the rate given by $\Gamma_d = g/c_p$, the lapse rate in an atmosphere with constant potential temperature. However, the difference between T and T' is usually small, so that the ratio of these absolute temperatures is close to one. Consequently

$$-\frac{dT}{dz} \approx \frac{g}{c_p} = \Gamma_d \tag{7.1}$$

for dry adiabatic processes.

In order to calculate the rate at which saturated air cools when it rises we utilize Eq. (4.10) for pseudoadiabatic processes of saturated air, with the further approximation that $p - e_s \approx p$:

$$-L\,dw_s = c_p\,dT - \frac{R^*T}{m_d}\frac{dp}{p} \tag{7.2}$$

From Eq. (4.2)

$$w_s \approx \frac{\epsilon e_s}{p}$$

Therefore

$$\frac{dw_s}{w_s} = \frac{de_s}{e_s} - \frac{dp}{p}$$

And from the hydrostatic equation

$$\frac{dp}{p} = -\frac{gm_d}{R^*T'}\,dz$$

Again we shall assume T/T' is sufficiently close to one. When we substitute these last two results into Eq. (7.2) we get

$$-Lw_s\left(\frac{de_s}{e_s}+\frac{gm_d}{R^*T}dz\right)=c_pdT+gdz$$

When we divide by dz, substitute for de_s/dz the equivalent expression

$$\frac{de_s}{dT}\frac{dT}{dz}$$

and solve for $-dT/dz$, we obtain

$$\Gamma_s\equiv-\frac{dT}{dz}=g\frac{1+\epsilon\dfrac{Le_s}{pR_dT}}{c_p+\epsilon\dfrac{L}{p}\dfrac{de_s}{dT}}$$

where Γ_s denotes the lapse rate for saturated pseudoadiabatic processes. We may eliminate e_s in favor of w_s by remembering that

$$\frac{\epsilon e_s}{p}\approx w_s \quad\text{and}\quad \frac{de_s}{dT}=\frac{m_vLe_s}{R^*T^2}=\frac{L}{R_d}\frac{w_sp}{T^2}$$

Then

$$\Gamma_s=\frac{g}{c_p}\left[\frac{1+\dfrac{L}{R_d}\dfrac{w_s}{T}}{1+\dfrac{\epsilon L^2}{c_pR_d}\dfrac{w_s}{T^2}}\right]\tag{7.3}$$

We see from this result that Γ_s is not a constant, but is equal to Γ_d multiplied by a factor which is a function of pressure and temperature. (Remember that w_s depends upon pressure and temperature.) Values of this saturated pseudoadiabatic lapse rate for various pressures and temperatures are given in Table 7.1. This moist adiabatic lapse rate is always less than Γ_d but approaches Γ_d as pressure increases or temperature decreases.

TABLE 7.1

VALUES OF THE SATURATED PSEUDOADIABATIC LAPSE RATE IN °C km^{-1}, CALCULATED FROM EQ. (7.3) FOR SATURATION WITH RESPECT TO WATER

T (°C) \ p (mb)	1000	700	500
−30	9.2	9.0	8.7
−20	8.6	8.2	7.8
−10	7.7	7.1	6.4
0	6.5	5.8	5.1
+10	5.3	4.6	4.0
+20	4.3	3.7	3.3

In order to take into account the effect of water vapor on the density one may think of Γ_d and Γ_s as lapse rates of virtual temperature.

7.3 The Parcel Method

In this very simple approach we shall consider the vertical motions of an individual parcel of air with the following simplifying assumptions:

(1) No compensating motions occur in the environment as the parcel moves.

(2) The parcel does not mix with its environment and so retains its identity.

Neither of these two assumptions is completely justifiable and we shall investigate later some of the consequences of dropping each of these assumptions. As for the first assumption, if a sample of air moves upward it cannot leave a vacuum behind. Therefore, some of the environmental air must sink to take its place. If the ascending element is small and isolated, however, then the subsidence in the large environmental area surrounding it will be small and may be neglected. With respect to the second assumption, in reality there are no physical barriers along the surface of a parcel to prevent mixing with the surrounding air. Observations indicate that such mixing does take place. Nevertheless we shall neglect mixing in order to obtain an initial simple result.

The stability criteria in this simple approach, which is called the *parcel method*, can be obtained without the formal use of mathematics. Consider an atmosphere in hydrostatic equilibrium with a certain lapse rate of virtual temperature, γ. Imagine a parcel of air which initially has the same temperature, pressure, and density as its surroundings. There will be no net vertical force on this parcel since it is in hydrostatic equilibrium. Such an element of air is "floating." Now suppose the parcel is given a small upward displacement by some external agency. If the parcel remains unsaturated it will expand and cool at the dry adiabatic rate, $\Gamma_d = g/c_p$. If the environmental lapse rate is less than the dry adiabatic value, $\gamma < \Gamma_d$, the parcel will be at a lower virtual temperature than its new surroundings. Since the pressure in the parcel very quickly becomes the same as the pressure in the environment, it follows that the parcel will be more dense than its surroundings and will not be buoyant, but will sink back to the original level.

These conditions represent the stable case. Indeed, one can see qualitatively that the parcel will accelerate downward because of the part of gravity which is not exactly cancelled by the vertical gradient of pressure. It will arrive at its original level at the same temperature and density with which it started, and will once more be in buoyant equilibrium with no net vertical force. However, it will have a certain downward momentum which will carry it beyond this point. As it sinks it will warm at the rate $\Gamma_d > \gamma$ and will find itself warmer and lighter than its new environment. Thus there will be a net upward force which will ultimately reverse its motion and send it upward, just as happens to a cork released under water. In this way, we see in the stable case that a displaced parcel should continue to oscillate about its original position until viscosity robs the oscillation of its energy.

In the case $\Gamma_d < \gamma$ a parcel displaced upward will find itself with a temperature greater than that of the new environment. Consequently it will be lighter than the surroundings and will be subject to a net upward force. In this case the parcel will continue to move upward and will not return to its original location. Similarly a parcel displaced downward will continue to move down—the unstable case.

Finally, when $\Gamma_d = \gamma$ a parcel displaced upward or downward will have the same temperature and density as its surroundings. Consequently it will be subject to no net force in either direction—the neutral equilibrium. In summary the stability criteria for unsaturated parcel displacements are

$$\Gamma_d > \gamma\text{: stable,}$$
$$\Gamma_d = \gamma\text{: neutral,}$$
$$\Gamma_d < \gamma\text{: unstable.}$$

Thus, as was pointed out earlier, the dry adiabatic lapse rate is the dividing line between mechanical stability and instability for dry air. Consequently the homogeneous atmosphere, which has a lapse rate far greater than Γ_d, is highly unstable and is found only in shallow layers near the ground where viscous and turbulent effects are dominant. We now see how useful it is to have dry adiabats on a thermodynamic diagram. When the results of a sounding are plotted we merely need to ascertain whether the potential temperature of the atmosphere decreases or increases with height in order to know whether the air is unstable or stable. In actuality a decrease of θ with height rarely can be maintained for long because the resulting instability causes mechanical mixing which tends to produce a constant θ throughout the mixed layer. Thus in practice a lapse rate equal to or near the dry adiabatic value is considered a sign of near instability which may produce convection shortly.

The preceding argument also can be made quantitative. Since the environment is assumed to be in hydrostatic equilibrium, there is no upward acceleration of the environmental air. This condition is expressed by the hydrostatic equation in the form

$$0 = -g - \alpha' \frac{dp'}{dz}$$

In general the parcel under consideration will have some vertical acceleration, $\ddot{z}$, where each dot represents one differentiation with respect to time. From Newton's second law of motion this acceleration must be equal to the algebraic sum of the two vertical forces per unit mass which are being considered, gravity and pressure-gradient force. As before, we denote properties of the environment by a prime and properties of the parcel by the absence of a prime. Since we assume the pressure of the parcel to be equal to the pressure of its environment we get

$$\ddot{z} = -g - \alpha \frac{dp'}{dz}$$

Elimination of dp'/dz between these two equations gives

$$\ddot{z} = g \frac{\alpha - \alpha'}{\alpha'}$$

Substituting for α and α' from the equation of state for moist unsaturated air we obtain

$$\ddot{z} = g \frac{T^* - T^{*\prime}}{T^{*\prime}} \tag{7.4}$$

We shall investigate small displacements of the parcel from its original location. Let $z = 0$ at this original location where the temperature is T_0^*. Then the temperature at any point z in the environment is

$$T^{*\prime} = T_0^* + \frac{dT^{*\prime}}{dz} z + \frac{1}{2} \frac{d^2T^{*\prime}}{dz^2} z^2 + \ldots$$

The above is a Taylor expansion of temperature as a function of height. If the displacements are sufficiently small the higher order terms involving powers of z will be small and may be neglected. We thus reduce to

$$T^{*\prime} \approx T_0^* + \frac{dT^{*\prime}}{dz} z$$

This will be an exact result if $T^{*\prime}$ varies linearly with height, for then the higher derivatives will be identically zero. In exactly the same way we may write for the temperature of the parcel

$$T^* = T_0^* + \frac{dT^*}{dz} z$$

Since $-dT^{*\prime}/dz = \gamma$, the environmental lapse rate, and $-dT^*/dz = \Gamma_d$, the dry adiabatic lapse rate, these two results may be written

$$T^{*\prime} = T_0^* - \gamma z$$
$$T^* = T_0^* - \Gamma_d z$$

When we substitute these equations in Eq. (7.4) we obtain

$$\ddot{z} = \frac{g}{T_0^* - \gamma z}(\gamma - \Gamma_d)\, z$$

But

$$\frac{1}{T_0^* - \gamma z} \approx \frac{1}{T_0^*}\left(1 + \frac{\gamma z}{T_0^*}\right)$$

since $\gamma z/T_0^*$ is a small quantity. This may be verified by performing the indicated division and neglecting all terms involving squares and higher powers of $\gamma z/T_0^*$. With this approximation the expression for the vertical acceleration becomes

$$\ddot{z} = \frac{g}{T_0^*}\left[(\gamma - \Gamma_d)\, z + \frac{(\gamma - \Gamma_d)\gamma}{T_0^*} z^2\right]$$

We once again neglect the term involving z^2 compared to the term involving z to obtain the final result

$$\ddot{z} + \frac{g}{T_0^*}(\Gamma_d - \gamma)\, z = 0 \tag{7.5}$$

This is a well-known standard differential equation. If the coefficient of z, $g(\Gamma_d - \gamma)/T_0^*$, is positive the solution for z is a sinusoidal function of time. That is, the parcel will oscillate about its original position with a period given by

$$\tau = \frac{2\pi}{\sqrt{\dfrac{g}{T_0^*}(\Gamma_d - \gamma)}}$$

This is the stable case.[1] If the coefficient of z is negative the solution of Eq. (7.2) is in terms of exponentials of time. Thus the displacement will increase indefinitely. This is the unstable case. Finally, if the coefficient of z is zero, $\ddot{z} = 0$ and a displaced parcel does not accelerate at all. This is the neutral case.

---

[1] It is easy to verify by substitution into Eq. (7.5) that its solution in this case is

$$z = A\cos\frac{2\pi}{\tau}t + B\sin\frac{2\pi}{\tau}t$$

where A and B are constants of integration.

Now the sign of the coefficient of z in Eq. (7.5) is determined solely by the relative sizes of Γ_d and γ. The criteria are then

$$\begin{aligned} \Gamma_d &> \gamma\text{: stable,} \\ \Gamma_d &= \gamma\text{: neutral,} \\ \Gamma_d &< \gamma\text{: unstable.} \end{aligned}$$

These are identical with the stability criteria arrived at earlier by qualitative reasoning.

So far we have discussed only the case of unsaturated air. When the displaced parcel is saturated its lapse rate is not the dry adiabatic value, Γ_d, but the moist adiabatic value, Γ_s. Thus it is clear that the same stability criteria apply in the saturated case except that one must compare the environmental lapse rate to the moist adiabatic value instead of the dry adiabatic value. Since the moist adiabatic lapse rate is smaller, it is easier to obtain instability for saturated than for unsaturated air. Furthermore, this means that we must consider a special case of instability.

Suppose the environmental lapse rate lies between the moist and dry adiabatic values. An initially unsaturated parcel forced to ascend will be stable since $\Gamma_d > \gamma$. But if the impulse forcing the air upward lasts long enough the parcel will reach the lifting condensation level and become saturated. Then the parcel lapse rate immediately becomes less than that of the environment and instability results, known as *conditional instability*. In this case, when the environmental lapse rate lies between Γ_d and Γ_s, a parcel is stable with respect to unsaturated lifting processes and unstable with respect to saturated lifting processes.

Figure 7.1 gives an example of conditional instability plotted on a

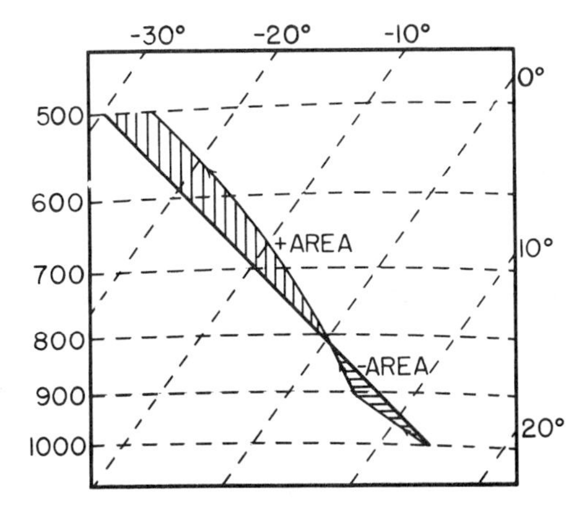

FIG. 7.1. Example of conditional instability portrayed on a tephigram

tephigram. A parcel forced to ascend dry adiabatically from 1000 mb at first is colder than the environment and is subject to negative or downward buoyancy. It reaches the LCL at 900 mb and ascends moist adiabatically thereafter. The parcel curve intersects the environmental curve at about 820 mb, thus defining an area on the tephigram which lies completely to the left of the sounding curve. This is called a *negative area* since it is proportional to the energy which must be supplied to the parcel in order to lift it this high. In this case the specific energy to be supplied is 0.5×10^6 ergs g^{-1}. Above 820 mb the parcel is warmer than the sounding and is subject to positive or upward buoyancy. From 820 mb to 500 mb, where the diagram has been terminated arbitrarily, the parcel curve and sounding curve define an area completely to the right of the sounding curve. This is called a *positive area*, since it is proportional to the energy which becomes available to the parcel from its environment. In this case the specific energy released is 3.8×10^6 ergs g^{-1} up to 500 mb. This is a very large energy, for if all of it were converted into upward motion the vertical speed at 500 mb would be nearly 28 m s^{-1}, or about 60 miles per hr. Such updrafts are quite extreme, even though the example given is quite realistic. Clearly the parcel method neglects significant retarding influences which greatly slow the ascending air.

In summary, five states can be recognized with respect to parcel displacement in an atmosphere of lapse rate γ. The atmosphere is said to be:

(1) *absolutely stable* if $\gamma < \Gamma_s$,
(2) *saturated neutral* if $\Gamma_s = \gamma$,
(3) *conditionally unstable* if $\Gamma_s < \gamma < \Gamma_d$,
(4) *dry neutral* if $\gamma = \Gamma_d$,
(5) *absolutely unstable* if $\gamma > \Gamma_d$.

7.4 Changes of Stability During Displacement of Layers

In the preceding section we found that the stability of a parcel depended upon the relationship of the lapse rate in the environment to Γ_d and Γ_s. There are meteorological situations, such as ascent or descent of a broad mountain barrier, in which an entire atmospheric layer may be lifted or lowered. Does this affect the environmental lapse rate and thereby the parcel stability? Throughout our discussion of this question we shall consider a layer of the atmosphere with a finite difference in pressure between its bottom and top. From the hydrostatic equation this pressure difference is directly proportional to the mass per unit area contained in the column. Provided no additional mass is brought into the layer by horizontal or vertical convergence of air, the vertical pressure increment between top and bottom of the layer must remain constant. This is the situation with which we shall deal.

Let us first consider unsaturated processes. We shall seek a relationship between the lapse rate of potential temperature, on the one hand, and the temperature lapse rate on the other hand. The relationship between T' and θ' is Poisson's equation which yields, when differentiated logarithmically with respect to height

$$\frac{1}{T'}\frac{\partial T'}{\partial z} - \frac{1}{\theta'}\frac{\partial \theta'}{\partial z} = \frac{\kappa}{p'}\frac{\partial p'}{\partial z}$$

When we substitute from the hydrostatic equation and solve for

$$\frac{1}{\theta'}\frac{\partial \theta'}{\partial z}$$

we obtain

$$\frac{1}{\theta'}\frac{\partial \theta'}{\partial z} = \frac{1}{T'}\left(\frac{\partial T'}{\partial z} + \frac{g}{c_p}\right)$$

or

$$\frac{1}{\theta'}\frac{\partial \theta'}{\partial z} = \frac{\Gamma_d - \gamma}{T'} \tag{7.6}$$

In order to take advantage of the constant pressure increment in our layer we wish to convert from a height derivative to a pressure derivative. Since

$$\frac{\partial \theta'}{\partial p'} = \frac{\partial \theta'}{\partial z}\frac{\partial z}{\partial p'}$$

and from the hydrostatic equation $\partial z/\partial p' = -1/(\rho' g)$, we may write Eq. (7.6) as

$$\frac{1}{\theta'}\frac{\partial \theta'}{\partial p'} = -\frac{\Gamma_d - \gamma}{g\rho' T'}$$

Finally we introduce the equation of state and find that

$$\frac{1}{\theta'}\frac{\partial \theta'}{\partial p'} = -\frac{R_d}{g}\frac{(\Gamma_d - \gamma)}{p'}$$

Since in a dry adiabatic process θ' is conserved, the difference in θ' from top to bottom of a layer is conserved also. Furthermore, we have confined ourselves to the case where the vertical pressure increment in the layer is constant. Therefore

$$\frac{1}{\theta'}\frac{\partial \theta'}{\partial p'}$$

is constant for the layer and

$$\Gamma_d - \gamma = \text{const}\, p'$$

That is, during lifting of the layer when the pressure decreases, the lapse rate becomes more nearly dry adiabatic. During sinking of the layer when the pressure increases, the lapse rate moves farther from Γ_d. Thus lifting always destabilizes a stable layer while sinking stabilizes it further with respect to parcel displacements.

When the lifting of an air layer is so appreciable as to cause saturation throughout the layer the result is completely different. The result is found most easily by means of graphical operations on a thermodynamic diagram. Figure 7.2 shows three different cases plotted on a

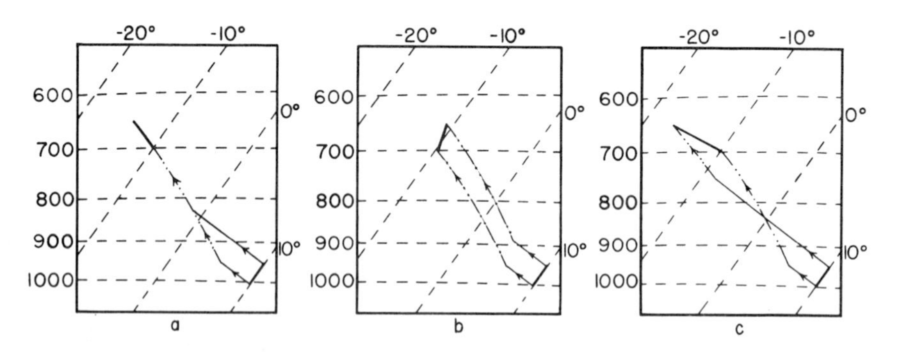

FIG. 7.2. Results of lifting an initially stable, unsaturated layer when

$$\text{(a)}\ \frac{\partial \theta_e}{\partial z} = 0, \quad \text{(b)}\ \frac{\partial \theta_e}{\partial z} > 0, \quad \text{(c)}\ \frac{\partial \theta_e}{\partial z} < 0$$

The layer lapse rates before and after lifting are given by heavy solid lines, dry adiabats by thin solid lines, and saturated adiabats by dash-dot lines.

tephigram. In all three cases the layer is 50 mb thick and is initially isothermal, so it is stable with respect to both dry and moist adiabatic parcel processes. In all three cases the layer is lifted 300 mb which suffices to saturate the entire layer. In Fig. 7.2*a* we assume that the equivalent potential temperature is constant in the layer. Thus every point in the layer, after a preliminary dry adiabatic expansion, reaches condensation along the same moist adiabatic line. Consequently, the layer lapse rate after lifting is exactly moist adiabatic and the layer has become neutral with respect to saturated parcel displacements.

In Fig. 7.2*b* we assume that θ_e increases with height. Thus the top of the layer reaches saturation along a moist adiabat which is to the right of the one along which the bottom of the layer reaches saturation.

Consequently the final lapse rate is less than Γ_s and the layer is stable with respect to saturated parcel displacements.

Finally, in Fig. 7.2*c*, we assume that θ_e decreases with height in the layer. Then the upper part of the layer reaches saturation along a moist adiabat which lies to the left of the one along which the bottom of the layer reaches saturation. Consequently the final lapse rate is steeper than the moist adiabats and is therefore unstable with respect to saturated parcel displacements.

Clearly these results are independent of the initial lapse rate and conditions chosen so we have a general result. The parcel stability of a layer of air which is lifted until it is completely saturated depends upon the lapse rate of equivalent potential temperature.

If $\frac{\partial \theta_e}{\partial z} > 0$ the saturated layer will be stable;

if $\frac{\partial \theta_e}{\partial z} = 0$ the saturated layer will be neutral;

if $\frac{\partial \theta_e}{\partial z} < 0$ the saturated layer will be unstable.

Because of the one-to-one correspondence between equivalent potential temperature and wet-bulb potential temperature, the latter may be substituted in these criteria. These results are called *convective stability*, *convective neutrality*, and *convective instability*, respectively.

7.5 The Slice Method

We now take into account compensating vertical motions in the environment as a parcel or column of air rises. The procedure we shall use was introduced by J. Bjerknes[2] and is called the *slice method*. Since we shall be concerned primarily with the formation of cumulus clouds we shall consider an initially horizontal layer of air which is saturated. Within such a layer there may be several regions in which the air is ascending and cooling moist adiabatically. Within the remainder of the layer there must be descent with warming at the dry adiabatic rate. Let the ascending air have total horizontal area A and let it be moving upward with speed w. The descending air will have area A' and vertical speed w'. We shall assume that the rate at which mass descends through a fixed reference level in the slice of originally saturated air is equal

[2] J. BJERKNES, Saturated Ascent of Air through a Dry-adiabatically Descending Environment. *Q. J. Roy. Meteor. Soc.*, 65, 1938.

to the rate at which mass ascends through the reference level. In a brief interval of time, dt, the masses transported upward and downward are

$$dM = \rho\, A\, w\, dt = \rho\, A\, dz = -A\, dp/g$$
$$dM' = \rho' A' w' dt = \rho' A' dz' = -A' dp'/g$$

where dz and dz' are the vertical distances through which the ascending and descending air move in the time dt, and dp and dp' are the corresponding hydrostatic changes in pressure. At the initial moment the slice is assumed to be horizontally homogeneous so $\rho = \rho'$. Since $dM = dM'$ we find at the start of the process

$$\frac{A'}{A} = \frac{w}{w'} = \frac{dz}{dz'} = \frac{dp}{dp'} \tag{7.7}$$

We shall assume further that local temperature changes are caused solely by vertical motion. That is, horizontal transport of cooler or warmer air will be neglected. In Fig. 7.3 are illustrated the reference

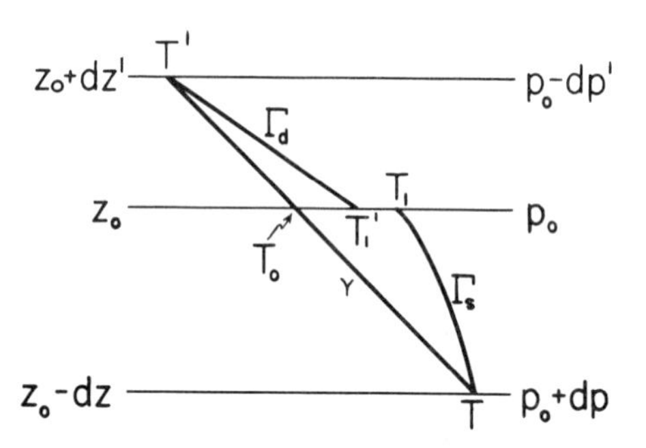

FIG. 7.3. Temperature changes in a conditionally unstable slice of air as a result of dry adiabatic descent and saturated adiabatic ascent to the reference level z_0

level z_0 where the pressure is p_0, above it the level from which air will sink to z_0 in time dt, and below it the level from which air will rise to z_0 in the same time. This is for an initially conditionally unstable atmosphere, which is the most interesting case. When the ascending air, which has a temperature T at the level $z - dz$, reaches the reference level it will have a temperature

$$T_1 = T - \Gamma_s\, dz$$

and when the descending air, which has a temperature T' at the level $z + dz'$, reaches the reference level it will have a temperature

$$T_1' = T' + \Gamma_d\, dz'$$

Just as in our treatment of the parcel method the establishment of instability requires positive buoyancy for the ascending air. That is, T_1, the temperature at z_0 of the rising air, must be greater than T_1', the temperature of the surrounding subsiding air at z_0. Therefore, for the unstable case

$$T - \Gamma_s\,dz > T' + \Gamma_d\,dz'$$

But $T = T_0 + \gamma dz$, and $T' = T_0 - \gamma dz'$, where T_0 is the initial uniform temperature at the reference level and γ is the initial lapse rate in the layer. Therefore

$$(\gamma - \Gamma_s)\,dz > (\Gamma_d - \gamma)\,dz'$$

But from Eq. (7.7), $A'/A = dz/dz'$, so

$$(\gamma - \Gamma_s)\,A' > (\Gamma_d - \gamma)\,A \text{ for instability;}$$
$$(\gamma - \Gamma_s)\,A' = (\Gamma_d - \gamma)\,A \text{ for neutrality;}$$
$$(\gamma - \Gamma_s)\,A' < (\Gamma_d - \gamma)\,A \text{ for stability.}$$

These results may be put in a compact form by using γ_n, the slice lapse rate for the middle case of neutral equilibrium, as a parameter:

$$\gamma_n = \frac{A\Gamma_d + A'\Gamma_s}{A' + A} \tag{7.8}$$

The stability criteria in terms of γ_n reduce to:

$$\gamma > \gamma_n\text{: unstable,}$$
$$\gamma = \gamma_n\text{: neutral,}$$
$$\gamma < \gamma_n\text{: stable.}$$

The above statements have the same form as those for parcel stability in the dry and saturated cases, except instead of making the comparison to Γ_d or Γ_s we compare to γ_n. We see from Eq. (7.8) that γ_n is a weighted mean of Γ_d and Γ_s, where the weighting factors are the areas of ascent and descent. Consequently, in the conditionally unstable case we are considering $\Gamma_d > \gamma_n > \Gamma_s$. Thus when one takes into account compensating motions in the environment it is easier to get instability than in the case of dry uncompensated parcel displacements, but harder to get instability than in the case of saturated uncompensated parcel displacements. In the comparison to dry parcel motions this result is due to the fact that we are dealing with initially saturated air, and even the parcel method indicates it is then easier to get instability. In the comparison to saturated parcel motions this result is due entirely to the effect of the compensating subsidence.

The slice method is undoubtedly a theoretical improvement over the parcel method because it eliminates one of the unrealistic assumptions upon which the parcel method is based. However, it is not easy to apply the slice method in practice, because Eq. (7.8) requires one to know the relative areas of ascent and descent, which are not known until after the cumulus clouds have actually formed. Thus it is difficult to forecast instability by this means. The greatest contribution of the slice method is the increased understanding which it gives us of the process of initiation of cumulus development. Equation (7.8) and the associated stability criteria indicate that the chances for development of slice instability are greatest when A' is large and A is small, for then γ_n is small and more easily exceeded. That is, small areas of ascent and large areas of descent are most favorable to the development of cumulus clouds. This makes physical sense, because if air is subsiding gently over most of a given region, there must be rapid upward compensating motion in the few narrow cumulus towers present. This is clearly conducive to their development and is a direct consequence of the introduction of compensating vertical motions in the theory.

It should be pointed out also that compensating subsidence in the surroundings helps to reduce the excessive buoyancy found in the parcel method. Since the descending air warms dry adiabatically, the temperature difference between a buoyant cloud and its subsiding environment is less than it would be for a resting environment. For example, in Fig. 7.1 the parcel method gives at 700 mb $T - T' = +2.5°$ C. But if we suppose that 25 per cent of the total horizontal area is covered by ascending air and 75 per cent by descending air then in the time a parcel could ascend from 1000 mb to 700 mb, the environment would descend from 600 mb to 700 mb. The dry adiabatic descent would produce in this case an environmental temperature 2.0° C warmer at 700 mb than it was originally. Thus the temperature excess of the cloud would be reduced to $+0.5°$ C. It should be clear from this example that buoyancy forces in the slice method can be several times smaller than in the parcel method. It is worth noting that the results of the slice method reduce to those of the parcel method as A, the area of ascent, approaches zero. For when $A \ll A'$, Eq. (7.8) reduces to $\gamma_n = \Gamma_s$ and the stability criteria become the same as those of the parcel method for saturated air.

7.6 Entrainment into Cumulus Clouds

This section describes some of the effects of mixing of environmental air into an ascending saturated cloud mass. A cumulus cloud may be looked upon as a narrow jet of rising air embedded in a relatively quiescent environment. Hydrodynamical theory and experiment indicate that when a current is injected into a fluid the motion of the jet draws into it some of the surrounding medium. This phenomenon has been called *entrainment*; it seems reasonable to expect a similar phenomenon in a cumulus. Our problem will be to relate the rate of entrainment to the measurable properties of the cloud and its environment. The discussion will be based upon the treatment by Austin and Fleischer.[3]

Consider a sample of saturated cloud of mass m, which ascends through a distance dz and experiences a pressure change dp. As the sample rises it entrains a mass dm of unsaturated environmental air. We shall consider the ascent and mixing to consist of (1) a saturated adiabatic cooling of the cloud mass, m; (2) isobaric cooling of m and warming of dm during the mixing; (3) evaporation of part of the liquid water content of m into dm so as to produce a final mixture which is saturated. We shall assume that the cloud has a uniform horizontal distribution of temperature, water vapor, and liquid water. A method for the graphical evaluation of these steps has been described.[4]

We shall assume that the cloud and the air it entrains are a thermodynamically isolated system, so that no external sources or sinks of heat need be considered. Thus heat gained by dm must be equal to heat lost by m except for work done during expansion. The heat required to warm the entrained environment from its original temperature, T', to the cloud temperature, T, is

$$dQ_1 = c_p (T - T')\, dm$$

As before, we ignore the heat changes of the vapor and the liquid water compared to the heat change of the dry air. In order to saturate dm heat is needed to evaporate some of the liquid water. This amount of heat is

$$dQ_2 = L (w_s - w')\, dm$$

where w_s is the saturation mixing ratio in the cloud and w' is the mixing ratio of the environmental air. Finally, as the sample ascends, a certain amount of vapor, dw_s, condenses. The heat released by this condensation is

$$dQ_3 = - mL dw_s$$

The cloud mass, m, thus loses amounts of heat given by dQ_1 and dQ_2, and gains dQ_3. Thus the first law of thermodynamics for the cloud mass may be written

$$- c_p (T - T')\, dm - L (w_s - w')\, dm - mL dw_s = m \left(c_p dT - \frac{R^* T}{m_d} \frac{dp}{p} \right) \qquad (7.9)$$

[3] J. M. AUSTIN, and A. FLEISCHER, A Thermodynamic Analysis of Cumulus Convection. *J. Meteor.*, 5, 1948, pp. 240-3.

[4] H. STOMMEL, Entrainment of Air into a Cumulus Cloud. *J. Meteor.*, 4, 1947, pp. 91-4.

Next we (1) divide by m, (2) substitute for dw_s from the approximate relationship

$$\frac{dw_s}{w_s} = \frac{de_s}{e_s} - \frac{dp}{p}$$

and (3) substitute for dp/p from the hydrostatic equation,

$$\frac{dp}{p} = -\frac{gm_d}{R^*T'}\,dz \approx -\frac{gm_d}{R^*T}\,dz$$

This gives

$$-\frac{dm}{m}\left[c_p(T - T') + L(w_s - w')\right] - L\left[\frac{w_s de_s}{e_s} + \frac{gm_d w_s}{R^*T}\,dz\right] = c_p dT + g dz$$

We now divide by dz, and substitute for de_s/dz the equivalent quantity

$$\frac{de_s}{dT}\frac{dT}{dz}$$

and solve for $-dT/dz$. The result is

$$-\frac{dT}{dz} = \frac{\dfrac{g}{c_p}\left[1 + \dfrac{L}{R_d}\dfrac{w_s}{T}\right] + \dfrac{1}{m}\dfrac{dm}{dz}\left[(T - T') + \dfrac{L}{c_p}(w_s - w')\right]}{1 + \dfrac{Lw_s}{c_p e_s}\dfrac{de_s}{dT}}$$

Finally we substitute for e_s from the Clausius-Clapeyron equation:

$$\frac{1}{e_s}\frac{de_s}{dT} = \frac{m_v L}{R^*T^2} = \frac{\epsilon L}{R_d T^2}$$

The result for the lapse rate in an entraining cloud is

$$-\frac{dT}{dz} = \frac{\dfrac{g}{c_p}\left[1 + \dfrac{L}{R_d}\dfrac{w_s}{T}\right] + \dfrac{1}{m}\dfrac{dm}{dz}\left[(T - T') + \dfrac{L}{c_p}(w_s - w')\right]}{1 + \dfrac{\epsilon L^2}{c_p R_d}\dfrac{w_s}{T^2}} \quad (7.10)$$

We first note that Eq. (7.10) reduces to Eq. (7.3), the expression for the moist adiabatic lapse rate, if $dm/dz = 0$—that is, if no entrainment takes place. Furthermore dm/dz is positive, and $T > T'$ if the cloud is to experience a buoyant lift, therefore the lapse rate in an entraining cloud is greater than Γ_s. For example, if at 700 mb, $T = 0°$ C, $T' = -1°$ C, relative humidity in the environment is 67 per cent, and an observable rate of entrainment of 25 per cent km^{-1} occurs; the cloud lapse rate from Eq. (7.10) is 6.6° C km^{-1} while $\Gamma_s = 5.8°$ C km^{-1}. Thus entrainment can make an appreciable difference in lapse rate.

Since the lapse rate in an entraining cloud is larger than the saturated adiabatic value it is less likely that such a cloud can achieve appreciable buoyant lift than a cloud which does not mix with its environment. Indeed airplane observations indicate very small differences in temperature inside and outside a cumulus—in other words, small buoyancy of the cloud. Thus entrainment also reduces the excessive buoyant forces which result from the parcel theory.

In order to explore the consequences of entrainment further we may take advantage of the observations which indicate that $T \approx T'$. This means, in Eq. (7.9), that $c_p(T - T')\, dm$ is small compared to $L(w_s - w')\, dm$, except in the rare case of a nearly saturated environment. With these simplifications Austin and Fleischer were able to integrate Eq. (7.9) numerically for various environmental lapse rates and relative humidities. Their results are given in Fig. 7.4, which indicates that so

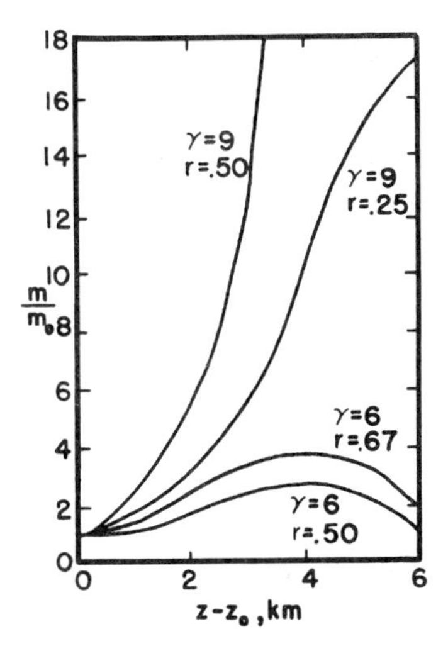

FIG. 7.4. Dilution of a cloud mass with height, where m/m_0 is the ratio of actual cloud mass m at height z to the mass m_0 originating at the cloud base z_0, γ is the lapse rate in °C km^{-1}, and r is relative humidity. The cloud base was taken at 900 mb and 20° C.

much air can be entrained that the cloud mass may easily double or triple in a few kilometers of ascent. Furthermore, they find that large lapse rates and high environmental relative humidities mean more rapid entrainment. The extraordinarily large entrainments shown for γ = 9° C km^{-1} are not realistic, but observations show that such large lapse rates in and around cumuli are not common. The curves for γ = 6° C km^{-1} are more realistic.

One difficulty with this entrainment theory can be seen immediately from Fig. 7.4. Since the clouds continue to entrain outside air for much of their ascent one would expect them to grow in width as they ascend. In fact, however, cumuli usually do not change much in diameter with height except near the top where they are often dome-shaped.

Austin and Fleischer have suggested that not all the air in such a cloud continues to ascend, but there may be downdrafts within the cloud which are not included in this theory. Others have suggested that the upward motion is larger for large entrainment rates, thus carrying the mass upward rather than allowing it to spread horizontally.

7.7 The Bubble Theory

There remains a number of serious deficiencies in our understanding of convection as represented by the entrainment theory. In the first place at least two physical forces which will tend to retard the ascent of a cumulus have been neglected. One of these is the downward force which is due to the weight of the condensed water being supported by the cloud air. These droplets exert a downward force per gram of cloud air equal to the liquid water content of each gram of air (w_l) times the acceleration of gravity: $w_l g$. The temperature excess of the cloud over its environment necessary to just balance this downward force can be obtained by equating the buoyancy force per unit mass,

$$g\frac{T - T'}{T'}$$

against the weight of liquid water per unit mass of air. This yields

$$w_l = \frac{T - T'}{T'}$$

If $T - T'$ has the value 1° C, which seems to be about right for many cumuli, and $T' \approx 273°$ K, then $w_l \approx 3.5$ g of liquid water kg^{-1} of dry air. This is a value which often occurs in large cumuli. Thus it appears that the weight of suspended liquid or solid water also makes a definite contribution towards reducing the excessive buoyancies predicted by the parcel method.

The other retarding influence that has been considered is an ordinary drag force exerted by the environment on an element of air rising within it owing to frictional and turbulent forces. The situation is one which we frequently take advantage of in meteorological measurements. When a balloon is filled with helium or hydrogen it is considerably less dense than air. In the absence of drag forces the balloon should accelerate upward. In actuality such balloons move upward with nearly constant speed, which is essential in single-theodolite wind measurements. The reason for this is that shortly after being released the balloon experiences downward frictional and turbulent forces essentially equal to the upward buoyancy. Malkus and Scorer[5] have considered a theory of this drag on cumuli and find it to be an important factor.

Another important factor which has been omitted is the fact that a cumulus has a fairly short life history of only several hours. This rapid growth and decay must be taken into account before a complete theory of convection will have been established. Observations of cumuli indicate that normally individual towers grow for a while, lose their impetus, and are succeeded by new towers. This led Scorer and Ludlam[6] to emphasize the individual bubble or "proton" of rising air as the essential ingredient to be studied. Furthermore, time-lapse photography, which produces "speeded-up" motion pictures of processes, shows clearly that the individual towers or bubbles are losing cloud material to the environment. That is, as the bubble rises through the environment its outer surface is being eroded by the relative descent of the outside air. This process seems like the opposite of entrainment, but we shall see the relationship between the two shortly.

Scorer and Ludlam view the life cycle of a convective bubble as follows: First, because of horizontal variations within a cumulus cloud mass, a portion finds itself buoyant with respect to the environment. This portion rises as a bubble with a roughly spherical top. As it rises it lifts the dry environmental air above it, which cools adiabatically and so begins to sink along the surface of the bubble. Second, this sinking environmental air erodes cloud material from the surface of the bubble and produces beneath the rising tower a "wake" of turbulent air which is a mixture of cloud and environment. Thus erosion is thought of as taking place along the surface of the bubble and entrainment in its wake. This process is illustrated in Fig. 7.5. Third, the bubble is

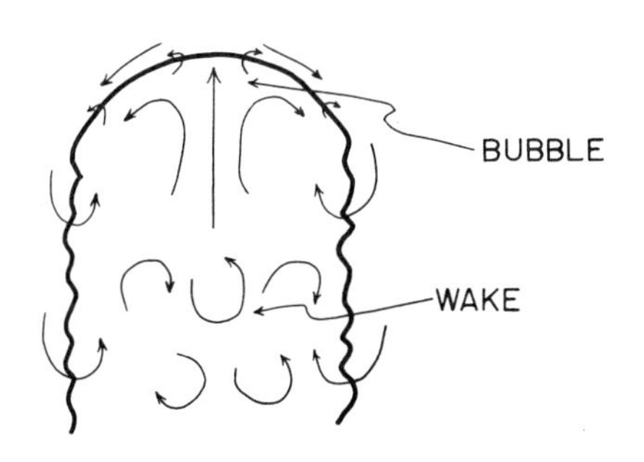

FIG. 7.5. Schematic representation of the ascent of a buoyant bubble through a relatively descending environment. Note erosion of the upper spherical bubble surface and the turbulent mixed wake.

[5] J. S. MALKUS, and R. S. SCORER, The Erosion of Cumulus Towers. *J. Meteor.*, 12, 1955, pp. 43-57.

[6] R. S. SCORER, and F. H. LUDLAM, Bubble Theory of Penetrative Convection. *Q. J. Roy. Meteor. Soc.*, 79, 1953, pp. 317-341.

eventually completely wasted away by this process, leaving only the turbulent, mixed wake. Because of the mixing with environmental air the wake is less buoyant than its parent bubble and may not be able to rise further. The net result has been an upward extension of the cloud mass.

This sequence may also be looked upon as a process of enrichment with water vapor of the environmental air above the original cloud mass. A moist environment is conducive to convection because less erosion can be accomplished by a moist environment. This cycle of growth and decay of a cloud bubble has therefore prepared the way for the development of another bubble which can penetrate farther than the first one. In this way the entire cloud mass grows upward through the action of successive rising bubbles, each of which alters the environment in such a way as to facilitate the ascent of subsequent bubbles.

A qualitative verification of one of the implications of the bubble theory of convection can be found by considering the case in which there is an increase of the horizontal wind speed with height (shear). As bubbles rise, the wake they produce should be carried downstream relative to the parent cloud mass by the higher winds above the original cloud. Accordingly, the environment immediately above the point at which the bubble originated will not be enriched with moisture, but will consist of fresh environmental air brought in by the wind. The enriched wake will be found downwind of the point of bubble origin. The bubble theory therefore predicts that the build-up of cumulus masses should occur on their downwind sides for wind increasing with height. The argument can easily be extended to the case of a decrease of wind speed with elevation, with the result that the build-up should occur on the upstream side. These two cases can be summarized by the single statement that *the greatest growth of a cumulus should be on its downshear side*. These two situations are portrayed in Fig. 7.6.

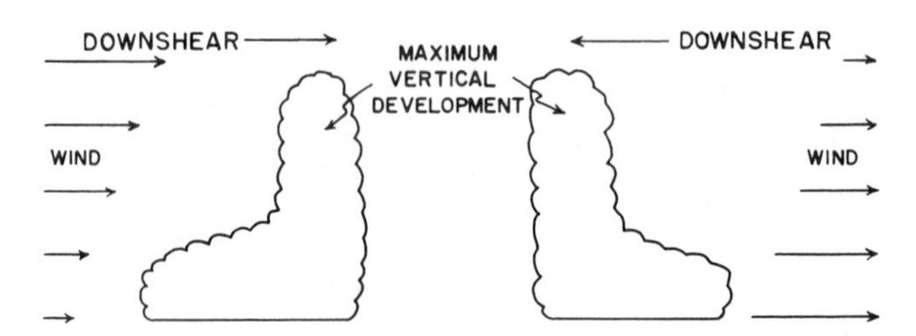

FIG. 7.6. Location of the region of maximum vertical development of a cumulus in the two cases of increase and decrease of wind speed with elevation

These implications of the bubble theory have been verified by direct observation, thus giving considerable support to the theory.

PROBLEMS

1. Calculate the numerical value of the saturated pseudoadiabatic lapse rate in °C km^{-1} for $p = 300$ mb, $T = -20°$ C.

2. A thick atmospheric layer is isothermal with temperature —3° C. A dry parcel is given an upward impetus and begins to oscillate about its original position. Calculate the period in seconds of the oscillation.

3. For the situation given in Problem 2 above, suppose the initial impetus is enough to cause the parcel to rise 1 km above its original level before it begins to sink again. Calculate the energy per unit mass which had to be imparted to the parcel to produce this displacement.

4. Show that $\Gamma_s \leq \Gamma_d$ in the meteorological range of temperatures.

CHAPTER 8

THE FUNDAMENTAL PHYSICS OF RADIATION

8.1 Nature of Radiation

Energy may be transferred from one place to another by any of three processes. *Conduction* involves transfer of the kinetic energy of atoms or molecules (heat) by what may be termed physical contact among molecules of varying speeds. Thus if one end of a metal bar is heated, the atoms at the hot end move about their mean positions faster than those at the cool end. As the fast atoms collide with the adjacent cooler (and slower) atoms they transfer part of their kinetic energy to them. The latter in turn transmit energy by contact to their neighbors farther along the bar, and consequently heat energy moves toward the cool end.

Energy may also be transmitted by physical displacement of matter. In gases and liquids, this transfer of energy takes place through a process known as *convection*. In its broadest sense convection applies not only to the upward transfer of heat when a gas or liquid is heated from below, but also to eddy diffusion and advection of heat in all directions.

Finally, energy may be transmitted from one body to another by means of electromagnetic waves with or without the presence of an intervening physical medium, through the process of *radiation*. Electromagnetic energy travels in straight lines and with a constant speed in a vacuum:

$$c \approx 3.00 \times 10^{10} \text{ cm s}^{-1}$$

One may characterize radiation by its *wave length*, λ, which is simply the distance from one crest of the wave to the next, or by the *frequency*, ν, which is the rate at which wave crests pass a fixed point. Clearly

$$\lambda\nu = c$$

and λ has dimensions $[L]$ while ν has dimensions $[T^{-1}]$. Sometimes wave number is used. This is simply the reciprocal of wave length and is clearly proportional to frequency.

The various kinds of electromagnetic waves which are generally recognized and named are given in Table 8.1 together with their approximate ranges of wave length and frequency. One should remember that the divisions are arbitrary and that the spectrum is really continuous.

TABLE 8.1

THE ELECTROMAGNETIC SPECTRUM

Type of radiation	*Range of wave length* (μ)	*Range of frequency* (s^{-1})
1. Cosmic rays, gamma rays, x-rays, etc.	up to 10^{-3}	3×10^{17} and up
2. Ultraviolet	10^{-3} to 4×10^{-1}	10^{15} to 3×10^{17}
3. Visible	4×10^{-1} to 8×10^{-1}	4×10^{14} to 10^{15}
4. Near-infrared	8×10^{-1} to 4	8×10^{13} to 4×10^{14}
5. Infrared	4 to 10^{2}	3×10^{12} to 8×10^{13}
6. Microwave	10^{2} to 10^{7}	3×10^{7} to 3×10^{12}
7. Radio	10^{7} and up	up to 3×10^{7}

8.2 Atomic and Molecular Spectra

Consider an elementary model of the atom consisting of a nucleus with a certain number of electrons moving in more-or-less fixed elliptical orbits around the nucleus. Each electron has a certain amount of energy (referred to as "energy level") associated with motion in its particular orbit. One might think that all geometrically possible orbits could be occupied by electrons, but the model comes into serious conflict with observed fact once this is allowed. Actually it is a postulate of the quantum theory that only certain orbits—and hence only certain energy levels—are permitted. These energy levels are separated by finite amounts, and orbits corresponding to intermediate energies are incapable of being occupied by electrons. Thus if an electron is to be elevated to a larger orbit a discrete amount of energy must be supplied to the electron.

Radiation must also be thought of as existing in discrete units or *quanta* in order to explain certain observations, although its wavelike character remains the same. It may be difficult to think of a discrete unit that is also a bundle of waves, yet this is precisely the view of quantum mechanics. The energy E of a quantum is related to the frequency by the expression

$$E = h\nu$$

where h is a universal constant called Planck's constant, having the value $h = 6.624 \times 10^{-27}$ erg s.

Now consider a beam of radiation containing a continuous range of frequencies and energies. If this beam falls upon matter a quantum of radiation will frequently come exceedingly close to an orbital electron. If the energy of the quantum is just equal to that required by the electron to elevate it to a higher permitted orbit, the electron may absorb this radiant energy and "jump" to the higher orbit. If the energy of the quantum does not match that needed for a specific transition, no absorption will take place and the quantum will pass on[1] or be transmitted. Since in any appreciable sample of matter there will be many atoms containing electrons, at some given energy level there will be many "captures" of quanta of the specific energy needed to raise these electrons to the next higher state. Consequently the incident radiation will suffer a selective depletion of the quanta of this specific energy. If the radiation passing through the sample is analyzed by means of a spectrometer it will be found that there has been a selective absorption at the frequency corresponding to E for the electronic transition that occurred, producing an *absorption line* in the spectrum.

Conversely, if a sample of matter contains electrons in high-energy levels because of previous excitation (absorption), there is a high probability that they will drop back into lower levels. The electrons will then emit quanta of light carrying exactly the difference in energy between the two levels of the transition—in other words, light of a fixed frequency will be emitted. When this emission occurs in enough atoms each second, a recognizable *emission line* will appear in the spectrum.

Such electronic transitions are not the only possible ones. We may think of the component atoms of polyatomic molecules as being physically separated from each other, yet bound together by electrical

[1] However, any quantum can be absorbed which has sufficient energy to separate an electron from the atom completely. Thus for each energy level there is a frequency above which continuous absorption can take place with the production of free electrons.

forces. A molecule may possess energy as a result of vibration of the component atoms about their mean positions in the molecule, or as a result of rotation of the molecule around its center of mass. It develops that we must think of these vibrational and rotational energies as being quantized also. That is, not all amplitudes and kinds of vibration are allowed, and not all rates of rotation are allowed. The permitted levels of vibration and rotation are also separated by finite amounts of energy, and a transition from one level to another requires absorption or emission of radiation of the correct frequency. Electronic, vibrational, and rotational transitions are illustrated schematically in Fig. 8.1.

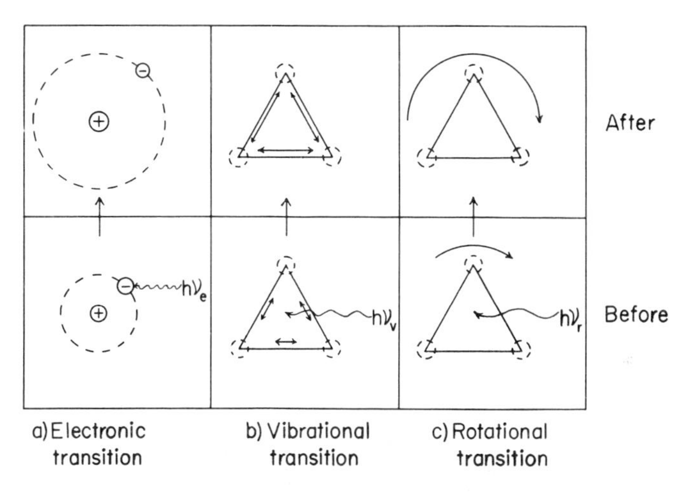

FIG. 8.1. Three kinds of atomic and molecular transitions that can occur upon absorption of a quantum of energy. The arrow lengths are proportional to the energy of vibration and rotation.

In actuality the energy absorbed or emitted in a given transition extends over a measurable range of frequencies rather than being confined to an infinitesimally small range, the reason being that the energy levels of the molecules of a sample of matter are not all precisely equal. For example, collision between the molecules of a gas will disturb the normal energy levels to varying degrees and produce a finite width to the absorption line or band.

In general, the greatest amounts of energy (highest frequency) are associated with electronic transitions; intermediate amounts of energy (medium frequency) with vibrational transitions; and the smallest energies (least frequency) with rotational transitions. Thus electronic absorptions are ordinarily found in the x-ray, ultraviolet, and visible regions of the spectrum; vibrational absorptions in the near-infrared and infrared; and rotational absorptions in the infrared and microwave regions.

The matter is further complicated by the possibility of having combinations of the three basic kinds of transitions. For example, suppose a collection of molecules is capable of undergoing a vibrational change which will produce an infrared absorption. At the same time, the molecules may undergo a number of rotational transitions. Some may jump from the lowest energy level (ground state) to the first energy

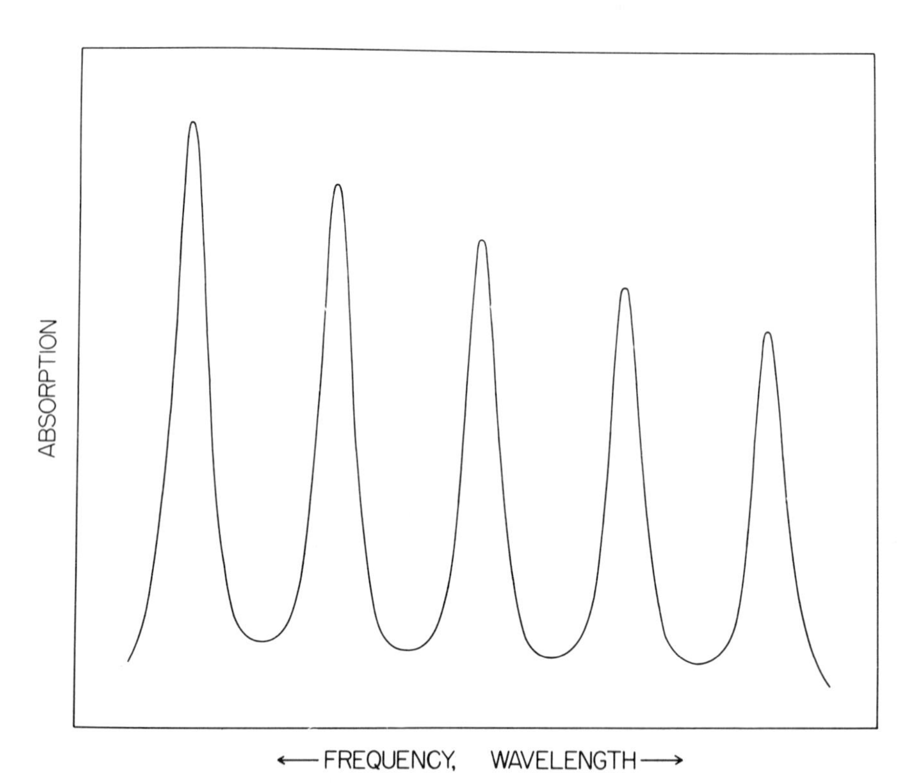

Fig. 8.2. Schematic representation of a rotation-vibration band

level of rotation; others may go from the ground state to the second energy level of rotation, and so on. All the rotational levels will share the same state of vibration. Thus a series of rotaticnal absorption lines will be superimposed upon the vibrational absorption, giving the absorption curve of Fig. 8.2. The ensemble is called a *rotation-vibration band* and may be observed with a spectrometer of sufficiently great resolving power. If, however, the spectrometer admits a fairly large range of wave lengths at once (low resolution) the line structure of the band may be indistinct in the resulting data, and the lines may not be separately visible. Figure 8.3 shows the absorption between 5.5 and 14μ by the atmosphere when relatively little water vapor is present. The measuring apparatus did not have sufficient resolving power to reveal the detailed line structure of the bands, although some hint of

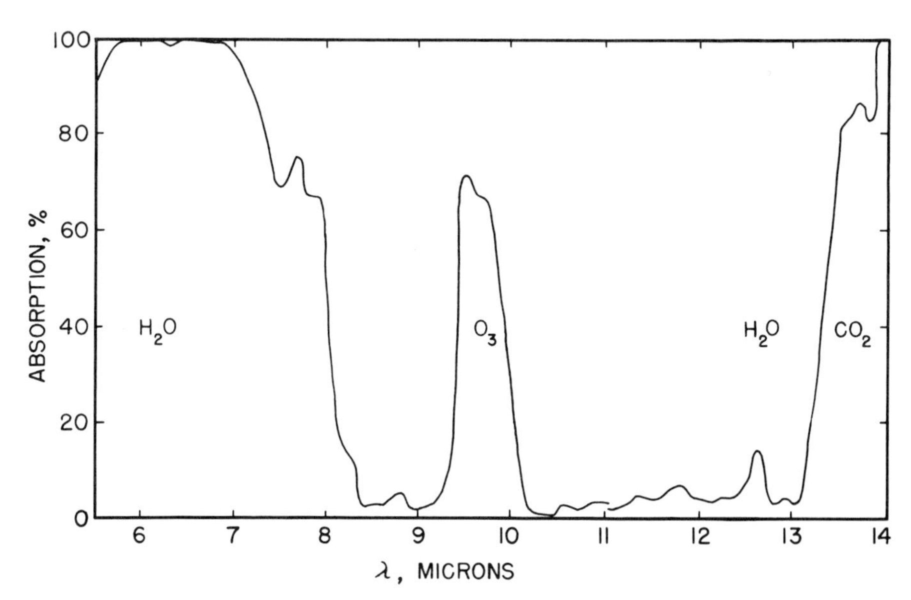

FIG. 8.3. Absorption by the atmosphere between 5.5 and 14 μ (after Adel). There were 2.1 mm of precipitable water present for the observations below 11 μ and 1.3 mm above 11 μ

the presence of lines can be seen. The bands shown are due to H_2O, O_3, and CO_2. Their great strength of absorption illustrates the overwhelming radiative importance of these minor constituents of the atmosphere.

8.3 Scattering

A beam of light passing through a medium may be depleted by *scattering* as well as by absorption. Scattering does not involve a net transfer of radiant energy into heat as does absorption, but merely a change in the direction of the radiant energy as a result of interaction with the scattering particles. Thus a parallel beam of solar radiation entering the atmosphere is partly scattered to the side and backward, so that upon reaching the surface the direct beam is less intense.

The amount and directional nature of scattering depends upon the ratio of the radius of the scattering particles to the wave length of the scattered light. If this ratio is small, as for air molecules and visible light, it is well known that the amount of scattering is proportional to λ^{-4}. This is the classical case called Rayleigh-Cabannes scattering. Since the shorter wave lengths are scattered more effectively than longer wave lengths, a beam of sunlight is somewhat depleted of blue light compared to yellow and red, which accounts for the blue of the sky and the redness of the sun near the horizon. Linke has calculated the transmission of solar radiation through the atmosphere, neglecting absorption and assuming dust-free conditions. Table 8.2 gives his values for a selection of wave lengths from which the dependence upon wave length is apparent.

TABLE 8.2

TRANSMISSION OF RADIATION DOWN THROUGH THE ATMOSPHERE TO SEA LEVEL IN THE ABSENCE OF ABSORPTION AND SCATTERING BY DUST (MOLECULAR SCATTERING ONLY)

Wave length, (Angstroms)	*Transmission, (%)*
3000	31.6
3500	55.1
4000	71.2
4500	81.2
5000	87.4
5500	91.3
6000	93.8
6500	95.5
7000	96.6
7500	97.4

If the scattering particles are not small compared to the wave length, the situation changes in a complex way. As the ratio of radius of particles to wave length increases the scattering does not remain proportional to λ^{-4}. The magnitude of the exponent of λ decreases progressively. That is, the scattering becomes less dependent upon wave length until, for sufficiently large particles, there is no wave length dependence—the scattering is neutral.[2] For water droplets the scattering becomes

[2] For more complete discussion of the complex subject of scattering see H. C. VAN DE HULST, *Light Scattering by Small Particles*, N. Y., Wiley, 1957.

neutral when the radius of the droplets is about equal to the wave length. Since this is sufficiently true for cloud droplets and visible light, clouds appear white; if the cloud droplets were appreciably smaller, clouds would look slightly blue.

The important scattering particles in the atmosphere are air molecules, water droplets, and dust.

8.4 Black-Body Radiation

When matter exists as a thin gas its molecules do not interact to any appreciable degree, and the line and band structure we have discussed can appear. In the liquid and solid state, however, the molecules are so close to each other that drastic changes occur in the energy levels of the molecules, and the variety of permitted energy levels becomes so great as to represent a continuous array. Thus, liquids and solids tend to emit and absorb in very extended continuous regions of the spectrum rather than in discrete lines and bands.

In studying this kind of emission and absorption we shall begin with *Kirchhoff's law*, which can be derived from the second law of thermodynamics. Consider a body that is emitting energy in a given direction and absorbing energy which falls upon it from that same direction. We define first the monochromatic *intensity of emission* (in the given direction), E_λ, as the radiant energy emitted at wave length λ per unit area, time, and wave length. It is often measured in units of cal cm^{-2} min^{-1} micron^{-1} and has dimensions $[ML^{-1}T^{-3}]$. Next, we define the monochromatic *fractional absorption*, a_λ, as the ratio of the incident radiation (from the given direction and at wave length λ) which is absorbed to the total which is incident.[3] The fractional absorption is nondimensional. Kirchhoff found that

$$\frac{E_\lambda}{a_\lambda} = f(\lambda, T) \tag{8.1}$$

That is, for *all* bodies the ratio of intensity of emission to fractional absorption is a function of wave length and temperature only. This ratio is called the *emissivity*. It follows from Eq. (8.1) that the emitted intensity becomes zero, at a given temperature above 0° K, only if the fractional

[3] The quantities E_λ and a_λ are often called *emissive power* and *absorptive power* respectively—poor teminology because they are not powers in the technical sense used in physics.

absorption also vanishes. That is, emission can occur only at wave lengths where absorption occurs. If the absorption varies with wave length, so also will the emission, and a body under these conditions is said to be a *selective emitter*. If the fractional absorption is constant with wave length, but less than one, the emission will be continuous but less than the maximum possible. An object with such properties is called a *gray body*. Finally, if absorption is complete ($a_\lambda = 1$ at all wave lengths) the emission will be the maximum possible:

$$E_\lambda = f(\lambda, T)$$

An object with this property of maximum emission absorbs all the radiation falling upon it and is called a *black body*.

The concepts of gray and black bodies are very convenient theoretically even though no real objects possess the required characteristics exactly. A close approximation to a black body can be achieved by constructing a hollow vessel which completely encloses a cavity except for a small hole through which radiation may enter and leave. The vessel must be completely insulated from its surroundings and allowed to come to a uniform temperature. If the interior of the vessel is constructed of a number of nonparallel surfaces of as black a material as possible, a beam of radiation entering through the hole will be completely absorbed. Such radiation will be reflected many times before it can emerge and a certain fraction of the energy will be absorbed at each reflection, so that very little can get out. As an example, suppose the interior material has a fractional absorption of 0.80, which is easily attained. Suppose further, only 5 reflections occur before the radiation comes out through the aperture. Since 0.20 of the energy is reflected each time, the proportion of the originally incident energy which will emerge is $0.20^5 = 0.0003$. That is, the fractional absorption for the whole apparatus will be 0.9997. Because such a device is so nearly a complete absorber, by Kirchhoff's law it must be a nearly perfect emitter. In other words, the energy emitted by the interior surfaces of the cavity will, when it finally escapes through the aperture, be essentially the maximum which any object at the temperature of the cavity can emit at each wave length.

At the end of the 19th century one of the most important unsolved problems of physics was to determine the explicit form of $f(\lambda, T)$, since this gives the characteristics of basic black-body radiation. The solution of this problem is a fascinating example of the step-by-step progress of science, and the interplay of experiment and theory.

The first results obtained dealt with specific properties of $f(\lambda, T)$

but did not give the functional form. In 1879 Stefan showed experimentally, and in 1894 Boltzmann proved from thermodynamic considerations, that the integral of $f(\lambda, T)$ over all directions (called the monochromatic flux) and over all wave lengths is

$$F = \sigma T^4 \tag{8.2}$$

where F is the total flux of energy (measured in cal cm^{-2} min^{-1}) and σ is a universal constant equal to 8.128×10^{-11} cal cm^{-2} °K^{-4} min^{-1}. The equation states the *Stefan-Boltzmann law.* Thus, if one makes a plot of $f(\lambda, T)$ against λ for a given temperature, this law tells us the area under the curve but does not tell us what the equation of the curve might be.

In the early 1890s it was known from observation that black-body radiation was a maximum at some wave length which depended upon temperature. In 1893 Wien developed a theory which gave a simple expression for this wave length of maximum emission, λ_m, as a function of the temperature of the emitting body:

$$\lambda_m = \frac{2897}{T} \tag{8.3}$$

where λ_m is measured in microns and T in °K. Equation (8.3) is called *Wien's displacement law.*

Next, in 1896, Wien suggested from experimental data that

$$E_\lambda d\lambda \sim \lambda^{-5} e^{-\frac{b}{\lambda T}} d\lambda$$

where b is a positive constant. This result, called *Wien's radiation law,* fits the observations of black-body cavities very well for most wave lengths but not for very long wave lengths. Later, in 1900, the *Rayleigh-Jeans radiation law* was proposed in which

$$E_\lambda \, d\lambda \sim \lambda^{-4} \, T d\lambda$$

This was found to fit the data for very long but not for short wave lengths. Thus functional forms of $f(\lambda, T)$ were found for the extremes of wave length, but the over-all synthesis into a single, all-inclusive, accurate equation remained elusive.

The culminating solution came in 1900 with the introduction of the quantum theory. Planck obtained a semi-empirical expression for $f(\lambda, T)$ that carried the theoretical implication that energy is quantized. He showed from the quantum theory that

$$E_\lambda d\lambda = \frac{c_1}{\lambda^5} \frac{1}{e^{\frac{c_2}{\lambda T}} - 1} d\lambda \tag{8.4}$$

Here $c_1 = 5.362 \times 10^5$ cal cm^{-2} min^{-1} micron^4, and $c_2 = 1.4385 \times 10^4$ micron °K. These are given in such units that if λ is expressed in microns and T in °K, E_λ will be in calories per cm^2 of emitting area per minute per micron of wave length. Equation (8.4) is called *Planck's law.* It fits the observational data for black bodies at all wave lengths and temperatures investigated, and all the preceding results may be derived from it by standard mathematical methods.

Figure 8.4 gives curves of black-body radiation, from Planck's law, as a function of wave length for several temperatures in the atmospheric

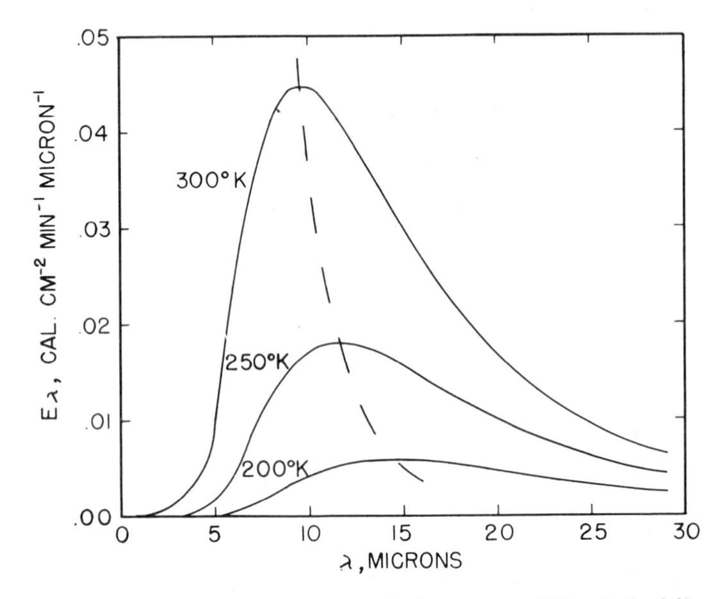

FIG. 8.4. Selected black-body radiation curves. The dashed line connects the points of maximum radiation of each curve.

range. Note that for each temperature the emission approaches zero for very small and very large wave lengths, and there is a maximum of emission at some intermediate wave length. The wave length of maximum emission increases with decreasing temperature, in agreement with Wien's displacement law, but the maximum intensity decreases with falling temperature. The E_λ curve for a warm black body lies above the curve for a cooler black body at each wave length.

Black-body radiation is extremely important for a number of reasons. First, the sun radiates very nearly like a black body with a temperature near 6000° K—and since virtually all of the energy affecting the atmosphere comes ultimately from the sun, its type of radiation is a vital matter. From Wien's displacement law the maximum intensity of sunlight comes at about 0.48 μ in the visual range. Second, the black-body radiation curves represent the upper limit of the emission rates of all materials. For example, a clean snow surface is a very poor radiator (good reflector) in visible light, but is a nearly perfect emitter (poor reflector) in the infrared above 1.5 — 2 μ. This combination of properties explains why temperatures above a snow surface can fall so very low. For with clear skies the incoming solar radiation during the day is largely reflected back to outer space, but at night the black-body emission from the snow, which is necessarily below 0° C, is a maximum between 10 μ and 14 μ (Fig. 8.4) where the atmosphere is nearly transparent (see Fig. 8.3). Thus strong radiant cooling occuring at night cannot be compensated by the weak absorption of sunlight until the surface is so cool it no longer radiates very intensely.

8.5 Radiative Transfer

When monochromatic radiation of intensity I_λ passes through a thin layer, *ds*, of gas, as in Fig. 8.5, a small increment dI_λ will be absorbed. By the definition of the fractional absorption,

$$dI_\lambda = -a_\lambda I_\lambda \tag{8.5}$$

We shall consider the case when the gas is so cool that its emission at the given wave length is negligibly small. Then measurements of radiation entering and leaving such a layer pertain only to the absorption. Measurements of this sort indicate that a_λ is proportional to the density of the gas and the distance *ds*:

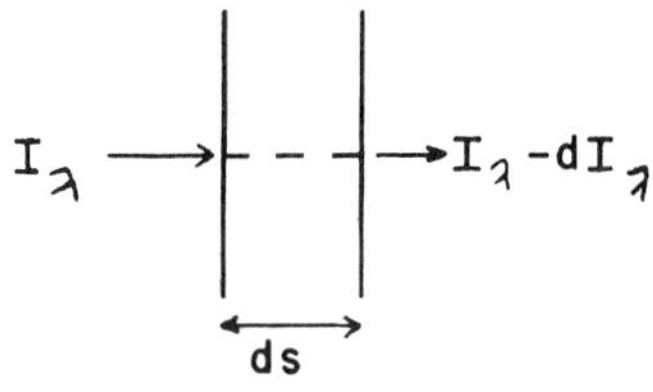

I_λ → $I_\lambda - dI_\lambda$

ds

FIG. 8.5. Absorption in an infinitesimal layer

$$a_\lambda = k_\lambda \rho ds$$

where k_λ is the constant of proportionality and is called the *absorption coefficient* of the gas $[L^2M^{-1}]$. This coefficient is generally a function of wave length. It is not necessarily constant at a given λ, but may vary with pressure and temperature.

Equation (8.5) may now be written

$$\frac{dI_\lambda}{I_\lambda} = -k_\lambda \rho ds \tag{8.6}$$

This says simply that the fractional decrease of intensity owing to absorption is proportional to the mass per unit area of the absorbing medium traversed. When we integrate Eq. (8.6) over a finite distance s we get

$$\int_{I_{\lambda 0}}^{I_{\lambda s}} \frac{dI_\lambda}{I_\lambda} = -\int_0^s k_\lambda \rho ds$$

or

$$I_{\lambda s} = I_{\lambda 0}\, e^{-\int_0^s k_\lambda \rho ds} \tag{8.7}$$

where I_0 is the incident intensity and I_s is the intensity after penetration to distance s. Equations (8.6) and (8.7) are differential and integral forms of *Beer's law* for absorption. Exactly the same equations hold for depletion of radiation by scattering, except k is then a scattering coefficient instead of an absorption coefficient and has quite a different wave length dependence.

The right side of Eq. (8.6) is often written as

$$k_\lambda \rho ds = d\tau$$

where τ is the optical thickness (nondimensional). Thus Eq. (8.7) may be written as

$$I_{\lambda s} = I_{\lambda 0} e^{-\tau}$$

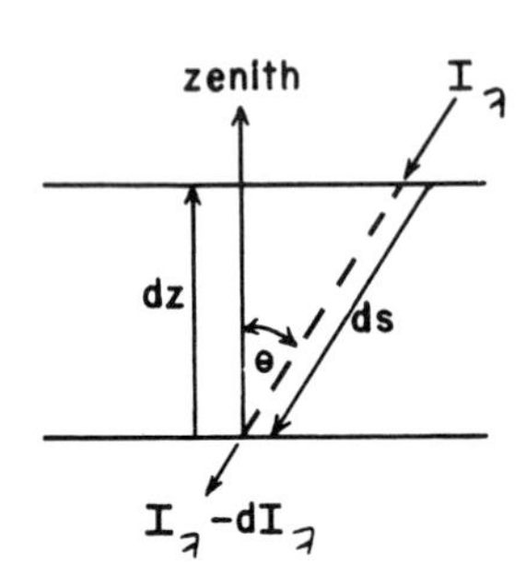

FIG. 8.6. Absorption from an oblique ray passing through an infinitesimal layer

Therefore, the significance of optical thickness is that when $\tau = 1$ the intensity of radiation has been reduced by the factor $1/e$.

In the important special case of radiation through the atmosphere, it is convenient to use the vertical distance, z, as a coordinate instead of distance along the radiation beam, s. Energy from the sun, for example, usually is not vertical but enters at a zenith angle, θ. The actual distance traversed by the rays through a thin atmospheric layer (see Fig. 8.6) is

$$ds = -dz \sec\theta$$

where the minus sign appears because z is counted positive upward while s is positive in the downward direction of the ray. Then Beer's law may be written as

$$dI_\lambda = k_\lambda I_\lambda \rho \sec\theta\, dz$$

or

$$I_{\lambda z} = I_{\lambda 0}\, e^{\int_z^\infty k_\lambda \rho \sec\theta\, dz} \tag{8.8}$$

In this development we have assumed that re-emission is negligible compared to the incident radiation. This assumption is valid for the important case of visible light passing through the atmosphere, for then the atmosphere is too cool to emit any detectable short-wave radiation. But when we consider the equally important case of transmission of infrared energy through the atmosphere, the re-emission is no longer negligible; air does radiate significantly in parts of the infrared spectrum. From Kirchhoff's law the emitted intensity is

$$E_\lambda = a_\lambda f(\lambda, T)$$

where $f(\lambda, T)$ is the Planck black-body emission rate. Thus when monochromatic radiation enters a layer of gas there is a loss by absorption and a gain by emission at the same wave length. The net loss is

$$dI_\lambda = - a_\lambda I_\lambda + a_\lambda f(\lambda, T)$$

or

$$dI_\lambda = a_\lambda [f(\lambda, T) - I_\lambda] \tag{8.9}$$

This is the *equation of radiative transfer*, also known as *Schwarzschild's equation*. Whenever absorption and re-emission are significant at a given wave length, Eq. (8.9) governs the change in radiation intensity.

PROBLEMS

1. Show from Planck's law that (a) Wien's radiation law holds approximately at short wave lengths; (b) The Rayleigh-Jeans radiation law holds approximately at long wave lengths.

2. Show from Planck's law that for a black body at some given temperature the product of the wave length of maximum emission with the temperature must be a constant (Wien's displacement law).

3. Consider an atmospheric constituent which has mixing ratio constant with height and whose absorption coefficient at wave length λ depends upon pressure according to the law

$$k_\lambda = k_{p_0} \left(\frac{p}{p_0}\right)^{1/2}$$

where p_0 is sea-level pressure. Show that sunlight which has intensity I_0 at λ and which enters the atmosphere at zenith angle θ arrives at sea level with intensity

$$I = I_0 e^{-\frac{2}{3}\left(\frac{k_{p_0} p_0 \sec \theta}{g}\right)}$$

4. Suppose for the conditions of Problem 3, $\theta = 0$ and the absorption by the total atmosphere is 75 per cent. Calculate the numerical value of the absorption coefficient at p_0 in metric units.

CHAPTER 9

SOLAR AND TERRESTRIAL RADIATION

9.1 The Nature of Solar Radiation

Radiation from the sun is the predominant source of energy for virtually all processes taking place on our planet. Some of the phenomena that are utterly dependent upon this source are the circulation of the atmosphere, all life processes, oceanic movements except tides, hydroelectric power, and the existence of fossil fuels. Comparatively tiny amounts of energy come from the moon and planets (reflected sunlight), the stars, gravitational influences (tides), volanoes, radioactive decay, and man-made nuclear fission and fusion. The amount of energy involved in these categories is quite negligible compared to the 3.7×10^{21} cal of solar energy that arrives at the earth every day.

The vast outpouring of energy from the sun requires a truly stupendous source. We now know that this source is the conversion, deep in the solar interior, of hydrogen to helium via the "carbon cycle." Four hydrogen nuclei have a mass nearly 1 per cent larger than the mass of the helium nucleus which results from their fusion. During this reaction the extra mass is converted into energy according to the famous Einstein equation, $E = mc^2$, where m is the mass converted into energy E, and c is the speed of light. Since c is a very large quantity, it is apparent that the production of energy is very great.

The heat produced in the interior by this process travels to the surface of the sun by radiation and convection and is radiated into space from the outer layers of the sun. The luminous surface of the sun which we see is called the *photosphere* which consists of very hot gases in various states of ionization. Although the photosphere appears smooth and featureless to the naked eye, telescopic examination reveals many details. Darker (and therefore cooler) regions called *sunspots* are common. A very fine mottling composed of so-called *rice grains* is normal. "Hot" spots called *faculae* and *flocculi* can be detected in monochromatic photographs taken at the wave lengths of emission of hydrogen and calcium. Above the photosphere lies the *reversing layer* of somewhat cooler gas which absorbs some of the energy coming up from below and which produces many of the well-known line absorptions in the solar spectrum called *Fraunhofer lines.*

If the image of the body of the sun is blocked out in a telescope equipped with an artificial eclipsing device one may study the outermost, tenuous layers. These are the reddish *chromosphere* consisting primarily of hydrogen and helium gases at low pressure, and the silvery *corona* consisting of extremely thin gases extending outward many millions of miles. Spectacular and explosive extensions of the chromosphere known as *prominences* are also observed. These seem to be associated with sudden localized bursts of high-temperature radiation called *flares.*

It is clear that our sun is an extremely complex body. Solar physics is still imperfectly understood and is an active field of research, but despite the complexities certain broad and important statements can be made.

(1) *Over much of the spectrum the sun radiates nearly as a black body.* Small departures occur in the Fraunhofer lines and over broader ranges of wave length in the ultraviolet. If we measure the total energy received from the sun and calculate from the Stefan-Boltzmann law the temperature of a black body which would radiate at this rate we obtain a solar surface temperature of 5750° K. On the other hand, we may measure the radiation as a function of wave length, determine the point at which the energy is a maximum, and calculate the temperature from Wien's displacement law. The maximum radiation is found at 4740A, which means a black body at 6108° K. Clearly the sun is not precisely a "black body" because these two temperatures do not agree. The discrepancy is due to absorption at short wave lengths in the outer layers of the sun which does not affect the position of the maximum at 4740A. However, the approximation to a black body is sufficiently close for many purposes.

(2) *The rate at which the sun radiates energy is very nearly constant.* Measurements of the intensity of solar radiation have been made for many years by S. P. Langley and especially by C. G. Abbot. The effect of the somewhat variable distance between earth and sun is easily taken into account, but the variable effects of absorption in our atmosphere are far more difficult to deal with. The fundamental method devised

by Abbot and his collaborators at the Astrophysical Observatory of the Smithsonian Institution makes use of Beer's law, Eq. (8.8). The intensity of solar radiation is measured at several wave lengths almost simultaneously, so that sec θ is the same for all λ. The measurements are repeated several times over a period of hours so that sec θ assumes different values as the sun moves in the sky. One assumes that the $k_{\lambda}\rho$ have not changed during the period of measurement; it is then possible to extrapolate to zero air mass in the path and obtain I_0, the intensity of sunlight outside the atmosphere at each wave length. This technique is used to determine I_0 in the range from 0.34 μ to 2.5 μ. A correction amounting to 3.4 per cent is added for the unobservable ultraviolet below 0.34 μ, and 2.0 per cent is added for the infrared above 2.5 μ. Since these corrections are small and cannot themselves be in error by more than a few parts per hundred, the value of the intensity of sunlight in space can be determined within one or two per cent.

This intensity is found to be 2.00 cal cm^{-2} min^{-1} falling upon a surface perpendicular to the sun's rays outside the atmosphere at the earth's average distance from the sun. This value is called the *solar constant.*[1] When Abbot began his work the value of the solar constant was very inadequately known. That we now know this number within only a few per cent is due to his remarkable patience, skill, and tenacity with these difficult observations.

A considerable controversy exists over whether the solar "constant" has any significant variations with time. There is no question that appreciable variation takes place in the ultraviolet, but changes in the visible region, where most of the energy is found, are debatable. Abbot feels that the solar constant deviates from its mean from time to time by a few per cent, and that these fluctuations are related to definite changes in weather. Others have strongly critized this theory. The preponderance of evidence seems to be that the total solar constant does not exhibit variations exceeding the limits of accuracy of measurement. In any case, percentage fluctuations in the reflective power of the earth-atmosphere system are much greater than any possible percentage variations in the total solar constant. Thus the energy absorbed changes with time, an effect that probably has more to do with certain weather phenomena than direct solar fluctuations. On the other hand, the percentage variations in the ultraviolet solar output are real and relatively large, the energy being absorbed at high levels in the atmosphere. If there are real and direct solar-terrestrial weather effects, the problem

[1] One cal cm^{-2} is frequently called a *langley* (abbreviated ly). Thus the solar constant is 2.00 ly min^{-1}.

remains to explain how the energy absorbed at high levels can produce appreciable changes in the denser lower atmosphere.

9.2 Geographical and Seasonal Distribution of Solar Radiation

The orbit of the earth about the sun is an ellipse with the center of the sun at one focus. The ellipse has only a small eccentricity (.017); thus the distance between the earth and sun varies only slightly during the year. The mean earth-sun distance is 1.4968×10^8 km (approximately 93,000,000 miles). The maximum deviations from the mean distance occur about January 1, when the earth is 1.67 per cent closer, and July 1, when it is 1.67 per cent farther away. Since the radiation intensity varies inversely as the square of the distance, the actual rate of receipt of solar energy can be 3.37 per cent higher or lower than the solar constant.

If initially we ignore the effect of the atmosphere it is possible to calculate the rate at which solar energy falls on a unit area of earth's surface (not perpendicular to the rays) per day at each latitude. The angle at which the sun's rays strike the earth's surface varies with latitude, season, and time of day, and since the length of a day varies with latitude and season the results depend only on season and latitude. Milankovitch has carried out the calculations from astronomical data and the results are presented in Fig. 9.1.

The rate of receipt of solar energy is nearly constant at the equator where it varies between 790 and 895 ly day^{-1}. This is because the sun is never more than $23\frac{1}{2}°$ away from the zenith at noon at this latitude, and the duration of sunlight is always almost exactly 12 hours. On the other hand, at the poles there is a large variation with time of year ranging from zero around the winter solstices to about 1100 ly day^{-1} around the summer solstices. This variation is due to the fact that during winter the sun is completely below the horizon (zero hours of sunlight per day) while during summer it is always above the horizon (24 hours of sunlight per day). Despite the low altitude of the sun at the poles at the summer solstices (zenith angle $66\frac{1}{2}°$) the relatively long duration of sunlight causes the daily receipt of energy to be larger at the poles in summer than at any other place or time on earth. A secondary maximum occurs near latitude 45° at the summer solstice of each hemisphere, the result of a combination of increasing duration with increasing latitude, and increasing intensity with decreasing latitude.

A definite asymmetry between the two hemispheres appears in Fig. 9.1. The Southern Hemisphere receives more daily radiation than does the Northern Hemisphere in their respective summers. For example,

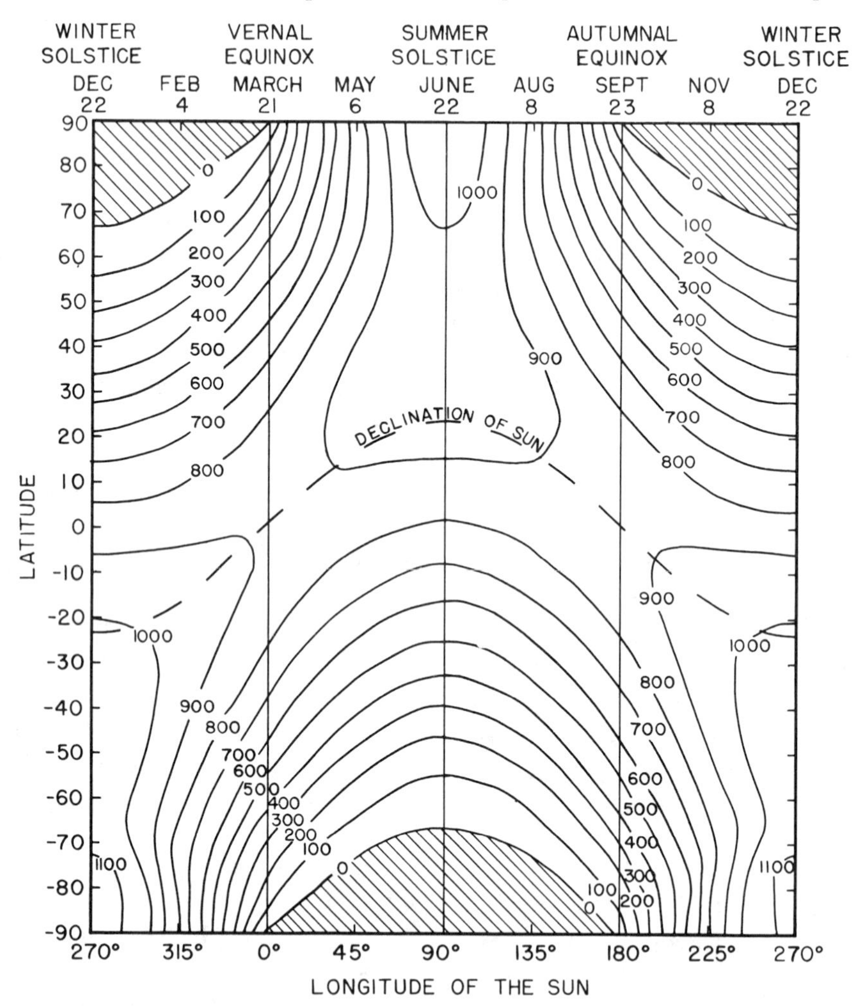

FIG. 9.1. Solar radiation in ly day^{-1} arriving at the earth's surface in the absence of an atmosphere

the maximum north polar radiation is 1077 ly day^{-1} while the maximum south polar radiation is 1149 ly day^{-1}, because earth is closest to the sun in Southern Hemisphere summer and farthest away in Northern Hemisphere summer.

This pattern undergoes a drastic change if we include a reasonable atmospheric transmission factor. We shall not take into account the influence of clouds at all and will assume one air mass transmits 70 per cent of the radiation entering it. This transmission will be assumed constant with wave length. The sec θ effect, which varies with latitude, season, and time of day, must then be included. The results, based upon Milankovitch's calculations are presented in Fig. 9.2.

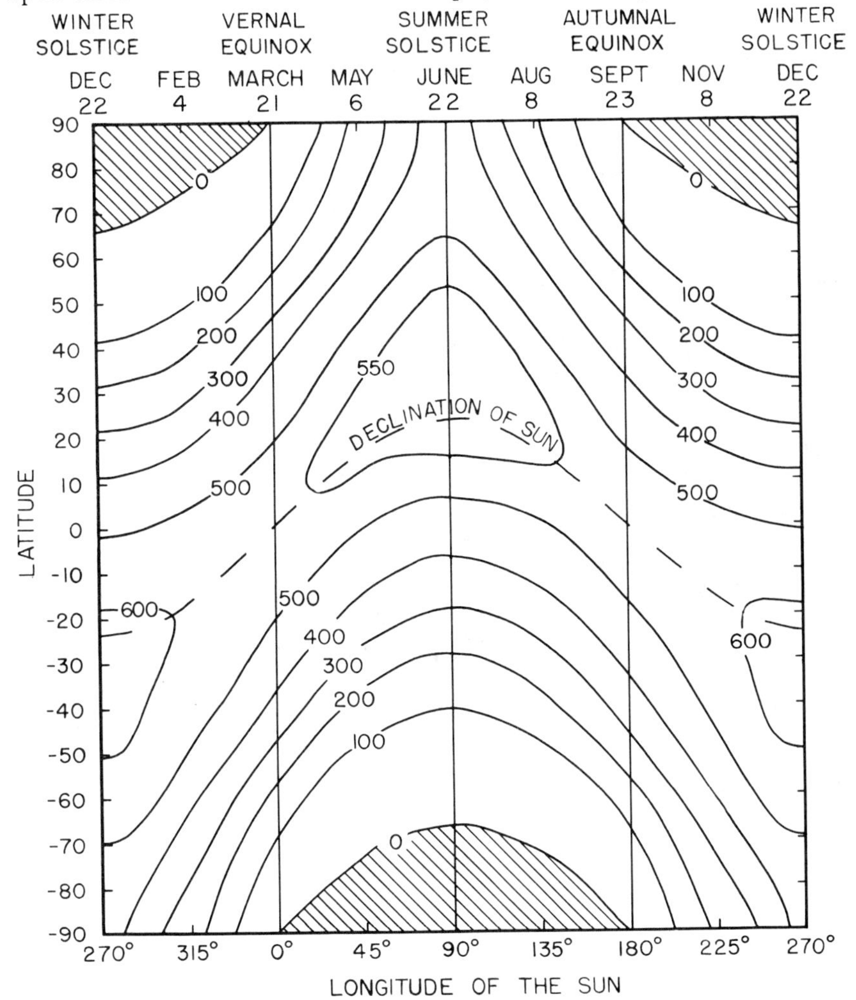

FIG. 9.2. Solar radiation in ly day^{-1} arriving at the earth's surface when the atmosphere transmits 0.7 of a vertical beam

As expected, the outstanding change introduced is a decrease in the rate of receipt of energy at the earth's surface. For example, at the equator at solar longitude 90°, 790 ly day^{-1} arrive but this atmosphere permits only 462 ly day^{-1} to penetrate to the surface. In polar regions the low elevation of the sun causes so much alteration of solar radiation that the summer polar maxima of Fig. 9.1 are wiped out. The maxima at the solstices are now near latitude 35°.

The presence of clouds interferes with the incoming beam in a manner that varies greatly according to cloud depth, cloud type, and proportion of the sky covered. Table 9.1 gives the annual mean cloud cover for the Northern Hemisphere as a function of latitude.

TABLE 9.1

ANNUAL MEAN CLOUDINESS FOR THE NORTHERN HEMISPHERE (AFTER LONDON)

Latitude, °N . . .	0-10	10-20	20-30	30-40	40-50	50-60	60-70	70-80	80-90
Cloudiness, % . .	51.2	43.8	40.8	47.2	57.2	63.5	63.5	61.2	54.8

The area-weighted, mean, hemispheric cloudiness is 51.1 per cent. London finds that the mean reflectivity of clouds is about 47 per cent. Therefore, on the average about 24 per cent of the incoming solar energy is reflected to space by clouds. Somewhat less than 24 per cent is lost in the relatively cloud-free subtropics and somewhat more than 24 per cent is lost in the cloudy region poleward of 40° N.

Of the remaining radiation which arrives at the surface of the earth only a part is absorbed, the rest being reflected. The reflectivity of earth's surface depends greatly upon the material composing the surface. It may range from 0.05 for green forests to about 0.9 for fresh snow. Furthermore, the reflectivity of the sea depends strongly upon the angle of incidence of the light. Houghton estimates that the average albedo (reflective factor) of the surface of the northern hemisphere is 0.100, with a range from .071 at 0° to 0.56 at 90° N. Thus except near the ice and snow fields of polar regions most of the sunlight reaching the surface is absorbed.

9.3 Terrestrial Radiation

Most materials comprising the surface of the planet earth radiate nearly as black bodies. Since the mean temperature near the surface is close to + 15° C, Wien's displacement law indicates that the maximum of radiation is at a wave length close to 10 μ. This value is rather far from the maximum of incoming solar radiation of nearly 0.5 μ. As a result, substances whose absorption is unimportant for short-wave solar radiation may become significant absorbers and emitters of the long-wave terrestrial radiation. The dominant constituents of the atmosphere in this respect are water vapor, carbon dioxide, and ozone. Figure 8.3 gives the observed absorption by unit air mass of a rather dry atmosphere in the infrared. Below about 5 μ (not shown in Fig. 8.3) important amounts of energy are not absorbed by the atmosphere, partly because the amounts of terrestrial radiation available for absorption are relatively small. From roughly 5 to 8 μ there is a strong absorption band of H_2O. Beyond 8 μ the absorption by H_2O becomes smaller up to about 13.5 μ except for a strong narrow band due to O_3 at 9.6 μ which masks a weaker band of CO_2 near 10 μ. This relatively transparent window in the atmospheric absorption spectrum falls in the wave-length regions where the earth's surface radiates most strongly.

The strong absorption beginning at 13.5 μ and extending to 17 μ is due to CO_2. At its center it absorbs terrestrial radiation completely. Although the atmosphere again becomes transparent beyond 17 μ until the rotation band of H_2O appears at 24 μ and beyond, the amount of terrestrial radiation at these wave lengths is secondary in importance and 14 μ is frequently taken to be the cut-off point beyond which the atmosphere may be thought of as completely opaque.

Data of this sort can, in principle, be combined with the radiation laws to study the flux of terrestrial radiation through the atmosphere to space. This flux is an essential part of the heat budget of the planet earth. We know from meteorological records that the mean temperature of our planet remains nearly constant over a period of years. Since the only major source is solar energy and the only way of losing heat is by flow of infrared radiation to space, it follows that these two must balance in the mean. We know fairly well the annual mean rate of receipt of energy as a function of latitude. Can we calculate from the known absorption characteristics of the atmosphere and the radiation laws the annual mean loss of energy and show that it balances the gain?

This is a considerable task because the spectrum of H_2O and CO_2 is complex and the radiation laws are quite involved. The very earliest attempts simplified matters by neglecting the band structure of atmospheric transmission and assumed a "gray" atmosphere—that is, one having a uniformly incomplete absorption at all wave lengths. It was not possible to account for the radiation balance with so simple a model. Then Simpson[2] attempted a less gross simplification which led to the first successful explanation of the radiation balance.

---

[2] G. C. Simpson, Further Studies in Terrestrial Radiation. *Mem. Roy. Meteor. Soc.*, 3, 1928, pp. 1-26.

Simpson dealt only with the effects of H_2O and CO_2, and from Hettner's measurements of the absorption spectrum of steam concluded that the complex transmission spectrum of thin layers of atmosphere could be smoothed considerably. He found that a layer containing 0.03 g cm^{-2} of water vapor could be characterized as:

(1) completely transparent below 4.0 μ
(2) partly transparent from 4.0 to 5.5 μ
(3) completely opaque from 5.5 to 7.0 μ (water vapor band)
(4) partly transparent from 7.0 to 8.5 μ
(5) completely transparent from 8.5 to 11.0 μ
(6) partly transparent from 11.0 to 14.0 μ
(7) completely opaque above 14.0 μ (carbon dioxide and water vapor bands)

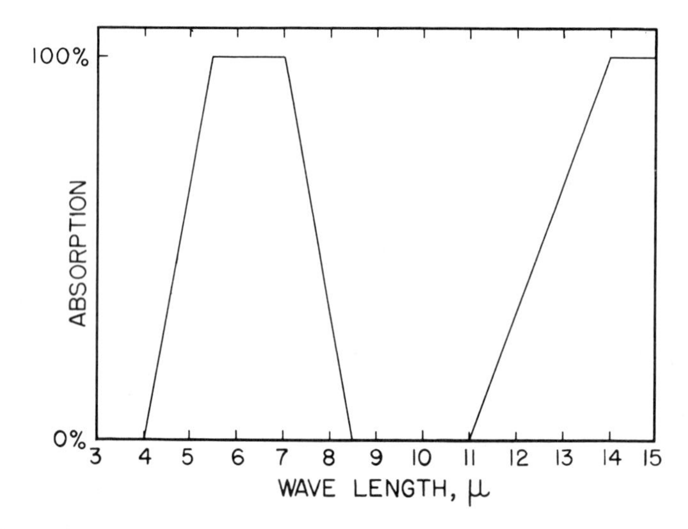

FIG. 9.3. Simpson's simplified infrared absorption spectrum for a layer of air containing 0.03 g cm^{-2} of water vapor

This simplified spectrum is shown schematically in Fig. 9.3. Hettner's absorptions are now known to be too high and ozone absorption is completely ignored, but the approach still has value as the first and simplest method to give acceptable results.

Simpson conceived of the atmosphere as divided into layers of varying thickness, each containing 0.03 g cm^{-2} of H_2O. The surface radiates upward very nearly as a black body. Part of the energy is absorbed completely by the H_2O and CO_2 in the layer of air immediately above the surface. The rest is partly or completely transmitted. This first layer then re-radiates at the same wave lengths and at a rate determined by its temperature. Half this radiation is downward, which merely compensates for part of the upward surface radiation, and half is directed upward. The upward beam is absorbed in the next higher layer and again re-radiated at a rate dictated by its temperature. In this way the radiation passes upward in the atmosphere until it reaches the last layer containing .03 g cm^{-2} of H_2O. Simpson took this to be the stratosphere. The upward radiation from this last layer then is lost to space.

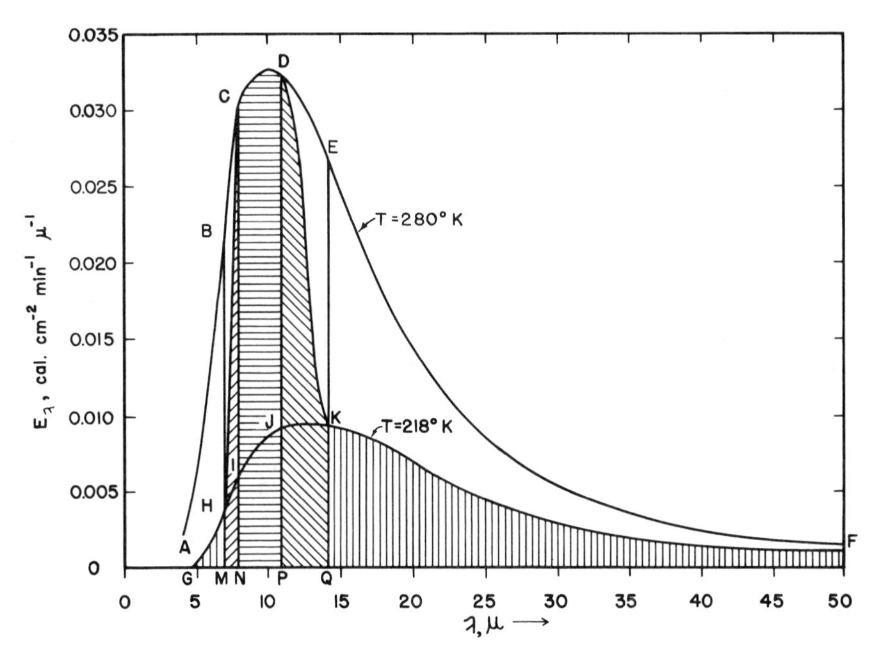

FIG. 9.4. Simpson's graphical calculation at latitude 50°. (Adopted by permission from *General Meteorology* by Horace R. Byers. Copyright 1944. McGraw Hill Book Co., Inc.)

Figure 9.4 illustrates the procedure. Here two black body curves are entered for mean conditions at latitude 50°. One is for the surface at 280° K (curve *ABCDEF*); the other is for the stratosphere at 218° K (curve *GHIJKF*). In the completely transparent window the surface radiation passes unimpeded to space. Thus the area NCDP is lost at once. In the two completely opaque regions the loss to space is at a rate dictated by the temperature of the stratosphere, given by areas *GHM* and *QKF*. In the semi-transparent parts of the spectrum the rate of loss is between the rates given by the two black body curves. Since the average absorption in the semi-transparent bands is about

50 per cent, Simpson arbitrarily divided these areas between the two Planck curves into equal halves with the lines *HC* and *DK*. Thus the loss to space through the semi-transparent region of the spectrum is given by the areas *MHCIN* and *PJDKQ*. Finally the sum of all these contributions (the shaded areas in Fig. 9.4) is the mean rate of radiation to space of the earth and atmosphere at latitude 50°. The procedure can be followed at all latitudes to yield the loss of radiation over the entire planet.

This determination neglected the presence of clouds. Since clouds absorb and radiate like black bodies in the infrared, the cloud top can be taken in place of the earth's surface in the calculation. This Simpson did, assuming 0.5 cloud cover and a cloud temperature of 261° K at all latitudes. So far as the over-all balance is concerned, he found an annual mean outgoing radiation of 0.271 cal cm^2 min^{-1}. This compares with an annual mean absorption of solar energy at the ground of about 0.24 cal cm^{-2} min^{-1}. The agreement is excellent in view of the difficulties of the problem. In Simpson's calculations by latitudes he found an excess of receipt over loss of energy below latitude 35° and a deficit above that latitude. Since the tropics are not warming and the polar regions are not cooling it is necessary that the atmosphere and the oceans transport the excess heat poleward. It is known that most of this transport is done by the atmosphere, and this latitudinal differential in the radiation budget is the basic cause of the circulation of the atmosphere. A more recent calculation of the heat balance will be discussed in detail in Chapter 10, Section 10.7.

PROBLEMS

1. Given that the solar constant is 2.0 cal cm^{-2} min^{-1}, the mean earth-sun distance is 1.5×10^8 km, and the radius of the sun is $6./9 \times 10^5$ km, calculate the black-body temperature of the sun.

2. Suppose the earth has albedo A for solar radiation but radiates as a black body in the infrared. Ignore absorption of infrared by the atmosphere. Show that the mean temperature of a rotating body of this sort in radiative equilibrium is given by:

$$T = \left[\frac{(1 - A)\, S}{4\sigma}\right]^{\frac{1}{4}}$$

where S is the solar constant. Calculate this black-body temperature if $A = 0.35$. Is this lower or higher than earth's actual mean temperature (288° K)? Why?

CHAPTER 10

APPLICATIONS TO RADIATION IN THE EARTH-ATMOSPHERE SYSTEM

10.1 The Basis of Elsasser's Method

In order to obtain more accurate and more general results than Simpson's it is necessary to deal more directly with the complexities of radiative transfer and the $H_2O—CO_2$ infrared spectrum. Beer's law for pure absorption, Eq. (8.6), holds true for a single wave length but may be quite inaccurate for a wide band of wave lengths. This suggests that one should deal with very narrow increments of λ. However, the fine structure of the spectrum of H_2O and CO_2 is so complex that the task of measuring and using the absorption coefficients over wide portions of the spectrum is overwhelming. Therefore, Elsasser[1] sought a way of averaging and smoothing the spectrum which would still permit him to derive a sufficiently simple expression for the transmission of moist air. He considered a model of an absorption band consisting of equally spaced and equally strong lines of the correct theoretical shape. The average transmission over a finite range of wave lengths in such a band is

$$\tau_I = \left(\frac{I}{I_0}\right)_{av}$$

where τ_I is called the transmission function. We shall define an *optical depth*

$$u = \int_0^s \rho ds$$

which is simply the mass of absorbing medium per unit area normal to the radiation. Then, from Beer's law

$$\tau_I(u, \lambda) = (e^{-k_\lambda u})_{av} \tag{10.1}$$

[1] W. M. Elsasser, Heat Transfer by Infra-red Radiation in the Atmosphere, *Harvard Meteor. Studies*, No. 6, 1942.

If in Elsasser's model of a band the spacing between lines is large compared to the half-width of a line, the averaging may be carried out mathematically. He found

$$\tau_I = 1 - \Phi\left(\sqrt{\frac{l_\lambda u}{2}}\right) \tag{10.2}$$

where Φ is the normal probability integral:

$$\Phi(x) = \frac{2}{\sqrt{\pi}} \int_0^x e^{-y^2}\, dy$$

which is tabulated numerically in statistics books. Its value varies from 0 at $x = 0$ to 1 at $x = \infty$. Thus Eq. (10.2) indicates that all the incident radiation is transmitted if $u = 0$ (no absorbing gas), and none is transmitted for $u = \infty$ (infinite amount of absorbing gas).

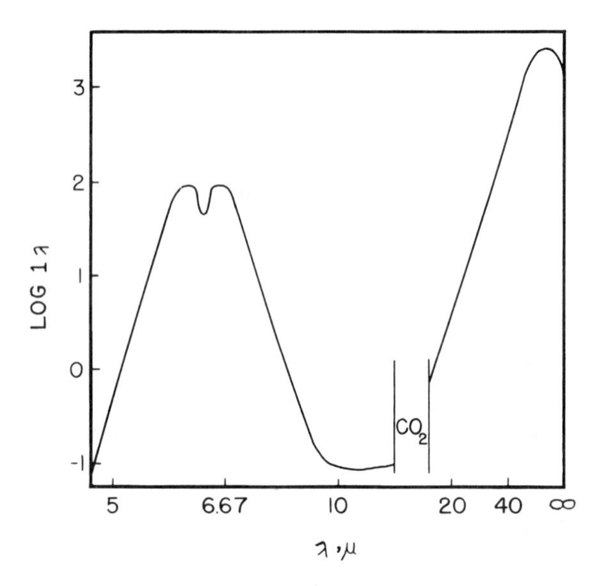

FIG. 10.1. Elsasser's smooth absorption curve for water vapor

The quantity l_λ is the smoothed absorption coefficient in which detailed line structure has been averaged out. It is shown graphically in Fig. 10.1 for water vapor. In the interrupted part of the curve, labelled CO_2, the absorption by CO_2 is very large and dominates that due to H_2O. There generalized coefficients for CO_2 are used and absorption by water vapor is negligible.

These results apply at a standard temperature and pressure. However, the absorption coefficients vary with temperature and pressure, so when we deal with the dominant case of vertical transmission of radiation the large variations of T and p may have to be considered. There is evidence that the dependence upon temperature is relatively small so this is neglected. Because the empirical data indicate that absorption varies as $p^{1/2}$ for constant u, Elsasser used

$$l_\lambda = \sqrt{\frac{p}{p_0}}\, l_{\lambda p_0} \tag{10.3}$$

where $l_{\lambda p_0}$ is the coefficient at the standard pressure p_0 (for example, 1000 mb).

When l_λ varies with p the combination $l_\lambda u$ in Eq. (10.2) must be replaced by $\int l_\lambda \, du$, where u varies with height and pressure. This can be related to the specific humidity q as follows:

$$du = q\rho dz$$

and from the hydrostatic equation

$$du = \frac{q}{g}\, dp$$

where sign has been disregarded. Thus

$$l_\lambda du = \frac{l_\lambda q}{g}\, dp \tag{10.4}$$

If we set $l_\lambda \, du = l_{\lambda p_0} \, du'$, where u' is a modified optical depth, then from Eq. (10.4)

$$du' = \frac{l_\lambda}{l_{\lambda p_0}} \frac{q}{g}\, dp$$

and from Eq. (10.3)

$$du' = \frac{q}{g} \sqrt{\frac{p}{p_0}}\, dp \tag{10.5}$$

The modified optical depth calculated in this way has the property that when multiplied by a standard absorption coefficient (independent of p) it may be used in Eq. (10.2) to calculate the transmission. Furthermore, only du' need be integrated with height since $l_{\lambda p_0}$ is independent of pressure. This integration must be performed numerically using the data supplied by a radiosonde. In such a numerical integration one replaces infinitesimal differentials by finite differences and sums over a finite number of layers instead of integrating. Thus Eq. (10.5) becomes

$$\Delta_i u' = \frac{q_i}{g} \sqrt{\frac{p_i}{p_0}}\, \Delta_i p$$

where Δ refers to a finite increment and the subscript i defines a value as belonging to layer number i. The total modified optical depth for any large column of the atmosphere is then

$$u' \approx \sum_{i=1}^{m} \Delta_i u' = \sum_{i=1}^{m} \frac{q_i}{g} \sqrt{\frac{p_i}{p_0}}\, \Delta_i p$$

where the summation is taken over the m layers contained in the column. Examples are given in Table 10.1 of an upward integration from the reference level 990 mb, and upward and downward integrations from the reference level 720 mb.

TABLE 10.1

COMPUTATION OF MODIFIED OPTICAL DEPTH FROM RADIOSONDE DATA

							Reference level 990 mb	*Reference level* 720 mb
p (mb)	T (°C)	q (g kg^{-1})	q_i (g kg^{-1})	$\Delta_i p$ (mb)	$\sqrt{\frac{p_i}{1000}}$	$\Delta_i u'$ (g cm^{-2})	u' (g cm^{-2})	u' (g cm^{-2})
990	15	6.5					0.00	1.03
			5.2	120	0.96	0.61		
870	6	4.0					0.61	0.42
			3.1	150	0.89	0.42		
720	−6	2.2					1.03	0.00
			1.6	130	0.81	0.17		
590	−17	0.9					1.20	0.17
			0.5	190	0.70	0.07		
400	−38	0.1					1.27	0.24
			0.1	118	0.58	0.01		
282	−52	0.1					1.28	0.25

Such a computation makes it possible to obtain a relationship between the observed T and the computed u'. We shall see in the next section how this can be used to determine infrared radiative flux in the atmosphere.

10.2 The Elsasser Diagram

The absorption of radiation by a layer of atmosphere does not depend upon the temperature to an appreciable extent, but does depend upon the modified optical depth, du', and the standard generalized absorption coefficient, $l_{\lambda p_0}$, as discussed above. On the other hand, the emission by this same layer depends upon the temperature as well as upon du' and $l_{\lambda p_0}$, from Kirchkoff's and Planck's laws. Thus the net transfer of radiation through the layer at a given wave length, determined from Schwarzschild's equation for radiative transfer (Eq. 8.9), is a function of T, du', and $l_{\lambda p_0}$. If one integrates over all wave lengths to obtain the total net flux of radiation the only variables left are T and du'. Elsasser showed how the complex integrals involved could be reduced to measurement of areas on a specially designed chart known as the *Elsasser diagram*. This is a plot of T as abscissa increasing to the left, and u' as ordinate increasing downward. The isotherms of the diagram are straight vertical lines and the lines of constant u' are gentle curves which converge to a point at 0° K. Thus the diagram is essentially

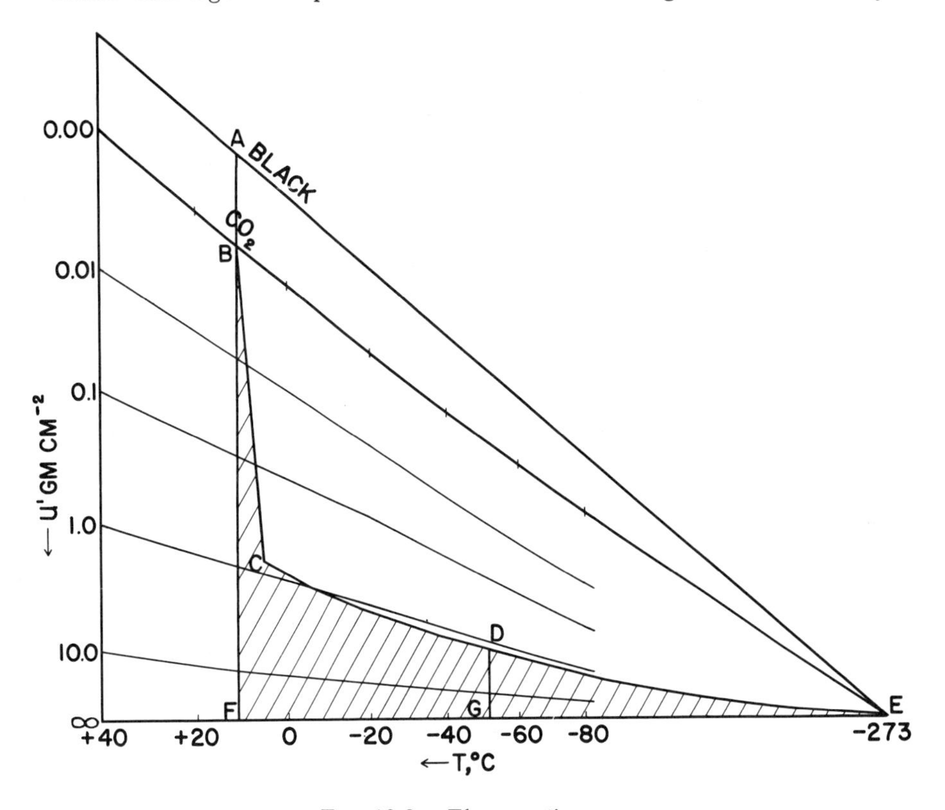

Fig. 10.2. Elsasser diagram

triangular as shown in Fig. 10.2. The diagram is so constructed that if one considers a vertical isotherm extending from the base (labelled $u' = \infty$) to the uppermost sloping line (labelled "black") the triangular area to the right of the isotherm is proportional to the energy emitted by a black body at the given temperature. The equivalence between energy flux in langlies per 3 hours and area is given on the diagram. Areas may be measured with a planimeter or by other means. The extreme right of the diagram is for temperatures below the meteorological range, so in practice the diagram is terminated at the $-80°$ C isotherm and the various areas so omitted are tabulated. We shall now give several examples of the determination of radiative flux from the sounding of Table 10.1.

Example 1—downward flux arriving at the surface: For this we use the values of u' calculated for the reference level 990 mb, that is, an upward integration of du'. For point B (Fig. 10.2), which is at the surface, there is no vapor below. Thus a point is plotted at the surface temperature and $u' = 0$. Successively higher in the sounding more and more water vapor is found below, defining the curve BCD. Above 282 mb (Point D) there is so little water vapor left that u' remains essentially constant. Therefore D is connected to the vertex, E, by an essentially straight line along a constant value of u'. The area $BCDEB$ then represents the downward flux of radiation from the moisture in the entire vertical column, but does not include the downward flux from CO_2. Since CO_2 absorbs essentially completely in the 15 μ band, which is the only important one, each layer of the atmosphere completely absorbs the CO_2 radiation from above and re-emits downward at its own temperature. Thus the radiation in the 15 μ band arriving at the surface is that from a thin layer immediately above the surface, and the CO_2 flux is a known fraction of the black-body radiation at the surface temperature. This relationship is represented on the diagram by the triangular area $ABEA$. Accordingly the total downward flux is given by the sum of the water vapor flux and the CO_2 flux, area $ABCDEA$.

Example 2—net loss of radiation by the surface (clear skies): Since the earth's surface radiates very nearly as a black body in the infrared, the upward radiation rate is given by the area to the right of the surface isotherm, $ABFEA$ (Fig. 10.2). If we subtract the downward flux calculated in Example 1, we find the net upward flux from the surface to be the hatched area $BFEDCB$. Note that the CO_2 radiation area cancels in the subtraction. The reason is that the surface air layer radiates as much to the ground as it receives in those parts of the spectrum where CO_2 is completely black. Only the water-vapor flux modifies the ground radiation and hence determines the nocturnal cooling rate.

Example 3—net loss of radiation from the surface (overcast skies): Suppose in Fig. 10.2 an overcast exists with its base at a level corresponding to point D. If a cloud layer is about 50 m or more thick it will absorb and radiate in the infrared as a black body. Therefore the downward flux arriving at the ground will be that due to the water vapor and CO_2 between the surface and the overcast, represented by the curve *ABCD*, and the additional black-body radiation from the clouds not absorbed by the vapor and CO_2 below it, represented by the line *DG*. The net downward flux is thus given by area *ABCDGEA*. When, as before, this area is subtracted from the upward black-body radiation of the surface we find the net radiative loss of the ground, area *BFGDCB*. This latter area is much smaller than the loss by the surface with clear skies given by the entire hatched area. The blanketing effect of clouds explains why surface temperatures fall to much lower values on clear nights than on cloudy nights, other things being equal.

Example 4—net flux at a selected level (*clear skies*): Net flux of radiation at a certain level is the difference between the downward and upward directed fluxes at that level. We have seen that downward flux can be determined from an upward integration of water vapor from a reference level; conversely, upward flux is obtained from a downward integration of du' from the reference level. In Fig. 10.3 are plotted the

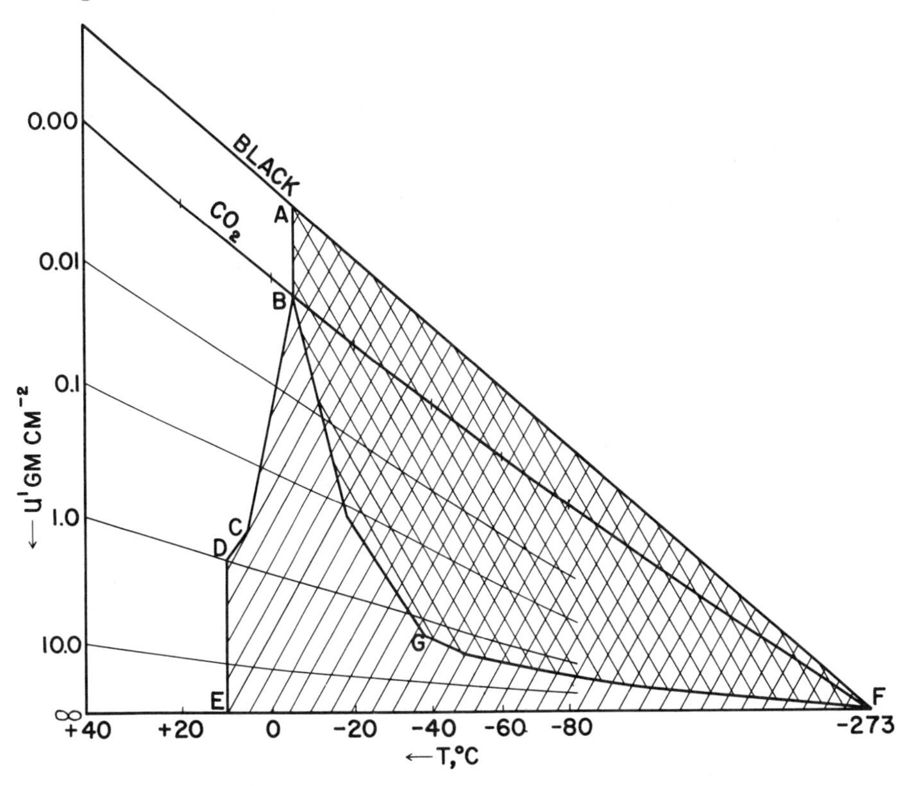

Fig. 10.3. Elsasser diagram

upward and downward integrations for the reference level 720 mb from Table 10.1. The curve *BCD* is the downward integrated curve from 720 mb to 990 mb. The area *DEF* to the right of the isothermal line *DE* is the water-vapor equivalent of the black-body radiation from the surface, so if we add the CO_2 flux then the upward flow of radiation through the reference level is given by the area *ABCDEFA*. The upward integrated water vapor curve is given by *BGF*; therefore the downward flow of radiation including the CO_2 flux, is given by the doubly hatched area *ABGFA*. The difference, represented by the singly hatched area *BCDEFGB*, is the net upward flux through the 720 mb level. Note that Elsasser's assumption about the nature of CO_2 radiation causes cancellation of its contribution to net flux. Once again, only the water vapor flux makes a significant contribution.

Example 5—radiative cooling and heating of an air layer: To determine radiative temperature changes for a layer one must compute the net upward flux at the bottom and top of the layer. The difference between them, called the *net flux divergence*, determines the rate of change of energy of the layer by radiation. If ΔQ represents the gain of energy per unit horizontal area by the layer in time Δt, and F_{NT} and F_{NB} represent the net upward fluxes through the top and bottom of the layer in question, then

$$\frac{\Delta Q}{\Delta t} = F_{NB} - F_{NT} \tag{10.6}$$

But from the first law of thermodynamics

$$\frac{\Delta Q}{\Delta M} = c_p \Delta T$$

if we neglect changes of pressure. Here ΔM is the mass of the layer per unit horizontal area. From the hydrostatic equation

$$\Delta M = \rho \Delta z = \frac{p_B - p_T}{g}$$

therefore Eq. (10.6) becomes

$$\frac{\Delta T}{\Delta t} = \frac{g}{c_p} \frac{F_{NB} - F_{NT}}{p_B - p_T}$$

or, in the limit as the pressure increment through the layer becomes infinitesimal,

$$\frac{\Delta T}{\Delta t} = \frac{g}{c_p} \frac{\partial F_N}{\partial p}$$

These results indicate that if $F_{NB} < F_{NT}$ (net flux divergence) cooling will result, while if $F_{NB} > F_{NT}$ (net flux convergence) warming will occur.

Since net radiative fluxes can be determined from the Elsasser diagram it can be used to calculate rates of radiative temperature change. Because both temperature and moisture usually decrease upward it is normal for more radiation to emerge from the top of an atmospheric layer than enters it from below. The average radiational cooling in the troposphere will be discussed in Section 10.4.

10.3 Radiative Heating and Cooling of Clouds

A sufficiently thick cloud acts as a black body. Therefore its base, which receives radiation from the generally warmer earth and air below, tends to rise in temperature. On the other hand, the cloud top loses part of its energy to space in the transparent windows of the spectrum and part to the water vapor and CO_2 above. Since the air above is usually colder than the cloud, less radiation comes back down to the cloud than it emits even in the infrared absorption bands. Thus cloud tops ordinarily lose energy in two ways, leading to cooling. The warming of cloud bases and cooling of cloud tops can proceed quite rapidly. It has been calculated that in a cloud of modest thickness the lapse rate can be converted by radiation from isothermal to the saturated adiabatic value in less than one hour.

Clearly such rapid destabilization has important synoptic consequences. For example, an originally stable stratiform cloud when sufficiently destabilized should break up into cumuliform cells, as is often observed. Other factors, however, may cancel this effect, as when continued cooling due to upward motion along a sloping frontal surface overcomes the radiative heating of the cloud base.

Another example which in part is due to this destabilizing effect is the nocturnal development of thunderstorms from clouds which could not develop beyong the swelling cumulus stage during the day. Not only is the surface heated by the sun during the day to produce instability, but once the cumuli are formed their tops are heated by absorption of sunlight, thus slowing the destabilization. When the sun sets the cumulus tops cool rapidly while their bases warm, and further growth of the cloud may result. Again, the process for formation of nocturnal thunderstorms is not solely a radiative one. For example, low-level convergence often plays a great role in this phenomenon.

10.4 Infrared Radiative Cooling of the Atmosphere

We have seen that the normal decrease of temperature and water vapor upward leads to radiative cooling of the atmosphere. London[2] has used the Elsasser diagram to evaluate this cooling in the mean for various seasons. Figures 10.4 and 10.5 show the results as a function of latitude and elevation for winter and summer for average cloudiness.

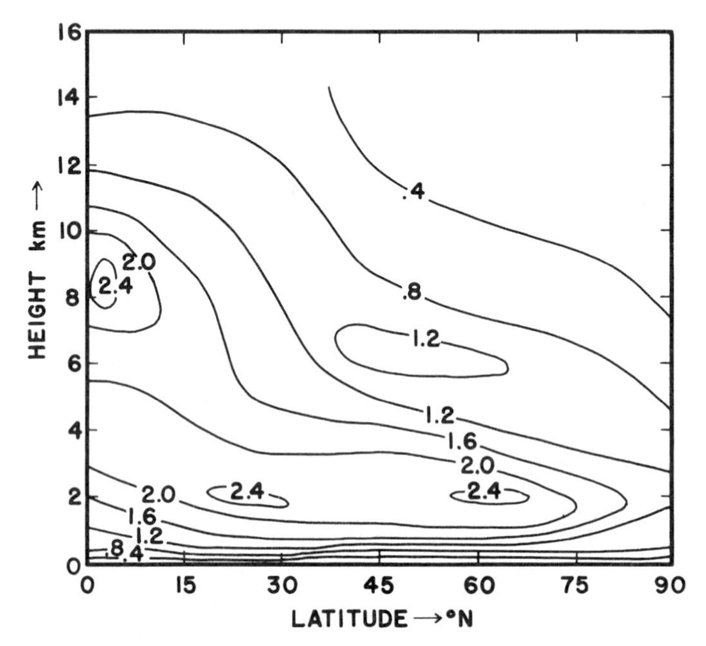

FIG. 10.4. Atmospheric infrared radiative cooling rate (°C day^{-1}) in winter, after London

In both seasons the cooling rate increases upward from the ground until a maximum of about 2.4° C per day is reached. This occurs at a nearly constant level of 2 km in winter and at a layer sloping downward from 5-6 km at the equator to 2 km near the pole in summer. The maximum cooling rate is associated, in general, with the top of the mean middle-cloud layer. Above this maximum the cooling rates decrease upward throughout the troposphere, except in the tropics. There a second maximum of cooling is found; a strong one at 8-9 km in winter and a weak one at 12 km in summer. This condition is pre-

[2] J. LONDON, A Study of the Atmospheric Heat Balance. Final Report, Contract No. AF 19(122)-165, New York University, 1957.

sumably a reflection of the radiative loss from the top of the deep moist layer produced by tropical convection.

Since the atmosphere maintains an essentially constant average temperature over a period of years, this radiative loss must be made up

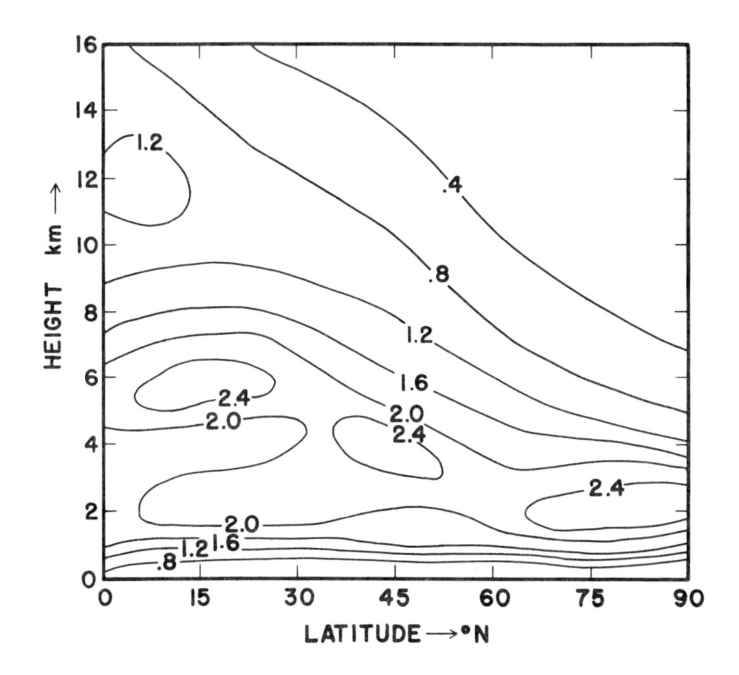

FIG. 10.5. Atmospheric infrared cooling rate (°C day^{-1}) in summer, after London

by other processes. The chief of these are release of latent heat of condensation, absorption of sunlight, and convective transport of heat upward from the ground (see Section 10.7).

10.5 Transformation of Maritime Polar Air into Continental Polar Air

In winter continental polar air (*cP*) is formed over the northern portion of the North American continent. This air mass is characterized by rather low surface temperatures (appreciably below 0° C in the source region), an increase of temperature upward for about one km, a nearly isothermal layer one or more km thick, and a normal rate of decrease of temperature thereafter. An example at Fairbanks in the interior of Alaska, is given in Fig. 10.6. Throughout the winter *cP* air drains southward in frequent cold waves and is replaced by maritime polar air (*mP*) from the adjacent oceans. The maritime air is characterized by surface temperatures near 0° C and a normal decrease upward throughout. An example at Cold Bay along the Alaskan coast, is given in Figure 10.6. By what processes is an *mP* air mass converted to *cP*?

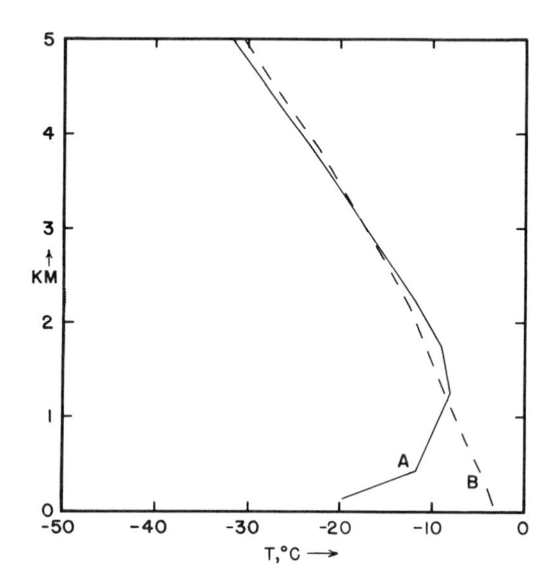

FIG. 10.6. Mean lapse rates in January 1958 at Fairbanks, Alaska (*A*) and Cold Bay, Alaska (*B*)

Wexler[3] has approached this problem from the radiative point of view using a modification of Simpson's method to determine the radiative cooling. Let us suppose that when an *mP* air mass moves over a snow-covered continent the surface temperature initially is 0° C and the lapse rate is as shown in Fig. 10.7 (Curve 1). Snow is almost exactly a black body in the infrared but has a high albedo in short wave lengths. Thus the feeble winter insolation is largely reflected during the day but the surface radiates copiously day and night. Since some of the surface radiation escapes directly to space through the windows in the infrared spectrum, the surface begins to cool very rapidly and a shallow inversion develops as shown by Curve 2.

[3] H. WEXLER, Cooling in the Lower Atmosphere and the Structure of Polar Continental Air. *Mon. Weather Rev.*, 64, 1936.

When the surface cooling has proceeded far enough the overlying atmosphere begins to lose energy, not only to space by upward flux but also to the cold underlying surface. Thus, a layer above the inversion begins to drop in temperature. Wexler's calculations show that this drop takes the form of development of an isothermal layer extending from the inversion to the original sounding curve. As the overlying air cools it returns energy to the snow surface less rapidly, thus the

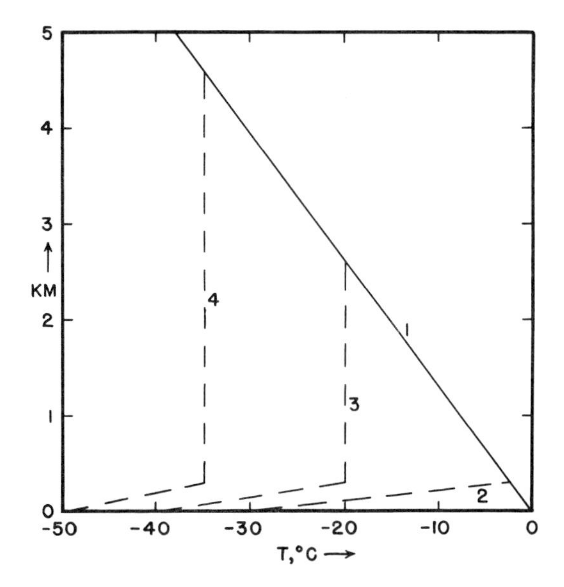

FIG. 10.7. Schematic representation of the conversion of *mP* air to *cP* air, after Wexler

surface temperature drops further and Curve 3 develops. If the air mass remains over the snow surface long enough, the process of cooling of the surface and deepening of the isothermal layer can continue to produce Curve 4 and still more extreme conditions. Note that above the isothermal region the lapse rate is unaffected and is still representative of *mP* air.

In Wexler's treatment the depth of the inversion is theoretically infinitesimal. However, molecular and eddy diffusion will deepen this layer to finite proportions. Indeed, if there is sufficient wind the mechanical stirring produced can wipe out the surface inversion completely. Thus this approach is valid only with calm or light winds. These are precisely the wind conditions found when *cP* air is being formed.

10.6 Radiative Equilibrium and the Stratosphere

Since the pioneering observations of Tiesserenc de Bort at the end of the 19th century it has been known that temperature decreases with elevation only up to a certain height. Above this level the atmosphere is nearly isothermal for a considerable distance. The lower region of temperature decrease is the *troposphere*, the upper isothermal region is the *stratosphere*, and the level of sharp change from one to the other is the *tropopause*. The elevation of the tropopause varies with latitude and time, but on the average is near 10 km.

Many attempts have been made to explain this structure theoretically, but with imperfect success. It appears that a complete explanation cannot be attained without including radiative, advective, turbulent, and other dynamic effects. Nevertheless, a helpful first approximation

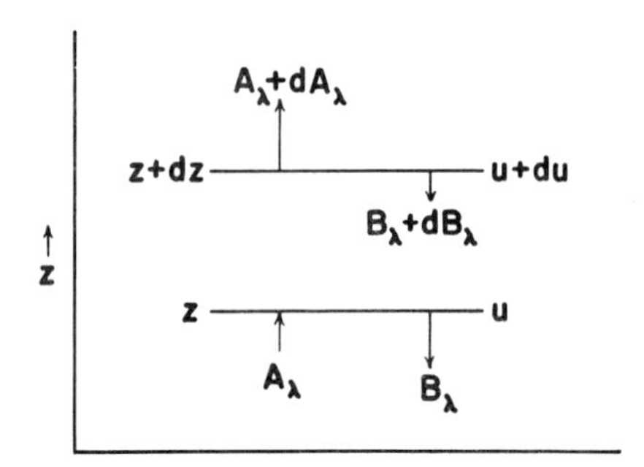

FIG. 10.8. Upward and downward radiative fluxes for an infinitesimal atmospheric layer

can be reached by considering radiation first and by assuming a state of radiative equilibrium. By this we mean that each horizontal layer of air exchanges energy by radiation alone and is losing at the same rate it is receiving radiant energy.

Consider a layer of air of infinitesimal optical thickness, as in Fig. 10.8. We shall denote the upward monochromatic flux of radiation by A_λ and the downward flux by B_λ. If A_λ and B_λ are the values at the bottom of the layer, then at the top of the layer we have $A_\lambda + dA_\lambda$ and $B_\lambda + dB_\lambda$. Each of these fluxes is governed by Schwarzchild's Eq. (8.9) which, since $a_\lambda = k_\lambda \, du$, may be written

$$\begin{aligned} \frac{dA_\lambda}{du} &= k_\lambda[f(\lambda, T) - A_\lambda] \\ -\frac{dB_\lambda}{du} &= k_\lambda[f(\lambda, T) - B_\lambda] \end{aligned} \tag{10.7}$$

where the minus sign in the second equation is inserted because B_λ is a downward flux. Since we assume radiative equilibrium, the fluxes into the layer must balance against the fluxes out of the layer:

$$A_\lambda + (B_\lambda + dB_\lambda) - (A_\lambda + dA_\lambda) - B_\lambda = 0$$

or

$$d(B_\lambda - A_\lambda) = 0$$

Thus from Eq. (10.7)

$$B_\lambda + A_\lambda - 2f(\lambda, T) = 0$$

Now the upward flux, A_λ, through a given level depends upon the temperature of the ground, the amount and distribution of water vapor below the level, and the temperature distribution below the level. The downward flux B_λ depends upon the intensity of effective solar radiation, the amount and distribution of water vapor above the level, and the temperature distribution above the level. Since we know the average ground temperature and the average effective insolation, it should be possible to find the radiative equilibrium lapse rate of temperature if the water vapor distribution is specified. Emden[4] solved this problem with the following assumptions:

1. Water vapor is the only radiation absorber, and its density decreases exponentially with height.

2. The water-vapor absorption spectrum is semi-grey. That is, one constant absorption coefficient is assumed for solar radiation and another constant value for terrestrial radiation.

With appropriate values for the various parameters Emden found the temperature curve given by the dashed line of Fig. 10.9. This radiative equilibrium lapse rate begins at $z = 0$ with a temperature 19° C lower than the ground temperature—that is, there is a temperature discontinuity at the surface. Immediately above $z = 0$ the lapse rate is considerably greater than the dry adiabatic rate. At higher elevations the lapse rate decreases progressively until above 8-9 km it becomes nearly isothermal with a temperature slightly warmer than −60° C. Thus Emden's model gives a good fit to the stratosphere both as to elevation and temperature but is quite unrealistic in the troposphere.

Emden gave a physically satisfying qualitative explanation of the low-level discrepancies. Clearly the lower layers of this model are mechanically unstable and must overturn. This means that heat will be transported upward by convection and this will predominate over the radiative transfer. A dry adiabatic lapse rate will be established near the

[4] R. EMDEN, Strahlungsgleichgewicht und atmosphärische Strahlung. *Sitz. K. Bayer. Akad. Wissensch.*, München, 1913.

ground, and if we assume a condensation level at 1 km a moist adiabatic lapse rate will be established above, as given by the solid line of Fig. 10.9. Emden made the reasonable assumption that this convective regime would give way to the radiative equilibrium regime some modest distance above the level where the temperature curves cross. In this way he arrived at a qualitative result in which the troposphere is the layer dominated by convective energy transport, the stratosphere is the layer dominated by radiative energy transport, and the tropopause is the region of rather abrupt transition from one to the other.

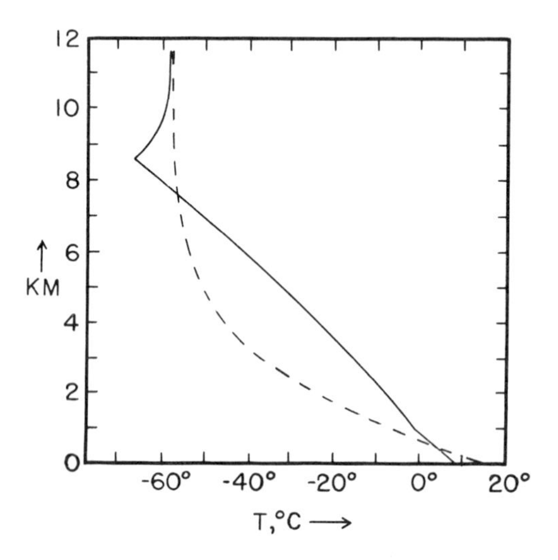

FIG. 10.9. Emden's curves for radiative equilibrium in the entire atmosphere (dashed), and convective equilibrium in the troposphere with radiative equilibrium in the stratosphere (solid)

This theory gives an essentially correct account of the most important factors influencing the vertical temperature distribution of the atmosphere below the ozone layer. However, its greatest defect is that the theoretical temperature of the stratosphere decreases with increasing latitude, in disagreement with observation. Goody[5] and others have considered somewhat different models which incorporate recent knowledge of radiative effects and which are more sophisticated. Goody assumed that the troposphere is in convective equilibrium with a lapse rate of 6.5° C km^{-1} and that the stratosphere is in radiative equilibrium with a nearly isothermal lapse rate. He further assumed a reasonable

[5] R. M. GOODY, The Thermal Equilibrium at the Tropopause and the Temperature of the lower Stratosphere, *Proc. R. Soc.* (*London*), *A*, 197, 1949, p. 487.

vertical distribution of CO_2, H_2O, and O_3 and a suitable pressure dependence of the absorption coefficients of these gases. He then solved the Schwarzschild equation for radiative equilibrium for the various infrared bands involved. With this approach Goody was able to ascertain the relative importance of the various gases in maintaining radiative equilibrium and to determine the latitudinal dependence of the stratospheric temperature.

Goody found that the balance is accomplished in middle and high latitudes by warming due to CO_2 and cooling by H_2O. Ozone makes no significant contribution here in the lower stratosphere. The cooling by H_2O is largely due to the rapid decrease in amount of vapor with height which was assumed. In low latitudes, where the tropopause is higher, H_2O vapor is negligibly small but ozone is more appreciable in amount. The balance here is between warming by O_3 and cooling by CO_2. This more inclusive model yields an increase of tropopause temperature with latitude, in agreement with reality.

10.7 The Mean Annual Heat Balance

The mean temperature of our planet undergoes oscillations covering a wide range of period. The annual cycle is well known, but in addition there are faster variations which occupy a few weeks and slower ones taking anywhere from a few years through a century and ranging up to the geological glacial variations requiring millions of years for their completion. The faster fluctuations can be smoothed out by averaging over a number of years, but the slower variations that remain imply a net gain or loss of energy during the period of averaging. Nevertheless, the remaining unsmoothed fluctuations are so slow that the net energy change per year is quite small. Thus we may neglect these energy changes without significant error and assume that conditions averaged over a period of 10-20 years represent a state of heat balance. In other words, the earth-atmosphere system must receive, to a high degree of accuracy, as much energy from the sun as it sends to space.

The study of this heat balance and determination of the heat budget of the earth-atmosphere system is one of the central problems of meteorology because all atmospheric phenomena ultimately derive their energy from this source. Consequently many investigators have looked into this matter and gradual improvement in our understanding of the energy balance has resulted from incorporation of advances in the theory of radiative transfer. We shall consider the most recent of these studies, carried out by London.[6]

In order to make accurate calculations of the major contributions to the heat budget by solar and terrestrial radiation it is necessary to know the geographical and vertical distributions of

1. temperature and pressure,
2. concentration of the absorbing and emitting gases, and
3. cloudiness.

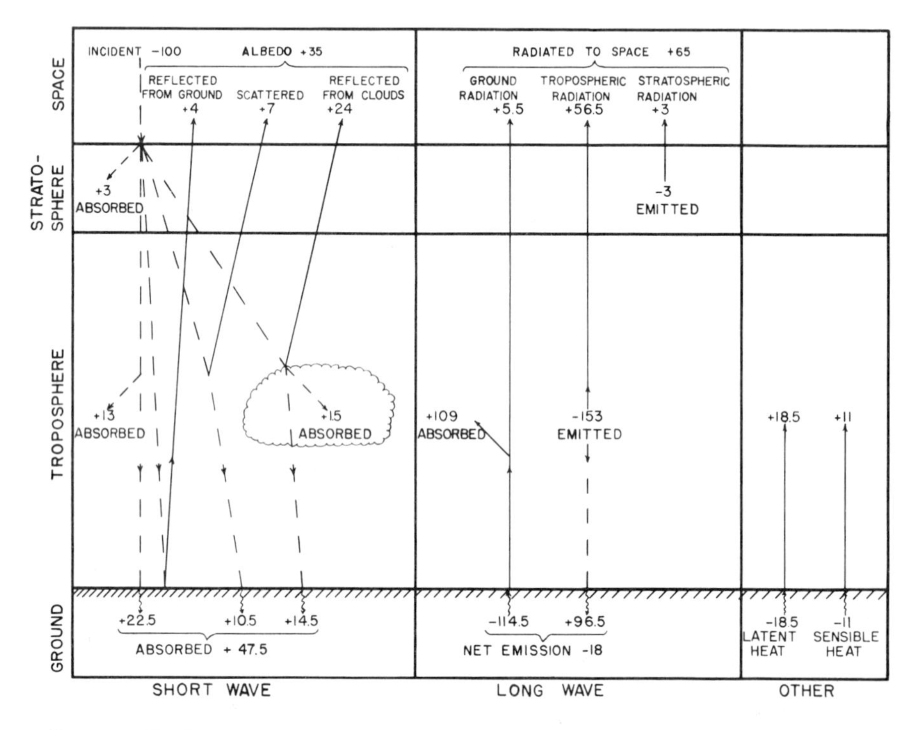

FIG. 10.10. The mean annual heat budget of the Northern Hemisphere, after London

The temperature and pressure distributions are known rather well from climatological data. The concentration of CO_2 is also well known since it is characterized by a constant mixing ratio, but the proportion of H_2O is known rather imperfectly. This is largely a deficiency of our radiosonde instruments, which do not measure water vapor satisfactorily above 5-6 km. London was obliged to extrapolate the observations above this level by judiciously fitting a relative humidity distribution to the known variations of pressure and temperature. Ozone values also

[6] *Op. cit.*, 1957.

leave something to be desired, especially in connection with their vertical variation, but since ozone plays only a minor role in the heat balance the existing data suffice. Our knowledge of the detailed mean distribution of clouds is also deficient, since more emphasis has been placed on reporting cloud amount than upon determining cloud type. Furthermore, the data for middle and high clouds are lower limits on their amounts because low clouds often conceal existing higher clouds. Again London made judicious corrections to the climatological data.

The tool used to determine infrared flux was the Elsasser diagram. The computations were carried out for mean data over the northern hemisphere for each of the four seasons and the results combined to obtain the annual mean. Figure 10.10 gives the results for short-wave solar radiation, long-wave terrestrial radiation, and nonradiative processes; separately for the earth's surface, the troposphere, the stratosphere, and space. We shall discuss each of these in detail.

1. *Short-wave radiation*:

(a) Solar energy arrives at the top of the atmosphere at the average rate of 0.500 ly min^{-1}, referred to in the diagram as 100 units. This is because the solar constant of 2.00 ly min^{-1} is intercepted by a disc of area πa^2 but is distributed over the area of a sphere, $4\pi a^2$. Thus the average rate of receipt of energy is one-fourth of the solar constant. The data are given to the nearest 0.5 unit in order to exhibit the balance, but this probably exceeds the accuracy of the computations and should not be taken too seriously.

(b) Three of these incident 100 units are absorbed in the stratosphere, primarily by ozone.

(c) Of the remaining direct beam, 13 units are absorbed in the troposphere, mostly by water vapor and dust, and 22.5 units are absorbed by the ground. Four units are reflected directly to space by the ground.

(d) The major contribution to the albedo of our planet comes from reflection of the solar beam to space by clouds. This reflection was found to amount to 24 units. Not all of the sunlight incident on clouds is reflected; 1.5 units are absorbed in the clouds and 14.5 units are sent downward as diffuse radiation which is absorbed by the surface.

(e) Finally, 17.5 units are scattered by the atmosphere, with 10.5 going downward to be absorbed by the ground and 7 going upward to escape into space.

Note that a total of 35 units of short-wave energy is returned directly to space, thus the mean albedo of our planet is 35 per cent. This is made up of reflection from the ground, reflection from clouds, and scattering.

2. *Long-wave radiation*:

(a) The surface is found to radiate 114.5 units upward. This is calculated from the Stefan-Boltzmann law for a mean ground temperature of 288° K. Of these 114.5 units, 109 are absorbed by H_2O and CO_2 in the troposphere and only 5.5 escape to space. Thus the atmosphere is about 95 per cent opaque to terrestrial radiation.

(b) The water vapor and carbon dioxide of the troposphere emit an average of 153 units. Of these 96.5 go downward and are absorbed by the essentially black ground, and 56.5 go upward and escape to space. Since the ground emits 114.5 units but gets 96.5 back from the atmosphere, the net ground emission is only 18 units.

(c) The stratosphere radiates 3 units to space in the ozone band at 10 μ and the carbon dioxide band at 15 μ.

3. *Nonradiative effects*:

(a) Some of the solar energy absorbed at the ground goes to evaporate water. This vapor is carried upward and horizontally and ultimately condenses, releasing its latent heat to the atmosphere. Only that part of the condensation which results in precipitation need be counted in this energy transfer because the droplets which do not fall out ultimately re-evaporate and take up their latent heat again. London used the world-wide data on mean rainfall and, assuming the condensation to occur at the middle-cloud level, found that 18.5 units are transferred to the atmosphere in this way.

(b) Turbulence in the atmosphere can transport sensible heat upward or downward, from or to the ground. Since this turbulent transport is extremely complex it is best not to attempt to compute it directly, but to determine its amount as the residue necessary to balance the heat budget. In this way London found that 11 units are transferred upward from the ground to the troposphere. Although our imperfect understanding of turbulent transfer makes it risky to compute this amount, we do know that the transfer must be upward, in agreement with the requirement of this balance computation. It is worth noting that some earlier evaluations of the heat balance gave a downward turbulent transfer—because the early estimates of the large radiative exchanges were so rough that the small remainder needed for balance was not even of the right sign.

We shall now verify the balance for each region: space, stratosphere, troposphere, and ground.

Space loses 100 units to the earth-atmosphere system (solar energy) and gets back 100 composed of 35 short-wave units (reflection and scattering) and 65 long-wave units (radiation from the ground, the troposphere, and the stratosphere). Of these losses to space the most important are tropospheric radiation and reflection by clouds.

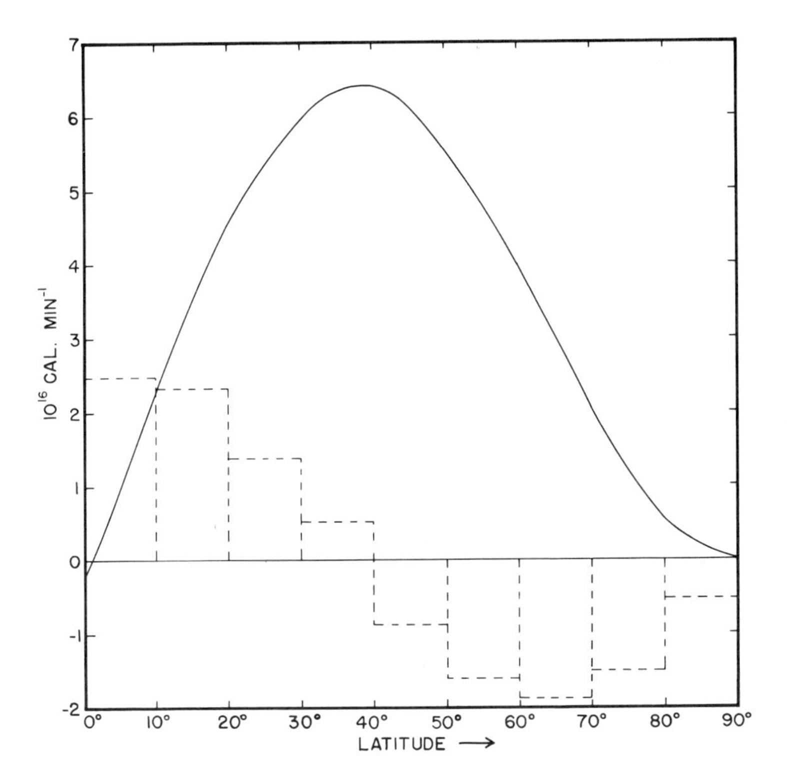

FIG. 10.11. Annual mean radiative excess or deficit (dashed blocks) and required latitudinal transport of energy (solid curve), after London

The stratosphere gains 3 units by absorption of sunlight and loses 3 units by long-wave radiation to space.

The troposphere absorbs 14.5 units of sunlight (atmosphere and clouds) and 109 units of ground radiation. It also receives 18.5 units of latent heat and 11 of sensible heat from the ground, for a total gain of 153 units. It loses these 153 units by long wave radiation upward and downward. Thus atmospheric absorption and re-emission of long waves are the dominant features of the tropospheric balance.

The ground gains 47.5 units by absorption of direct and diffuse sunlight, and 96.5 units by absorption of tropospheric radiation, for a total gain of 144 units. It loses precisely 144 by black body radiation (114.5), transfer of latent heat (18.5), and transfer of sensible heat (11). The dominant process is the exchange of long-wave radiation between the ground and the troposphere.

At first glance it seems strange that with an input of 100 units the ground and the troposphere can absorb and emit more than 100 units of energy flux. This transfer can occur because of the greenhouse effect of the atmosphere which, by largely blocking ground radiation, forces the surface temperature to rise sufficiently above the value it would have in the absence of an atmosphere to produce the needed flux to space. The mean temperature of the ground is raised 33° C in this way.

London's calculations were performed for 10° latitude strips from the equator to the pole and then summed up over all latitudes to get the over-all balance. There is no reason, however, to assume that each latitude band balances by means of vertical flux of energy. The data actually show that for an annual mean there is a surplus of energy received over that lost to space from the equator to about latitude 40° N and a deficit from 40° N to the pole (dashed blocks, Fig. 10.11). Since all latitudes maintain essentially the same annual mean temperature, this heat must be transported from low towards high latitudes. To calculate this transport we start at the pole (across which no heat can flow) and integrate the surplus-deficit amounts with latitude. This gives the solid curve of Fig. 10.11. At all latitudes (except 0° and 90° N) there is a transport of heat northward and the maximum is between 6 and 7 $\times$ 10^{16} cal min^{-1} at latitude 40° N.

This poleward flow of heat can be accomplished only by the atmosphere and the oceans, and the atmosphere has been estimated to perform 70-90 per cent of the task. The atmospheric poleward flux is produced by air currents which transport both latent heat (in the form of water vapor) and sensible heat. Indeed, this energy surplus in low latitudes and deficit in high latitudes is the fundamental cause of the general circulation of the atmosphere.

PROBLEM

1. The Elsasser diagram has the property that the triangular area bounded by the temperature axis, the line labelled "black", and any isotherm is proportional to the rate of emission of a black body at the isotherm temperature. Show from this that the abscissa of the diagram is proportional to T^2.

CHAPTER 11

THE EQUATIONS OF MOTION ON A ROTATING EARTH

11.1 Inertial versus Noninertial Coordinate Systems

Newton's second law of motion will be taken as the fundamental statement to be elaborated in the present chapter. However, it is necessary to realize that this law is applicable only in certain kinds of coordinates called *inertial coordinate systems*; all other systems are *noninertial.*

A first appreciation of the importance of the coordinate system can be gained by considering an application of Newton's law to experiments inside a closed elevator car. When the car is stationary (or moving with constant velocity) a weight dropped within the car will fall to the floor with a measurable, constant acceleration. This is explained by saying that a constant force (gravity) has been applied to a constant mass and is associated with the observed constant acceleration: $\mathbf{F} = m\mathbf{a}$. Note that the acceleration is measured relative to the elevator car which constitutes our frame of coordinates. The coordinate system in this case is inertial.

If now the car is allowed to fall freely in the elevator shaft and the experiment is repeated it will be found that the weight dropped will remain suspended at a constant elevation above the floor. Measured relative to the same set of coordinates as before (the car) the weight has zero acceleration even though we know quite well that it is being acted upon by the force of gravity. There is an apparent failure of Newton's second law in this set of coordinates.

The essential characteristic of the coordinate frame in the second experiment is that the system is accelerating. Thus an observer outside the car could determine that the weight is "really" accelerating downward *relative* to the earth's surface, as it should. He would conclude that the floor of the car is accelerating at the same rate so that the *relative* acceleration of weight to floor is zero. Thus the validity of Newton's law depends utterly upon the nature of the coordinate system relative to which we attempt to verify it. These hypothetical experiments indicate that noninertial coordinate systems are those which are accelerating significantly. Inertial coordinates are those which are accelerating not at all, or so little that the effect is negligible.[1]

Let us now consider the geophysical coordinate system most used on earth. This is simply our grid of latitude, longitude, and height above mean sea level. This system rotates with the earth about its axis, and as a result is an accelerating frame in which departures from Newton's second law are to be expected. One may ask how Newton was able to arrive at a law describing inertial events when he lived in a noninertial system. As the analysis to follow will show, he was able to do so because the effects of the acceleration of our reference frame are negligible for small-scale motions such as those observed in laboratories and in everyday life. When, however, one considers large-scale motions such as those occuring in the major atmospheric and oceanic currents the noninertial effects become important.

11.2 The Dynamical Equations in a Rotating Coordinate System

Let us consider a fixed, right-handed, Cartesian coordinate system χ', η', ζ' as in Fig. 11.1. Consider also a coordinate frame χ, η, ζ with the same origin as the primed system but which is rotating at a constant angular rate, Ω $[T^{-1}]$ with respect to the fixed system. We shall take the axes ζ' and ζ to be identical with each other and with the axis of rotation. We know how to write Newton's second law for a particle of constant mass in the fixed system:

$$\ddot{\chi}' = \frac{F_{\chi}'}{m}; \quad \ddot{\eta}' = \frac{F_{\eta}'}{m}; \quad \ddot{\zeta}' = \frac{F_{\zeta}'}{m} \tag{11.1}$$

[1] These remarks may seem to imply the existence of some fixed universal inertial system against which the acceleration of all others may be measured. This assumption is unnecessary if we establish in advance the accuracy with which we are required to meet Newton's law. Measurements in the chosen system of coordinates will then reveal whether the system is inertial within the tolerable error or not, and no external reference will be needed.

where each dot means one differentiation with respect to time. The problem is to find the form of this law when we transform to the rotating system.

At a given time, t, the angle between corresponding axes of the two systems will be Ωt. Thus the position of a point P in Fig. 11.1 in the primed system can be related to its position in the unprimed system by the following relationships :

$$\begin{aligned} \chi' &= \chi \cos \Omega t - \eta \sin \Omega t \\ \eta' &= \chi \sin \Omega t + \eta \cos \Omega t \\ \zeta' &= \zeta \end{aligned} \tag{11.2}$$

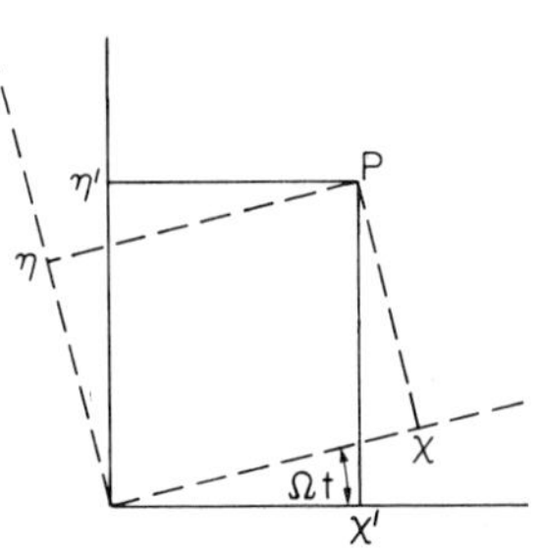

FIG. 11.1. Fixed and rotating Cartesian coordinate systems

Similarly, forces measured in the primed system can be related to the same forces measured in the unprimed system by:

$$\begin{aligned} F_{\chi'} &= F_\chi \cos \Omega t - F_\eta \sin \Omega t \\ F_{\eta'} &= F_\chi \sin \Omega t + F_\eta \cos \Omega t \\ F_{\zeta'} &= F_\zeta \end{aligned} \tag{11.3}$$

If we differentiate Eq. (11.2) twice with respect to time and substitute the result, together with Eq. (11.3), into Eq. (11.1) we get

$$(\ddot{\chi} - 2\dot{\eta}\Omega - \chi\Omega^2) \cos \Omega t - (\ddot{\eta} + 2\dot{\chi}\Omega - \eta\Omega^2) \sin \Omega t = \frac{F_\chi}{m} \cos \Omega t - \frac{F_\eta}{m} \sin \Omega t$$

$$(\ddot{\chi} - 2\dot{\eta}\Omega - \chi\Omega^2) \sin \Omega t + (\ddot{\eta} + 2\dot{\chi}\Omega - \eta\Omega^2) \cos \Omega t = \frac{F_\chi}{m} \sin \Omega t + \frac{F_\eta}{m} \cos \Omega t$$

$$\ddot{\zeta} = \frac{F_\zeta}{m} \tag{11.4}$$

Let us now remove all reference to sin Ωt and cos Ωt by elimination between the first and second of Eq. (11.4). The result may be written

$$\begin{aligned} \ddot{\chi} - \frac{F_\chi}{m} &= \chi\Omega^2 + 2\dot{\eta}\Omega \\ \ddot{\eta} - \frac{F_\eta}{m} &= \eta\Omega^2 - 2\dot{\chi}\Omega \\ \ddot{\zeta} - \frac{F_\zeta}{m} &= 0 \end{aligned} \tag{11.5}$$

Equations (11.5) are Newton's second law transformed to a rotating coordinate system. In a fixed frame the difference between the acceleration and the force per unit mass is zero. In the rotating system this difference is zero only along the axis of rotation. The first of the additional terms on the right in Eq. (11.5) are accelerations pointing radially outward from the rotation axis, of combined magnitude $\sqrt{\chi^2 + \eta^2}\ \Omega^2$. This is the same as $R\Omega^2$, where R is the distance of the particle from the axis of rotation. Such an effect is a centrifugal acceleration whose existence in a rotating coordinate system is already familiar from elementary physics.

The second terms on the right of Eq. (11.5) represent accelerations owing to the combined effects of rotation of the coordinates and motion of a particle relative to the rotating system. These are called *Coriolis accelerations* after the physicist who first discussed them.[2] They have the following important properties which may be verified directly:

(1) They involve only those components of the total motion, measured relative to a rotating frame, which lie in a plane perpendicular to the axis of rotation (equatorial plane).

(2) They are directed to the right of these velocity components in an equatorial plane with magnitude $2\Omega\sqrt{\dot{\chi}^2 + \dot{\eta}^2} = 2\Omega V$, where V is the part of the velocity which lies in an equatorial plane.

(3) Since the Coriolis acceleration is always perpendicular to the velocity it can never change the speed of a particle, but can change only its direction of motion. For this reason it is sometimes referred to as a *deflecting force*.

The centrifugal and Coriolis accelerations are departures from Newton's second law which arise from the noninertial character of the frame of reference. We may, however, continue to think of dynamics in such a system in Newtonian terms if we adopt the convenient fiction that two additional forces per unit mass are acting, namely centrifugal and Coriolis forces per unit mass. If this is done we may still say, even for a rotating system, that the acceleration of a particle is equal to the sum of all the forces per unit mass which are acting, including these two additions. It is to be remembered that these forces do not exist in an inertial system.

We shall now apply these results to events on a spherical earth by placing the origin of the coordinate system at the center of the earth, by identifying the ζ axis with the earth's axis, and by making Ω the angular velocity of the earth. We ordinarily prefer to use a coordinate system tangent to the surface of the earth at some latitude ϕ, in which x is distance along the eastward tangent, y is distance along the northward tangent, and z is distance upward, as in Fig. 11.2. We wish to write Eq. (11.5) in the x, y, z system instead of the χ, η, ζ system. We recognize, first, that the terms on the left of Eq. (11.5) transform immediately to

$$\ddot{x} - \frac{1}{m}F_x; \qquad \ddot{y} - \frac{1}{m}F_y; \qquad \ddot{z} - \frac{1}{m}F_z$$

[2] G. G. Coriolis, *Traité de la Mécanique de Corps Solides*, Paris, 1844.

To grasp this one should think of the acceleration as being a single three-dimensional vector with components $\ddot{\chi}$, $\ddot{\eta}$, $\ddot{\zeta}$ in the coordinate system of Eq. (11.5). When we shift to the tangent plane coordinates the acceleration vector has unique components in that system which we shall call $\ddot{x}$, $\ddot{y}$, $\ddot{z}$ regardless of their magnitudes. The same argument applies to the forces per unit mass. Such simple results are valid because we are transferring between equivalent Cartesian systems.

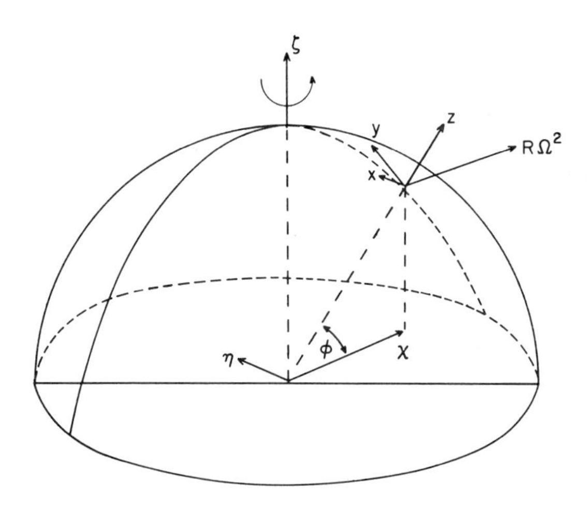

Fig. 11.2. The tangent-plane coordinate system

Next, consider the centrifugal terms. We know them to be directed outward perpendicular to the earth's axis. Since $R = a \cos \phi$, where a is the distance from the earth's center, the centrifugal force per unit mass, $R\Omega^2$, clearly has the components at latitude ϕ which are listed in Table 11.1.

TABLE 11.1

Components of Centrifugal and Coriolis Forces per Unit Mass in the Tangent Plane Coordinates

Direction	*Centrifugal*	*Coriolis*		
		due to eastward motion	*due to northward motion*	*due to upward motion*
eastward	none	none	$2\,\Omega v \sin \phi$	$-2\,\Omega w \cos \phi$
northward	$-a\Omega^2 \cos \phi \sin \phi$	$-2\,\Omega u \sin \phi$	none	none
upward	$a\Omega^2 \cos^2 \phi$	$2\,\Omega u \cos \phi$	none	none

Next we take up the Coriolis terms at the same latitude. We shall consider separately the effect of motion in each of the three coordinate directions. The total Coriolis effect will be the sum of these three effects. If a particle is moving eastward with speed u the Coriolis acceleration will be $2\Omega u$, since u is already parallel to the equatorial plane, and is directed to the right or outward. Therefore, just as above, this acceleration has the components listed in Table 11.1. If a particle moves northward with speed v, the component of this motion in the equatorial plane is $v \sin \phi$. The acceleration due to this motion has the components listed in Table 11.1. Finally, if a particle moves upward with speed w it will have a component of velocity in the equatorial plane equal to $w \cos \phi$, and the components of the Coriolis acceleration on this motion are as listed in the last column of Table 11.1.

If we now add all these components, Newton's second law of motion on a rotating earth may be written:

$$\begin{aligned} \ddot{x} &= 2\Omega(v \sin \phi - w \cos \phi) + \frac{1}{m} F_x \\ \ddot{y} &= -2\Omega u \sin \phi - a\Omega^2 \cos \phi \sin \phi + \frac{1}{m} F_y \\ \ddot{z} &= 2\Omega u \cos \phi + a\Omega^2 \cos^2 \phi + \frac{1}{m} F_z \end{aligned} \tag{11.6}$$

These are referred to as the *equations of motion.*[3]

11.3 Gravitation versus Gravity

Let us consider a body at rest on the rotating earth and subject to no other external force per unit mass than that of gravitation, $G\ [LT^{-2}]$.

[3] These equations are for a Cartesian system tangent to the earth's surface at the latitude ϕ. If one redefines x, y and u, v to be distances and velocities along the curved surface of the earth, instead of tangent to the surface, additional terms are introduced. These arise from transformation of the accelerations in the Cartesian system of Eq. (11.5) to this non-Cartesian system. In such a curvilinear frame the equations of motion are:

$$\begin{aligned} \ddot{x} &= 2\Omega\,(v \sin \phi - w \cos \phi) + \frac{1}{m} F_x + \frac{uv}{a} \tan \phi - \frac{uw}{a} \\ \ddot{y} &= -2\Omega u \sin \phi - a\Omega^2 \cos \phi \sin \phi + \frac{1}{m} F_y - \frac{u^2}{a} \tan \phi - \frac{vw}{a} \\ \ddot{z} &= 2\Omega u \cos \phi + a\Omega^2 \cos^2 \phi + \frac{1}{m} F_z + \frac{u^2}{a} + \frac{v^2}{a} \end{aligned} \tag{11.7}$$

These clearly reduce to Eq. (11.6) if the radius of the earth is so large that the additional curvature terms become negligibly small.

We shall mean by the term gravitation solely the attractive force between bodies which is described in Newton's law of universal gravitation. Equations (11.6) become

$$\begin{aligned} \ddot{x} &= 0 \\ \ddot{y} &= -a\Omega^2 \cos\phi \sin\phi \\ \ddot{z} &= a\Omega^2 \cos^2\phi - G \end{aligned} \tag{11.8}$$

Thus it is clear that the initial state of rest cannot continue since the accelerations do not all vanish. The primary effect from the second of Eq. (11.8) will be to accelerate all bodies towards the earth's equator. This effect is simply the influence of the centrifugal force which accelerates all bodies outward. Since the centrifugal effect is quite insufficient actually to throw objects off a spherical earth against the attraction of gravity, it will slide them equatorward instead and thus cause them to accelerate away from the axis of rotation. A secondary effect, from the third of Eq. (11.8), is to counteract partially the force of gravitation at all latitudes but the poles.

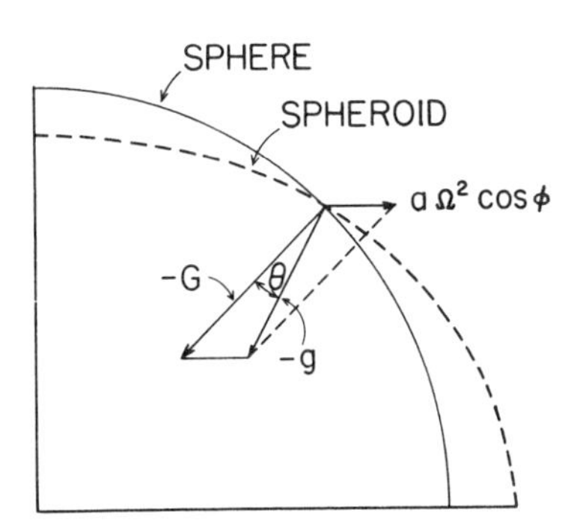

FIG. 11.3. Gravitational and centrifugal forces on an equilibrium oblate spheroid

It is clear, then, that a spherical rotating earth is not an equilibrium configuration, since all particles on it tend to slide equatorward. When this sliding process has gone on long enough, such a hypothetical earth will have become an equilibrium *oblate spheroid.* The essential property of this configuration is that its surface is perpendicular to the vector sum of centrifugal and gravitational forces per unit mass. This is shown on an exaggerated scale in Fig. 11.3. An observer on the rotating earth can measure only the *combined* gravitational and centrifugal effects—he has no direct way of separating them. We call this combined effect *gravity.* Gravity is perpendicular to all equilibrium surfaces, which are therefore called *level surfaces.* All the level surfaces for the earth are oblate, their oblateness being due to the centrifugal term. Since this term becomes larger as one goes outward, while gravitation becomes smaller, it is apparent that the level surfaces become more and more oblate as distance from the earth's axis increases, as illustrated in Fig. 11.4. It is seen that the distance separating successive level surfaces is greater near the equator than near the poles.

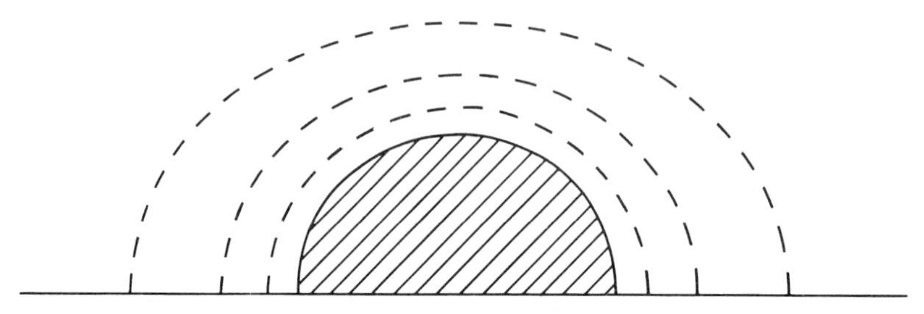

FIG. 11.4. Level surfaces in the atmosphere

The fact that a particle supported on one of these level surfaces is not subject to any component of gravity along the surface means that the surfaces are also levels of constant potential energy in the gravity force field. If this were not so, a particle would roll towards a point of minimum potential energy on the surface. Accordingly these levels are also called *equipotential surfaces*. Aerological data are commonly referred to such equipotential surfaces rather than to surfaces of constant geometric height above mean sea level.[4] This is done to insure that tangential components of gravity will not enter into the consideration.

In Fig. 11.3 the angle θ between the simple gravitation vector and the combined gravity vector is quite small, reaching a maximum of about 0.1° at latitude 45° and diminishing to zero at the equator and poles. As a result we may take the oblateness of the earth into account by considering the z axis to be along the gravity vector instead of the gravitation vector, and in all other respects neglect the nonsphericity

[4] A certain amount of confusion may arise in the terminology of equipotential measurements. The potential energy per unit mass at an elevation, z, above mean sea level is $\psi = \int_0^z g\, dz$. This has dimensions $[L^2T^{-2}]$. Since g is about 9.80 m s^{-2}, ψ is numerically about ten times the geometric height in meters. This led V. Bjerknes to define a *geodynamic height* (or *dynamic height*) by the expression: geodyn. hgt $= \psi/10$. When ψ is measured in units of m^2 s^{-2} this expression gives *geodynamic meters*, which are numerically about 2 per cent smaller than the geometric height. Another unit which is internationally agreed upon is the *geopotential meter* which is simply 0.98 of a geodynamic meter. Thus, while the work needed to lift a unit mass through one geometric meter is about 0.98 geodynamic meters, it is also closely 1.00 geopotential meters. For a detailed discussion see R. J. LIST, ed., *Smithsonian Meteorological Tables*, Smithson. Inst., Washington, 1951, pp. 217-218.

of the earth. The error so introduced will be negligible, since the rotation about the x axis is small, and we can then combine gravitation and centrifugal effects together so that their sum will be along the z axis only. The equations of motion become:

$$\begin{aligned}
\ddot{x} &= 2\Omega(v \sin\phi - w \cos\phi) + \frac{1}{m} F_x \\
\ddot{y} &= -\ 2\Omega u \sin\phi + \frac{1}{m} F_y \\
\ddot{z} &= 2\Omega u \cos\phi - g + \frac{1}{m} F_z
\end{aligned} \tag{11.9}$$

Here F_z represents all external forces in the new z direction except gravity, and no centrifugal effects appear explicitly.

11.4 The Pressure-Gradient Force

Another important force in fluid systems is that due to the pressure. Consider a small fluid cube whose sides are dx, dy, dz as in Fig. 11.5. Let us calculate the pressure force on this fluid element in the x direction.

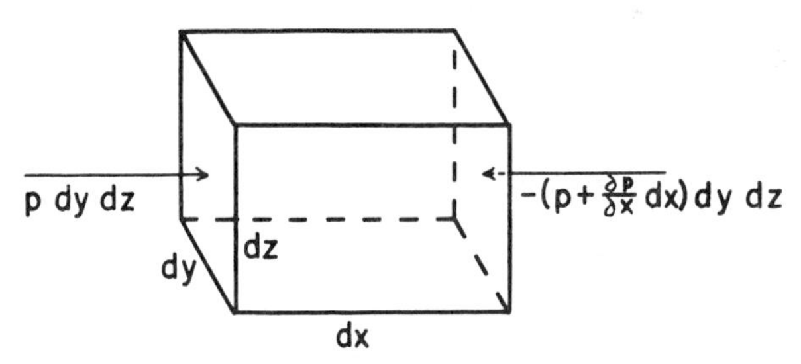

FIG. 11.5. Calculation of the pressure-gradient force

The force due to the surrounding fluid on the left-hand vertical face is $+p\ dy\ dz$, where the positive sign indicates that the force is directed to the right. The force on the right-hand vertical face is

$$-\left(p + \frac{\partial p}{\partial x}\,dx\right) dy\ dz$$

where the additional term $(\partial p/\partial x)\ dx$ takes into account the variation of pressure through the small distance dx. The minus sign expresses the fact that this force is directed to the left. The net pressure force in the x direction is the sum of these two:

$$\text{pressure force} = -\ (\partial p/\partial x)\ dx\ dy\ dz$$

Since the equations of motion (11.9) contain forces per unit mass we divide by dM, the mass of this infinitesimal cube, and obtain:

$$\text{pressure force per unit mass} = -\frac{1}{\rho}\frac{\partial p}{\partial x}$$

Here ρ is the mean density of the cube and is equal to $dM/(dx\,dy\,dz)$.

This result means that the force is due to a gradient of pressure and is directed from high towards low pressure. As a result it is referred to as the *pressure-gradient force*. Similar calculations may be performed for the y and z directions with analogous results.

We may now rewrite the equations of motion to include explicitly the components of the pressure-gradient force:

$$\begin{aligned} \ddot{x} &= 2\Omega(v \sin\phi - w\cos\phi) - \frac{1}{\rho}\frac{\partial p}{\partial x} + \frac{1}{m}F_x \\ \ddot{y} &= -2\Omega u \sin\phi - \frac{1}{\rho}\frac{\partial p}{\partial y} + \frac{1}{m}F_y \\ \ddot{z} &= 2\Omega u \cos\phi - g - \frac{1}{\rho}\frac{\partial p}{\partial z} + \frac{1}{m}F_z \end{aligned} \qquad (11.10)$$

where F_x, F_y, F_z now represent additional forces excluding pressure-gradient forces.

11.5 Inertia Motion

A first approach to an appreciation of the role which the Coriolis force plays in meteorology may be gained by considering the case when all other forces are absent and when the motion is purely horizontal. The latter assumption is justified by the fact that most large-scale atmospheric and oceanic motions are quasi-horizontal. That is, the horizontal velocity components usually exceed the vertical component by two or more factors of ten.[5] The first assumption means, among other things, that friction is assumed to vanish and a particle under consideration remains in motion because of inertia. The resultant phenomenon is therefore known as *inertia motion*.

Under these conditions the equations of motion become

$$du/dt = f\,dy/dt$$
$$dv/dt = -f\,dx/dt$$

where $f = 2\,\Omega \sin\phi$, the *Coriolis parameter*. Upon integration, assuming

[5] J. E. Miller, Studies of Large-Scale Vertical Motions of the Atmosphere, *New York Univ., Meteor. Papers, 1*, 1948, No. 2, 49 pp.

f to be constant, we get equations relating the velocity components to the position of the particle:

$$\begin{aligned} u &= fy + A \\ v &= -fx + B \end{aligned} \tag{11.11}$$

where A and B are constants of integration. However, we know that the Coriolis force is directed at right angles to the velocity and can change only the direction of motion but not the speed. Therefore $u^2 + v^2 = c^2$, where c is the constant speed. Substitution of Eq. (11.11) into this expression and rearrangement yields:

$$\left[x - \left(\frac{B}{f}\right)\right]^2 + \left[y + \left(\frac{A}{f}\right)\right]^2 = \left(\frac{c}{f}\right)^2$$

This is the equation of a circle of radius $r = c/f$, with its center at $x_o = B/f$, $y_0 = -A/f$. It is referred to as the *inertia circle*. Since the only force acting is the horizontal Coriolis force, it must point always towards the center of the circle, which means that the particle goes around in a counterclockwise sense in the Southern Hemisphere, clockwise in the Northern Hemisphere.

The period, T, of motion in the inertia circle is given by the circumference divided by the speed:

$$T = \frac{2\,\pi c/f}{c} = \frac{\pi}{\Omega \sin\phi}$$

Since a larger speed increases the distance to be covered in one revolution as well as the rate at which distance is covered, the period does not depend upon the particle's speed at all, but only upon the mean latitude of the event.

The period of the inertia oscillation is often spoken of as being one-half of a *pendulum day*, which is the time required for a Foucault pendulum to complete one revolution and is equal to a sidereal day divided by $\sin\phi$. Since a sidereal day is the time required for the earth to make one revolution on its axis relative to the fixed stars, it is equal to $2\pi/\Omega$. Thus a pendulum day is $2\pi/(\Omega \sin\phi)$, or twice the period of the inertia circle.

We may now consider, qualitatively, the effect of the increase in the Coriolis parameter with latitude. Since the radius of curvature of the path varies inversely as f, a true inertia path in middle and high latitudes will have a smaller radius on its poleward side than its equatorward side. Thus the path will not really be closed but the particle will move westward while describing almost circular paths. This effect becomes of great importance in low latitudes where f varies rapidly with ϕ, and where inertia trajectories differ greatly from circles.

These properties of inertia paths bring out the intrinsic deflecting nature of the Coriolis force. They also show that motions in which the earth's rotation plays a role have a natural period of vibration which is the inertia period, one-half a pendulum day. We shall see cases later in which this natural period emerges as a by-product in theoretical developments.

A widely quoted example of an inertia oscillation was observed in the Baltic Sea by Gustafson and Kullenberg. Under conditions of well-defined vertical stratification of the water, which permitted a nearly frictionless gliding of one layer over the next, they found the current at a depth of 14 meters turned continually to the right with a period of 14 hours. The inertia period at this latitude is $14^{h}\ 08^{m}$.

One must juxtapose with this example the fact that observations of this kind of motion are quite rare. Despite the frequent appearance of inertial components in theoretical results from particle dynamics the natural period is quite difficult to find in nature. The reason for this appears to be twofold. First, since the period is quite long it seems as though frictional and turbulent effects may dissipate much of the energy of the inertia oscillation before it has gone through enough cycles to be recognizable among all the other phenomena which may be occurring. Second, it has been shown theoretically that when a localized fluid acceleration excites an inertial period the energy of this motion is dispersed rapidly into the surrounding fluid.[6] Thus the amplitude of the vibration is reduced to presumably unrecognizable proportions.

11.6 Individual versus Local and Convective Derivatives

The equations of motion involve terms like $\dot{x} = dx/dt = u$. The physical meaning of such a term is: an increment in the x position of a fluid particle divided by the increment in time needed for it to move that distance, evaluated in the limit as the time increment approaches zero. Such a derivative is called an *individual derivative*, because it is taken following an individual particle in its motion; it is also called a *total derivative*.

We must deal with a number of other kinds of derivatives. To illustrate these, we consider some fluid property like the specific volume,

[6] A. CAHN, An Investigation of the Free Oscillations of a Simple Current System, *J. Meteor.*, *2*, 1945, pp. 113-119.

α, which is a function of x, y, z, and t. From the calculus we know that

$$d\alpha = dt\,\frac{\partial \alpha}{\partial t} + dx\,\frac{\partial \alpha}{\partial x} + dy\,\frac{\partial \alpha}{\partial y} + dz\,\frac{\partial \alpha}{\partial z}$$

where a partial differentiation has the usual meaning that all other independent variables are held constant while the partial derivative is being evaluated. It follows, upon dividing by dt, that

$$\frac{d\alpha}{dt} = \frac{\partial \alpha}{\partial t} + u\,\frac{\partial \alpha}{\partial x} + v\,\frac{\partial \alpha}{\partial y} + w\,\frac{\partial \alpha}{\partial z} \tag{11.12}$$

This equation means that the individual rate of change of α is due to two main contributions. The first, represented by $\partial\alpha/\partial t$, is the rate of change in α the particle being followed would experience if it could remain in a fixed location. This is the *local derivative* with respect to time. A local derivative of α could exist, for example, because of expansion or compression of a gas, or because of heating or cooling. The second kind of contribution is represented by the last three terms on the right of Eq. (11.12). These together give the rate of change in α which a particle would experience if α did not vary with time at any fixed point, but if the particle moved through a fluid in which specific volume varied with x, y, and z. This rate of change is called the *convective derivative* with respect to time. Physically as well as mathematically the total derivative is made up of the sum of the local and convective derivatives.

It is possible to set up a special kind of total derivative which represents the rate of change of a fluid property at a point moving with some arbitrary velocity different from the velocity of the fluid. If we denote such a total derivative operator by $\mathfrak{D}/\mathfrak{D}t$ and the velocity components of the moving point by C_x, C_y, C_z, then

$$\frac{\mathfrak{D}\alpha}{\mathfrak{D}t} = \frac{\partial \alpha}{\partial t} + C_x\,\frac{\partial \alpha}{\partial x} + C_y\,\frac{\partial \alpha}{\partial y} + C_z\,\frac{\partial \alpha}{\partial z}$$

Such derivatives are useful when one is dealing with waves which move at a speed different from that of the fluid. Then one may wish to form total derivatives following the waves rather than the particles. It is important to realize which kind of derivative is intended whenever a total differentiation is performed.

The interpretations given above have been applied to specific volume as an example, but the analysis applies to any variable that is a continuous function of time and position. An important special case occurs when the local derivative of a property is everywhere zero. In this case a *steady state* is said to exist with respect to that property.

PROBLEMS

1. Consider a plane tangent to a point on the earth's surface. Show that h, the height of this plane above the earth's surface at a reasonably small distance from the point of tangency, is given approximately by $h = d^2/2a$, where d is distance from the point of tangency and a is the radius of the earth. What is the value of this height when $d = 100$ miles?

2. Show that at the surface of the earth the angle between a line to the earth's center and a plumb bob should be approximately $\theta = 350'' \sin 2\phi$, if one neglects the difference between the Newtonian gravitation of a sphere and a spheroid.

3. What is the magnitude of the total Coriolis force acting on a bullet of mass m, fired with speed c from latitude ϕ, whose path lies in a plane containing the earth's axis,

(a) at angles ϕ from the zenith?

(b) at angles $90° - \phi$ from the zenith?

4. Calculate the magnitude of the additional curvature terms in the equations of motion compared to $2\,\Omega v \sin\phi$ for the terms in the x equation, $2\,\Omega u \sin\phi$ for terms in the y equation, and $2\,\Omega u \cos\phi$ for terms in the z equation. Use the following reasonable meteorological values: $u = 20$ m/s, $v = 10$ m/s, $w = 5$ cm/s, $\phi = 60°$ N. Omit terms which are negligible within an accuracy of 5 per cent and compare the result with the tangent-plane equations of motion, Eq. (11.6). In what regions of the earth would this result be misleading?

5. A particle is thrown vertically upward in a vacuum with initial speed w_0 at latitude ϕ. Show that when it returns to its original elevation it will be at a distance $\Omega \cos\phi\, 4\, w_0^3/3g^2$ west of its starting point. (*Hint*: use the tangent-plane equations and neglect $2\,\Omega u \cos\phi$ compared to g). Compute this displacement for $w_0 = 10^3$ cm/s at latitude 45°.

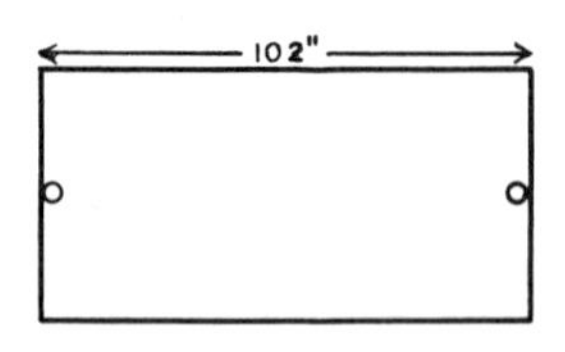

FIG. 11.6. Plan view of a billiard table

6. A billiard table is 102 inches long. Two billiard balls, 2 in. in diameter, are placed at opposite ends of the table as shown in Fig. 11.6. Neglecting friction, how fast does one ball have to be propelled directly at the other at the initial moment in order that it shall just barely miss striking the second ball due to the Coriolis force? The table is located at 43° N ($f = 10^{-4}$ s^{-1}).

CHAPTER 12

HORIZONTAL MOTION UNDER BALANCED FORCES

12.1 Equilibrium Motion

It has been observed that when a physical system is subject to a number of stresses it frequently tends towards a state of balance or equilibrium among the various stresses. Further, if the system reaches equilibrium and one of the stresses is then altered, the state of the system frequently will change so as to relieve the added stress and restore the balance. This concept is referred to in physics and chemistry as Le Chatelier's principle. If this idea is of sufficiently broad application in geophysical phenomena it will be worth while to study various possible equilibria since they represent states towards which the geophysical system in question will tend. Moreover it is to be expected that the solution to such equilibrium problems will be far simpler than solutions of the transient unbalanced cases.

It should be remembered, however, that not all physical systems exhibit this stable reversion to equilibrium. We have no guarantee that this approach will lead to usable results in any given case. In this, as in all other theoretical work, the validity of the approach will have to be judged ultimately by the extent of the agreement between theoretical results and appropriate observations.

12.2 Geostrophic Flow

An important geophysical example of this urge towards equilibrium may be found in the case of a parcel of air on the rotating earth subject to both Coriolis and pressure gradient forces. Let the speed of the parcel relative to the earth be zero at the initial moment, and let the subsequent motion be horizontal. At the initial moment the Coriolis force will be zero, since the speed is zero, but there will be an acceleration towards the low pressure. As the parcel picks up speed across the isobars the Coriolis force will begin to act to the right of the direction of motion (in the Northern Hemisphere) and will begin to turn the parcel away from perpendicularity with the isobars (Fig. 12.1). As this action goes on

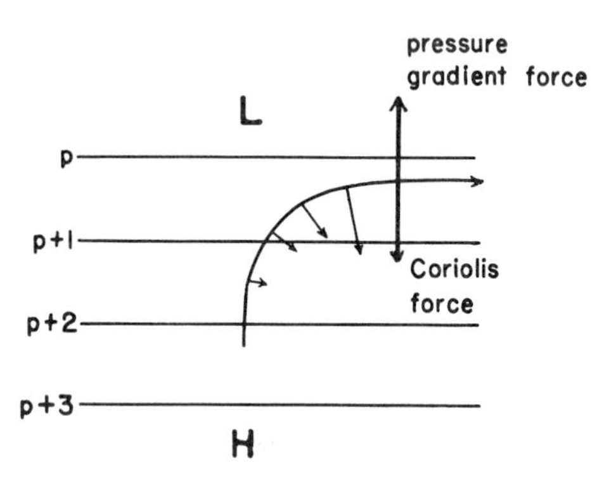

FIG. 12.1. Approach to geostrophic equilibrium by a parcel initially at rest

the Coriolis force increases, since the speed increases, and it is to be expected that ultimately the motion will be parallel to the isobars with the Coriolis and pressure gradient forces equal in magnitude but oppositely directed. It is this final equilibrium[1] which we shall express quantitatively below.

The sequence just described exemplifies the so-called "Geodynamic Paradox." This paradox states that on the rotating earth a particle subject to a constant force does not move parallel to the force with constant acceleration as expected, but ultimately will move perpendicular to the force with constant speed. The physical reasons for this seeming contradiction should now be clear.

The final state then, will be one in which the net acceleration is zero, the motion is horizontal, and the only forces present are those owing to the pressure gradient, gravity, and the earth's rotation. The equations of motion (11.9) become:

---

[1] A detailed analysis of the approach to this equilibrium shows that the parcel will overshoot the balance point and oscillate about equilibrium, with the inertia period. If the energy of the oscillation is dissipated the balance discussed above will be established.

$$fv = \frac{1}{\rho}\frac{\partial p}{\partial x}$$
$$fu = -\frac{1}{\rho}\frac{\partial p}{\partial y} \tag{12.1}$$

$$g = -\frac{1}{\rho}\frac{\partial p}{\partial z} + 2\Omega u \cos\phi \tag{12.2}$$

If one neglects the small Coriolis term, Eq. (12.2) is the familiar hydrostatic equation, which expresses the balance between gravity and the vertical pressure-gradient force per unit mass. Equations (12.1) are the *geostrophic wind equations*, which express the balance between horizontal Coriolis and pressure-gradient forces. The chief practical utility of these expressions lies in the fact that they enable us to delineate the horizontal wind field by analyzing the pressure field instead. It has usually proved more feasible to determine the distribution of pressure rather than the direct distribution of wind. The direction of the wind is then given by the orientation of the isobars, and its speed by the lateral spacing of the isobars.

It was recognized early that the special situation described by Eq. (12.1) was one which frequently occurred to an approximate extent in the atmosphere. Thus, these equations require that the geostrophic wind blow parallel to the isobars with low pressure on the left in the Northern Hemisphere and on the right in the Southern Hemisphere (where f becomes negative). A similar behavior of the atmosphere was noted over one hundred years ago and is known today as Buys-Ballot's law. Furthermore, it is widely observed that in middle and high latitudes the actual wind is stronger the more closely packed the isobars on an equipotential surface. Thus there is an approximate agreement between geostrophic theory and fact, the extent of which will be elaborated further below.

It should be clear from the assumptions made that at low latitudes, where f is small, it is more difficult to establish this kind of balance. Theory and experience both indicate that the geostrophic approximation should not be applied in equatorial regions.

The property of geostrophic flow which requires the pressure to decrease to the left of the motion may be taken advantage of to compress the two equations into one. Since the direction of motion and the normal to it are the basic orientations we shall define a ***natural coordinate system*** which has an s axis tangent to the flow, along which s increases in the direction of flow, and an n axis perpendicular to the flow, along which n increases to the left of the direction of motion. The geostrophic wind equations become:

$$fc = -\frac{1}{\rho}\frac{\partial p}{\partial n}$$

where c is the total horizontal wind speed, which is always positive. This simplification is possible because the geostrophic wind relationships do not change when the coordinate axes are rotated about the vertical.

Since the wind components in geostrophic flow must be constant for an individual particle ($du/dt = dv/dt = 0$), and since the pressure gradient must always be in balance with the horizontal Coriolis force, it follows that the value of the pressure gradient must be constant along the isobars (if variations of f and ρ are neglected) and constant with time. Thus, if the wind is exactly geostrophic the isobars must be straight parallel lines that are fixed in position for all time. If this were so it would be unnecessary to forecast atmospheric flow patterns. Despite the fact that in reality isobars are frequently curved, nonparallel, and move with time, it may be that the actual wind is sufficiently well approximated by the geostrophic value. It is therefore not inconsistent to speak of the geostrophic wind in a nonuniform pressure-gradient field, so long as we realize that we are speaking approximately.

Consider now the case of flow of air from a region of large pressure gradient to one of smaller pressure gradient as in Fig. 12.2. A given parcel carrying its initially high momentum into a region of more widely separated isobars will always find the Coriolis force overbalancing the pressure-gradient force. Such *supergradient winds* will therefore turn and blow across the isobars towards high pressure. The converse case of *subgradient winds* will have pressure-gradient force exceeding Coriolis force and winds blowing across the isobars towards low pressure. These are cases of disturbance of equilibrium, and it is worth while to note how the reaction of the system (cross-isobar flow) is such as to alleviate the stress which upset the balance. For in the supergradient case the wind removes air from the region of low pressure and transports it to the area of high pressure, a process that tends to increase the pressure gradient and restore the balance. A similar argument can be offered in the subgradient case.

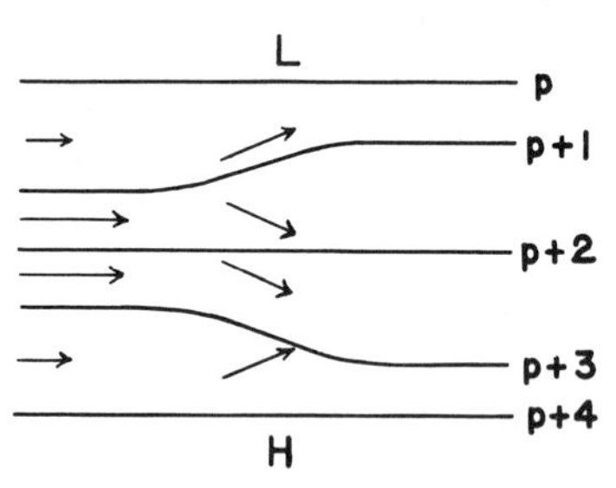

FIG. 12.2. Cross-isobar flow resulting from a downstream change in pressure gradient

12.3 The Effect of Friction

Let us now complicate the geostrophic case by introducing friction against the earth's surface. If we assume that friction acts exactly opposite to the direction of motion, then a balance of the three forces becomes possible as shown in Fig. 12.3. The pressure-gradient force will not be balanced by the Coriolis term alone, but by the vectorial sum of the frictional force and the Coriolis force. This requires the wind to blow at an angle α to the isobars towards low pressure.

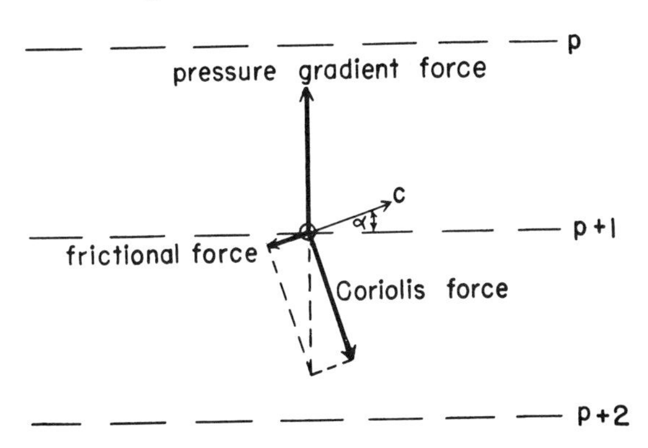

FIG. 12.3. Balance among pressure-gradient, Coriolis, and frictional forces

These conditions may be expressed quantitatively under the additional assumption that friction is not only opposite to the motion but proportional to the speed. Then the external force terms in the equations of motion (11.10) become $F_x/m = -ku$, $F_y/m = -kv$, where k is a positive constant.[2] If the isobars are oriented east-west and, in parallel to the geostrophic case, we consider horizontal motion with zero accelerations, the equations of motion become:

$$0 = fv - ku$$
$$0 = -fu - kv - \frac{1}{\rho}\frac{\partial p}{\partial y}$$

By elimination we obtain

$$u = -\frac{f}{f^2 + k^2}\left(\frac{1}{\rho}\frac{\partial p}{\partial y}\right)$$
$$v = -\frac{k}{f^2 + k^2}\left(\frac{1}{\rho}\frac{\partial p}{\partial y}\right)$$

When $k = 0$ (no friction) these reduce to $v = 0$ and the geostrophic equation for east-west flow. When k differs from zero the east-west flow

[2] C. M. GULDBERG and H. MOHN, *Études sur les Mouvements de l'Atmosphère, I*, 1876, A. W. Brögger, Oslo.

is reduced below the geostrophic value, and a north-south flow towards low pressure exists. The total wind speed is

$$c = \sqrt{u^2 + v^2} = \frac{1}{\sqrt{f^2 + k^2}} \left(\frac{1}{\rho} \frac{\partial p}{\partial y} \right)$$

which, when friction is present, is below the geostrophic value. The angle α between the isobars and the wind is given by $\tan \alpha = v/u = k/f$. This angle is zero if k vanishes.

The assumption of a frictional force proportional to the wind speed is reasonable only close to the earth's surface, and even there it is only an approximation. However, this artifice suffices to elaborate the physics of the problem and gives results which are in qualitative agreement with observations on any extratropical sea-level weather map.

12.4 Gradient Flow

Now that we have gained some insight into the effects of friction, we shall return to the ideal frictionless situation in order to consider the effects of curvature of the flow. It is to be expected that the results to follow will be modified by interaction with the ground in the same way, qualitatively, that the geostrophic case was modified. That is, we expect the wind to be reduced below its frictionless value, and the flow to have a component across the isobars towards low pressure.

We shall consider horizontal motion, in the steady state, when Coriolis and pressure-gradient forces are present. The isobars on an equipotential surface will be assumed to be circular and the motion may be curved. Strictly speaking, centripetal accelerations are present when the motion is curved so that a net force must act and equilibrium does not exist. The convenience of dealing with balanced forces may be retained, however, by equating the net excess of force, which is centrally oriented, to the centrifugal force, which is merely the reaction to the centripetal action. The horizontal equations of motion are then

$$\begin{aligned} u \frac{\partial u}{\partial x} + v \frac{\partial u}{\partial y} &= fv - \frac{1}{\rho} \frac{\partial p}{\partial x} \\ u \frac{\partial v}{\partial x} + v \frac{\partial v}{\partial y} &= -fu - \frac{1}{\rho} \frac{\partial p}{\partial y} \end{aligned} \tag{12.4}$$

We shall introduce polar coordinates:

$$x = r \cos \theta \qquad y = r \sin \theta$$

or

$$r = (x^2 + y^2)^{1/2} \qquad \theta = \tan^{-1} \frac{y}{x}$$

If one differentiates with respect to time, remembering that both r and θ may vary for a moving particle, the results are

$$u = v_r \cos\theta - v_\theta \sin\theta$$
$$v = v_r \sin\theta + v_\theta \cos\theta$$

where the speed along the radius is v_r (radial velocity) and the speed perpendicular to the radius is v_θ (tangential velocity).

These equations can be used to replace u and v by v_r and v_θ. We must now replace all derivatives with respect to x and y in Eq. (12.4) by derivatives with respect to r and θ. Let F be some function of x and y or r and θ. Since x is a function of r and θ, we know from the calculus that

$$\frac{\partial F}{\partial x} = \frac{\partial r}{\partial x}\frac{\partial F}{\partial r} + \frac{\partial \theta}{\partial x}\frac{\partial F}{\partial \theta} = \cos\theta \frac{\partial F}{\partial r} - \frac{\sin\theta}{r}\frac{\partial F}{\partial \theta}$$

Similarly

$$\frac{\partial F}{\partial y} = \frac{\partial r}{\partial y}\frac{\partial F}{\partial r} + \frac{\partial \theta}{\partial y}\frac{\partial F}{\partial \theta} = \sin\theta \frac{\partial F}{\partial r} + \frac{\cos\theta}{r}\frac{\partial F}{\partial \theta}$$

When we combine all the results we find that the horizontal convective derivative is

$$u\frac{\partial F}{\partial x} + v\frac{\partial F}{\partial y} = v_r \frac{\partial F}{\partial r} + \frac{v_\theta}{r}\frac{\partial F}{\partial \theta}$$

When we apply these results to u and v instead of to F, Eq. (12.4) become

$$\begin{aligned} v_r \frac{\partial v_r}{\partial r} + v_\theta \frac{\partial v_r}{r\partial \theta} - \frac{v_\theta^2}{r} &= f v_\theta - \frac{1}{\rho}\frac{\partial p}{\partial r} \\ v_r \frac{\partial v_\theta}{\partial r} + v_\theta \frac{\partial v_\theta}{r\partial \theta} + \frac{v_\theta v_r}{r} &= -f v_r - \frac{1}{\rho}\frac{\partial p}{r\partial \theta} \end{aligned} \qquad (12.5)$$

These are the equations of steady horizontal motion in polar coordinates.

We shall now particularize to the case of circular concentric isobars with their centers at $r = 0$. Then $\partial p/\partial\theta = 0$. If we also require that the velocity distribution have circular symmetry, then $\partial v_r/\partial\theta = \partial v_\theta/\partial\theta = 0$. Under these symmetric conditions it is clear that one way of satisfying the second of Eq. (12.5) is by taking $v_r = 0$.[3] The first of Eq. (12.5) may then be written, if v_θ is replaced by c, as

$$\frac{c^2}{r} + fc - \frac{1}{\rho}\frac{\partial p}{\partial r} = 0 \qquad (12.6)$$

[3] If v_r is not zero then the second of Eq. (12.5) becomes

$$f + \frac{\partial v_\theta}{\partial r} + \frac{v_\theta}{r} = 0$$

It will be shown later that the above relationship means the absolute vorticity about the vertical must be zero. This proves to be a very special condition, and it is reasonable to reject it in favor of taking $v_r = 0$.

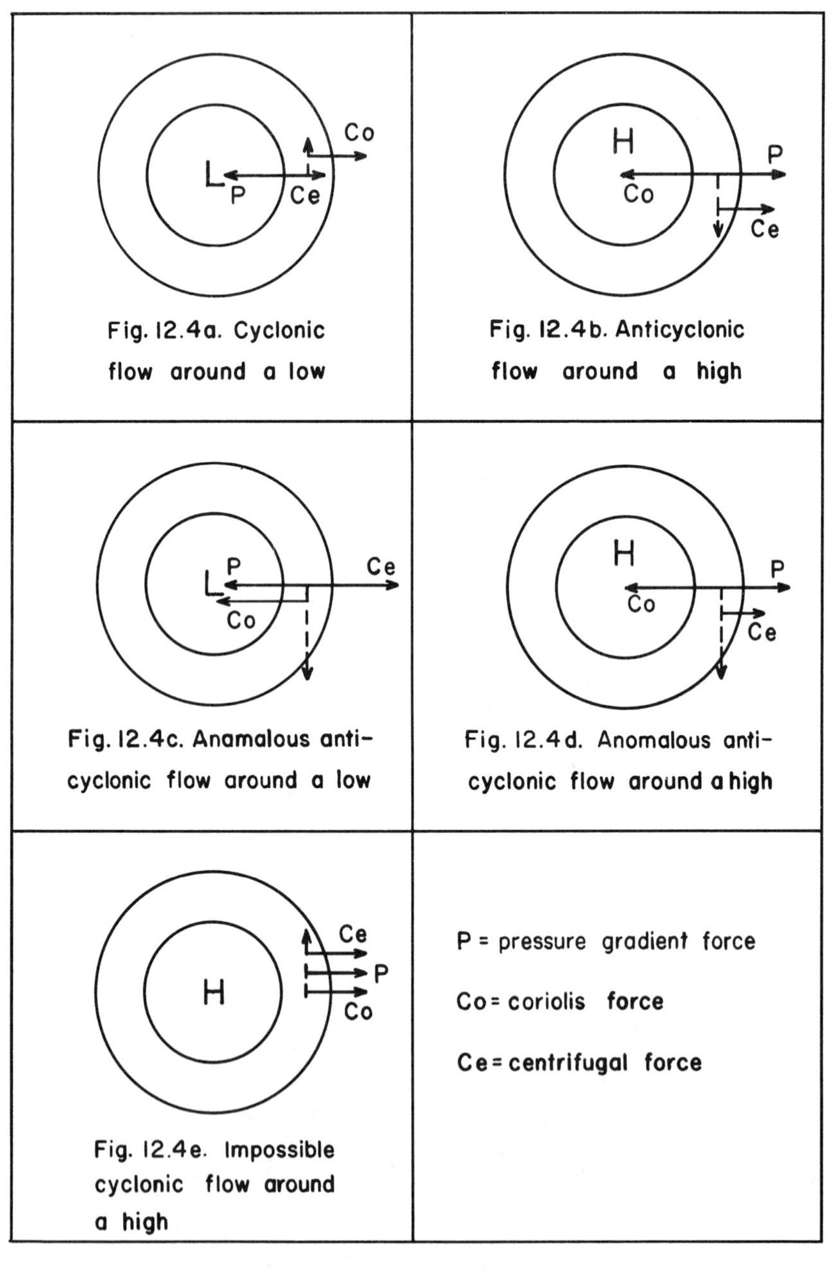

FIG. 12.4. Various cases of gradient balance

This is the equation that governs circular motion, parallel to the isobars with a balance among centrifugal, Coriolis, and pressure-gradient forces per unit mass. A wind blowing under these conditions is called a *gradient wind*. Equation (12.6) could have been set down immediately from the condition that these three forces must balance.

Since the pressure gradient may point inward or outward, and since in each of these cases this quadratic equation for c has two roots, we recognize four cases that are mathematically possible, as represented in Fig. 12.4. Still a fifth case is that of counterclockwise motion about a high, but here it is impossible to have a balance of forces, and Eq. (12.6) does not apply (Fig. 12.4*e*).

The solutions to Eq. (12.6) are

$$c = -\frac{fr}{2} \pm \sqrt{\frac{f^2r^2}{4} + \frac{r}{\rho}\frac{\partial p}{\partial r}} \tag{12.7}$$

In agreement with the usual polar coordinate convention, a positive value of c means counterclockwise motion (cyclonic in the Northern Hemisphere), and a negative value of c means clockwise motion (anti-cyclonic in the Northern Hemisphere). First, consider the positive root of Eq. (12.7). As $\partial p/\partial r$ approaches zero, c will also approach zero—exactly what we would expect from the results obtained in geostrophic flow. Indeed, this root describes the normally observed cases of cyclonic flow around a low-pressure center (Fig. 12.4*a*), and anticyclonic flow around a high-pressure center (Fig. 12.4*b*), because when $\partial p/\partial r > 0$ (a low), the square root exceeds $fr/2$ in magnitude and c is positive (cyclone). When $\partial p/\partial r < 0$ (a high) the square root is smaller than $fr/2$ in magnitude and c is negative (anticyclone).

Consider next the negative root of Eq. (12.7). As $\partial p/\partial r$ approaches zero, c approaches $-fr$. This represents anticyclonic motion with no pressure gradient in the limit, and will be recognized as motion in the inertia circle. When the pressure gradient differs from zero, c will remain negative since both terms of Eq. (12.7) will be negative. Thus when $\partial p/\partial r > 0$ (a low) the motion will still be anticyclonic (Fig. 12.4*c*). Although mathematically possible, this anticyclonic rotation about a low has not been observed on a large scale. When $\partial p/\partial r < 0$ (a high) the negative root gives an additional case of anticyclonic motion about a high (Fig. 12.4*d*). In this case the wind speed is high because the two terms of Eq. (12.7) act together, while in the normal motion around a high (Fig. 12.4*b*) the two terms are opposite in sign and the speed is low. The motions described by the negative root may be termed anomalous, because they are not normally observed. Anomalous anticyclonic flow has been reported[4] with the suggestion that it may be more frequent than has hitherto been supposed.

The reason for the relative rarity of actual motions corresponding to the negative root of Eq. (12.7) lies in the high wind speeds—and therefore high kinetic energy—required for this kind of flow.

Let us consider only the more common case:

$$c = -\frac{fr}{2} + \sqrt{\frac{f^2r^2}{4} + \frac{r}{\rho}\frac{\partial p}{\partial r}}$$

It is clear that when $\partial p/\partial r$ is positive (a low) the square root can never become imaginary so that all values of pressure gradient are permissible. There is no theoretical restriction on the magnitude of the pressure gradient for a low. However, when $\partial p/\partial r$ is negative (a high) it is possible for the square root to become imaginary. The resultant complex value for c has no physical meaning and must be excluded from consideration. The condition that c shall be real is

$$\left|\frac{\partial p}{\partial r}\right| \leq \frac{\rho f^2 r}{4} \tag{12.8}$$

That is, in a high the magnitude of the pressure gradient may not exceed a certain value determined largely by the latitude and distance from center. This is the explanation for the observed fact that near the center of a high the pressure gradient is small and the winds are light, for as r becomes small the pressure gradient must also become quite small in order to satisfy this criterion. In a low there is no such restriction and the pressure gradient and winds can be, and often are, very large near the center.

This restriction for a high may be reversed and the conclusion drawn that for a given pressure gradient the radius of curvature of the flow must not fall below a minimum value given by inequality (Eq. 12.8). J. Bjerknes has taken advantage of this condition to explain the developments that follow the establishment of a ridge of sharply curved isobars aloft. If, as in Fig. 12.5, the isobars in the anticyclone have a radius of curvature smaller than permitted by Eq. (12.8), an air parcel cannot move parallel to the isobars but must depart from gradient flow and cross isobars from high towards low pressure. The work done by the pressure-gradient force in this process accelerates the air and increases the Coriolis force acting to the right of the parcel. This action ultimately increases the anticyclonic curvature and may cause the air to cross

[4] A. F. Gustafson, On Anomalous Winds in the Free Atmosphere. *Bull. Am. Meteor. Soc.*, 34, 1953, p. 196.

isobars towards high pressure farther downstream, which produces a deceleration, predominance of the pressure-gradient force, and a turn to the left in a cyclonic path. The net result is the initiation of a pronounced cyclonic flow to the south of the original sharp anticyclonic area. Such upper-air cyclones form frequently in winter and are very important synoptic features. The chain of events leading to their

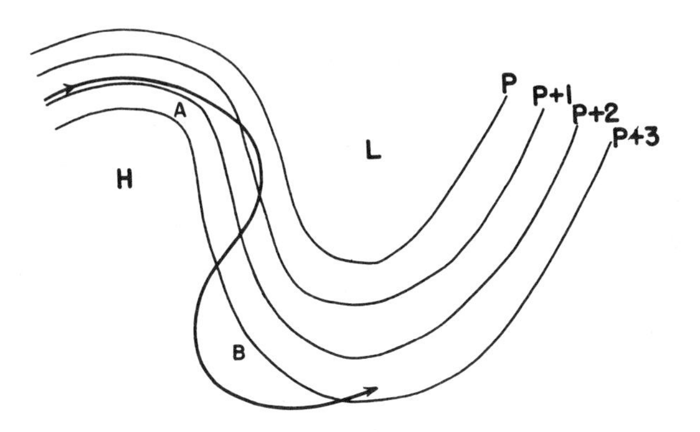

FIG. 12.5. Trajectory of air (heavy line) through a sharply curved ridge of high pressure when gradient balance cannot be maintained

appearance begins with the departure from gradient balance required by Eq. (12.8) when the isobars in a high-pressure ridge have so small a radius that it is dynamically impossible for the air to conform to the sharp curvature.

12.5 Comparison of Geostrophic and Gradient Wind Values

We may look upon the value of the geostrophic wind as a measure of the pressure gradient, for, in polar coordinates,

$$c_g = \frac{1}{\rho f}\frac{\partial p}{\partial r}$$

where c_g represent the geostrophic wind. We may therefore write Eq. (12.7) as

$$c = -\frac{fr}{2} + \sqrt{\frac{f^2r^2}{4} + frc_g}$$

where only the normal positive root is considered. This expression enables us to compare the geostrophic estimate of the wind speed, which neglects curvature of the flow, with the gradient estimate of the wind speed, which includes curvature of the flow. Upon squaring both sides and solving for c_g we get

$$c_g = c + \frac{c^2}{fr}$$

In normal flow around a low, c is cyclonic and therefore positive. Both r and f are also positive (in the Northern Hemisphere). Thus, in a low, $c_g > c$, and the geostrophic value is an overestimate of the gradient wind. In normal flow around a high both c and c_g are negative. If we introduce absolute magnitudes we get

$$-|c_g| = -|c| + \frac{|c|^2}{fr}$$

or

$$|c_g| = |c| - \frac{|c|^2}{fr}$$

Hence, in a high, $|c_g| < |c|$, or the geostrophic value is an underestimate of the gradient wind speed.

These results may be understood physically from the balance of forces in Fig. 12.4*a* and 12.4*b*. In a cyclone the Coriolis and centrifugal terms act together so only a low speed is needed to balance a given pressure-gradient force. In an anticyclone, the Coriolis force alone must balance the centrifugal term as well as the pressure-gradient force so a higher speed is needed for the same pressure gradient. The geostrophic wind must have a Coriolis effect which balances only the pressure-gradient term and so its speed lies between the above two values.

In actual practice on weather maps, the difference between the geostrophic and gradient winds is difficult to detect except when the wind speed is high. Then the term c^2/r, which is the source of the difference, becomes quite appreciable. As a result of an experiment[5] designed to compare the geostrophic and gradient winds to each other and to the observed winds at 700 mb, it was concluded that most of the time the geostrophic wind is a better approximation to the observed value than the gradient wind. Some of the important factors in this result are: the errors in pilot balloon measurement of the true wind, errors in radiosonde measurements, insufficiency of the observing network to uniquely define the field of pressure gradient, and departures of the atmosphere from the state of equilibrium assumed in both the geostrophic and the gradient theory.

---

[5] M. Neiburger, L. Sherman, W. W. Kellogg, A. F. Gustafson, On the Computation of Wind from Pressure Data. *J. Meteor.*, 5, 1948, p. 87. The situations studied in this paper had weak winds so that the centrifugal term was small.

12.6 Cyclostrophic Flow

In small-scale systems the radius of rotation may be of the order of hundreds of feet as compared with the usual systems portrayed on weather maps in which the radius of rotation may be measured in hundreds of miles. For this reason the centrifugal term becomes much larger in such small systems. Because the centrifugal term varies as the square of the speed and the Coriolis term as the first power of the speed, the Coriolis force is negligible even for high speeds. In such situations a balance of forces is to be sought between pressure-gradient and centrifugal effects alone:

$$\frac{c^2}{r} = \frac{1}{\rho}\frac{\partial p}{\partial r}$$

The flow will always be around a low-pressure center and since the Coriolis effect is negligible, the sense of rotation may be either cyclonic or anticyclonic. This type of motion is called *cyclostrophic* and is to be expected in low latitudes where f is small. Hurricanes obey this law near the center. Other examples are dust devils, waterspouts, and tornadoes.

12.7 Representation of the Pressure Gradient on Other Than Horizontal Surfaces[6]

For various reasons we often wish to represent atmospheric flow patterns on special surfaces which are not level. In order to apply the geostrophic or gradient approximations on such surfaces we shall have to derive an expression for the horizontal pressure gradient determined point-by-point along an arbitrarily defined sloping surface.

We shall denote the gradient of pressure along a level surface by terms like $(\partial p/\partial x)_z$, where the subscript z means "holding z constant." Similarly the gradient of pressure along a surface characterized by a constant value of some property q will be denoted by terms like $(\partial p/\partial x)_q$. We wish to relate these two quantities to each other. One may expect, in advance, that this relation will involve the slope of the q surface in some way.

[6] D. T. Perkins and A. F. Gustafson, Differential Relations between Quantities Projected from Meteorological Surfaces. *J. Meteor.*, 8, 1951, pp. 418-420.

Consider a cross section in the x, z plane as in Fig. 12.6. We can see that the following equality is correct:

$$\frac{p_3 - p_1}{\Delta x} = \frac{p_2 - p_1}{\Delta x} - \frac{p_2 - p_3}{\Delta z}\frac{\Delta z}{\Delta x}$$

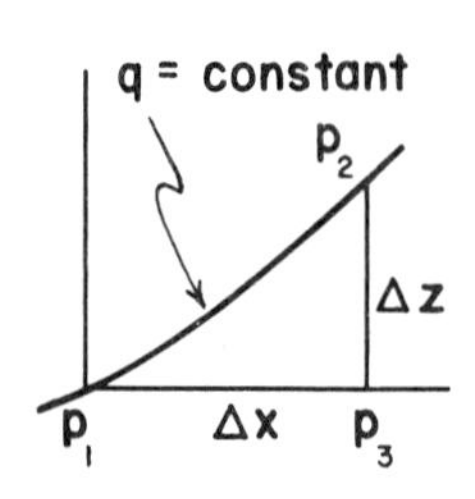

FIG. 12.6. Cross-section in the x,z plane

If we pass to the limit as the increments in distance become infinitesimally small we get

$$\left(\frac{\partial p}{\partial x}\right)_z = \left(\frac{\partial p}{\partial x}\right)_q - \left(\frac{\partial p}{\partial z}\right)\left(\frac{\partial z}{\partial x}\right)_q$$

If we divide both sides by the density and substitute on the right from the hydrostatic equation we obtain

$$\frac{1}{\rho}\left(\frac{\partial p}{\partial x}\right)_z = \frac{1}{\rho}\left(\frac{\partial p}{\partial x}\right)_q + g\left(\frac{\partial z}{\partial x}\right)_q \qquad (12.9)$$

A similar equation may be derived for the y, z plane in which x is replaced by y. These equations relate the horizontal pressure-gradient force per unit mass to the pressure-gradient force per unit mass along the q surface. The last term of Eq. (12.9) does involve the slope of the q surface, as expected. When Eq. (12.9) and the corresponding equation in the y direction are substituted into the geostrophic and gradient wind equations, the result is the appropriate form of those equations on a surface of constant q.

We shall now consider two specific examples:

Case (*a*)—Let $q = p$, so that we are dealing with an isobaric surface. Then Eq. (12.9) becomes

$$\frac{1}{\rho}\left(\frac{\partial p}{\partial x}\right)_z = g\left(\frac{\partial z}{\partial x}\right)_p$$

Thus the geostrophic wind equations on an isobaric surface are:

$$v = \frac{g}{f}\left(\frac{\partial z}{\partial x}\right)_p; \qquad u = -\frac{g}{f}\left(\frac{\partial z}{\partial y}\right)_p \qquad (12.10)$$

Thus on an isobaric surface the variable to be mapped is the height, z, of the surface—or better yet, its geopotential gz. The horizontal gradient of this quantity measured on an isobaric surface is proportional to the horizontal gradient of the pressure measured along an equipotential surface.

It is important to note that Eq. (12.10) do not involve the density. This is one of the advantages gained by the usual practice of analysis of isobaric charts rather than constant-level charts. On constant-level charts any device for measuring geostrophic winds must allow for the variation of density with elevation. This usually results in the need for a different geostrophic wind scale for each level analyzed. In analysis of the height contours of isobaric surfaces, on the other hand, one wind scale applies to all elevations.

Case (*b*).—Let $q = \theta$, the potential temperature, so that we are dealing with an isentropic surface. This case is of physical interest because we have seen in our study of meteorological thermodynamics that, under adiabatic conditions, isentropic surfaces are made up of the same particles day after day. The potential temperature is given by Poisson's equation, $T/\theta = (p/1000)^{\kappa}$. If we differentiate logarithmically along a surface of constant θ we get

$$\frac{1}{T}\left(\frac{\partial T}{\partial x}\right)_\theta = \frac{\kappa}{p}\left(\frac{\partial p}{\partial x}\right)_\theta$$

Therefore, from Eq. (12.9)

$$\frac{1}{\rho}\left(\frac{\partial p}{\partial x}\right)_z = \frac{p}{\kappa\rho T}\left(\frac{\partial T}{\partial x}\right)_\theta + \frac{\partial}{\partial x}(gz)_\theta$$

But since $\kappa = R/c_p$ and $p/(\rho T) = R$,

$$\frac{1}{\rho}\left(\frac{\partial p}{\partial x}\right)_z = \frac{\partial}{\partial x}(c_pT + gz)_\theta$$

and a similar result is true for the y direction. The geostrophic wind equations for an isentropic surface are then

$$v = \frac{1}{f}\frac{\partial}{\partial x}(c_pT + gz)_\theta; \qquad u = -\frac{1}{f}\frac{\partial}{\partial y}(c_pT + gz)_\theta$$

Thus the geostrophic wind field on an isentropic surface may be portrayed by drawing lines of constant value of the quantity $c_pT + gz$ measured on the chosen isentropic surface. This quantity is called the *isentropic stream function.*[7]

12.8 The Thermal Wind Equations

We now have at our disposal certain fairly good approximations to the true state of the atmosphere in middle and high latitudes, namely the geostrophic wind equations. We shall now proceed further with these relationships to see what additional information may be gotten from them.

---

[7] R. B. MONTGOMERY, A Suggested Method for Representing Gradient Flow in Isentropic Surfaces. *Bull. Am. Meteor. Soc.*, 18, 1937.

It is possible to show, without mathematics, that the vertical shear of the geostrophic wind is related to the horizontal temperature gradient. Consider two surfaces of constant pressure in a vertical cross section, as in Fig. 12.7. We have learned that the geostrophic wind is proportional to the slopes of the isobaric surfaces. Thus, in Fig. 12.7 the geostrophic wind increases with elevation because the upper pressure surface is more inclined than the lower. The cause of the greater slope of one surface over another is to be sought in the finite-difference form of the hydrostatic equation:

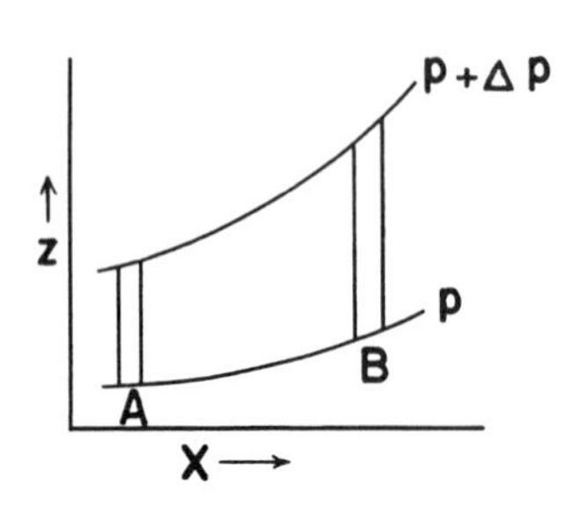

FIG. 12.7. Separation of isobaric surfaces

$$\Delta z \approx -\frac{\Delta p}{\rho g}$$

Now the pressure difference between the two surfaces is the same for columns A and B, so the difference in separation between them is to be attributed to a decrease in density in going from A to B. Since the average pressure is the same in the two columns this means column B has a higher average temperature than column A. Thus a shear of the geostrophic wind with elevation must be associated with a quasi-horizontal temperature gradient.

In actuality the temperature gradient to be considered is not a purely horizontal one but must be measured along an isobaric surface. Thus one may expect that the vertical wind shear could be related to the horizontal gradient of temperature plus a correction term. The correction term should involve the slope of the isobaric surface and the temperature variation in the vertical. Since the slopes of isobaric surfaces are generally quite small (of the order of 1/5000), the correction term may well be small also.

The equations we shall need to express these ideas quantitatively are the geostrophic wind equations:

$$fv = \frac{1}{\rho}\frac{\partial p}{\partial x} \qquad \text{and} \qquad fu = -\frac{1}{\rho}\frac{\partial p}{\partial y}$$

together with the hydrostatic equation and the equation of state:

$$g = -\frac{1}{\rho}\frac{\partial p}{\partial z} \qquad \text{and} \qquad \rho = \frac{p}{RT}$$

Note that there are five dependent variables u, v, ρ, p, T and only four equations relating them. Therefore we cannot expect to obtain a complete solution giving all these variables as functions of position. We shall be able to obtain a useful interrelationship, however.

If we eliminate reference to ρ in the geostrophic and hydrostatic expressions by means of the equation of state, we obtain:

$$\frac{fv}{T} = R\frac{\partial \ln p}{\partial x}; \qquad \frac{fu}{T} = -R\frac{\partial \ln p}{\partial y}; \qquad \frac{g}{T} = -R\frac{\partial \ln p}{\partial z}$$

Next we differentiate the first of these equations with respect to z, differentiate the last of these equations with respect to x, and eliminate the derivatives of pressure between the resulting expressions. This is called *cross-differentiation.* We then cross-differentiate between the second and third equations to eliminate all reference to pressure again. The results are:

$$\frac{\partial}{\partial z}\left(\frac{fv}{T}\right) = -\frac{\partial}{\partial x}\left(\frac{g}{T}\right); \qquad \frac{\partial}{\partial z}\left(\frac{fu}{T}\right) = \frac{\partial}{\partial y}\left(\frac{g}{T}\right)$$

When we complete the differentiation and rearrange we get

$$\begin{aligned} \frac{\partial v}{\partial z} &= \frac{g}{fT}\frac{\partial T}{\partial x} + \frac{v}{T}\frac{\partial T}{\partial z} \\ \frac{\partial u}{\partial z} &= -\frac{g}{fT}\frac{\partial T}{\partial y} + \frac{u}{T}\frac{\partial T}{\partial z} \end{aligned} \tag{12.11}$$

These are the *thermal wind equations* in differential form. They have the characteristics which were physically anticipated since the first terms on the right are the contribution of the horizontal temperature gradient, and the second terms on the right are correction terms involving the slopes of the isobaric surface (u, v) and the vertical temperature gradients. The correction terms on the right are indeed relatively small. This can be seen by dividing the second of Eq. (12.11) by u so that we have an expression for the fractional rate of change of u with height. The correction term is then $(1/T)(\partial T/\partial z)$, which, even if the lapse rate is as high as the dry adiabatic, is about 4 per cent per kilometer. The normal shear of west wind with height in the troposphere in middle latitudes is approximately 25 per cent per kilometer. Thus it is apparent that the additional term in question does not usually make a major contribution to the vertical wind shear, although there are times and places where it should be taken into account. This point will be elaborated further later.

In order to simplify matters we shall neglect these terms and write

$$\frac{\partial v}{\partial z} \approx \frac{g}{fT}\frac{\partial T}{\partial x}; \qquad \frac{\partial u}{\partial z} \approx -\frac{g}{fT}\frac{\partial T}{\partial y} \tag{12.12}$$

Equations (12.12) require that for v to increase with height temperature must increase to the east, and for u to increase with height temperature must increase to the south, in the Northern Hemisphere. This may be restated as follows: The vertical shear of the geostrophic wind is a vector which lies parallel to the isotherms on a level surface with low temperatures on the left in the Northern Hemisphere, and low temperatures on the right in the Southern Hemisphere. The fact that actual westerly winds in middle latitudes normally increase in strength going upward through the troposphere is to be explained, therefore, as a result of the normal decrease of temperature towards the poles in the troposphere.

These properties of the thermal wind may be used to show the relationship between the turning of wind with elevation and the horizontal temperature gradient. If we express equations (12.12) in approximate finite difference form we have

$$\Delta v \approx \frac{g}{fT}\frac{\partial T}{\partial x}\Delta z$$

$$\Delta u \approx -\frac{g}{fT}\frac{\partial T}{\partial y}\Delta z$$

where Δu and Δv represent the change of the wind through a finite increment of height Δz, and T and its derivatives are now to be thought of as mean values in the layer under consideration. If the mean isotherms

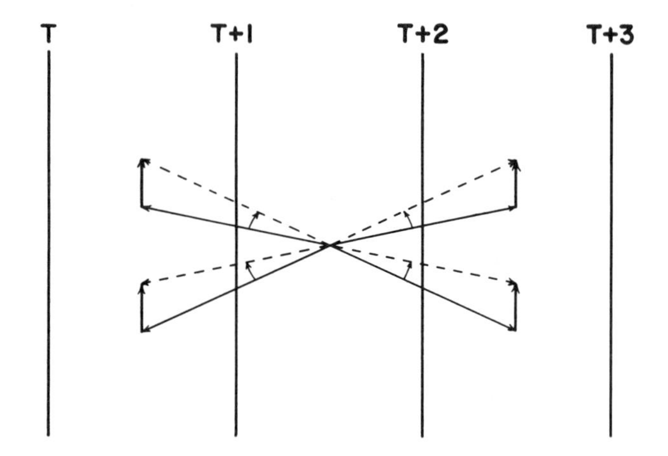

FIG. 12.8. Veering and backing of wind in relationship to horizontal temperature advection

are as shown in Fig. 12.8 then the wind shear vector, whose components are Δu and Δv, will be parallel to the mean isotherms with cold air on the left (Northern Hemisphere). This means that a shear vector so oriented, and of magnitude determined by the mean temperature gradient, must be added to the wind at the bottom of the layer to get the wind at the top. Then each of the four winds shown must turn with elevation as indicated. Inspection of Fig. 12.8 indicates that the wind turns anticyclonically with height (*veers*) whenever there is a wind component from warm towards cold air, and the wind turns cyclonically with height (*backs*) whenever there is a wind component from cold towards warm air. Since such flow will tend to change the temperature distribution by horizontal advection of cold or warm air we may formulate the following rule: Veering of the geostrophic wind with height occurs with warm air advection, backing of the geostrophic wind with height occurs with cold air advection. With the aid of this rule it is possible to estimate the orientation, spacing, and advection of the mean isotherms in a given layer from a pilot-balloon observation alone.

We consider next the vertical wind shear in a barotropic and a baroclinic atmosphere. A *barotropic atmosphere* is one in which the density at each point is determined solely by the pressure at that point. A *baroclinic atmosphere* is one in which no such simplifying condition exists. The importance of these two cases in atmospheric dynamics will be demonstrated later. Here we shall concern ourselves only with the thermal wind equations for each of these two states.

The condition for barotropy is $\rho = \rho(p)$, or, via the equation of state, $T = T(p)$. Then

$$\frac{\partial T}{\partial y} = \frac{dT}{dp}\frac{\partial p}{\partial y} = -\rho f u \frac{dT}{dp}$$

Similarly

$$\frac{\partial T}{\partial z} = \frac{dT}{dp}\frac{\partial p}{\partial z} = -\rho g \frac{dT}{dp}$$

Substituting these two results in the second of Eq. (12.11) we obtain

$$\frac{\partial u}{\partial z} = \frac{\rho g u}{T}\frac{dT}{dp} - \frac{\rho g u}{T}\frac{dT}{dp} = 0$$

with a similar result for the other thermal wind equation. Thus *in a barotropic atmosphere there can be no increase of geostrophic wind with height.* This is because the barotropic condition is precisely the one which makes the vertical-temperature-gradient term of the complete thermal wind equations no longer negligible, but equal and opposite to the horizontal-temperature-gradient term. Our previous calculation that the vertical gradient term seldom makes a major contribution to the wind shear means, accordingly, that the atmosphere is normally significantly baroclinic. In a barocline atmosphere the vertical wind shear depends mainly upon the horizontal temperature gradient.

Barotropy manifests itself on constant-level charts, in part, by parallelism of isobars and isotherms. This condition is more easily recognized on isobaric charts since these must then have constant temperature also, and no isotherms may be drawn. In the normal baroclinic case there is a gradient of temperature along an isobaric surface.

We next consider the thermal wind equations on isobaric surfaces. We shall illustrate with the equation for shear of the west wind with height:

$$\frac{\partial u}{\partial z} = -\frac{g}{fT}\left(\frac{\partial T}{\partial y}\right)_z + \frac{u}{T}\frac{\partial T}{\partial z}$$

A similar derivation may be given for the other thermal wind equation. The slope of an isobaric surface passing through the point under co-sideration is $(\partial z/\partial y)_p = -fu/g$. Upon substitution into the thermal wind equation we get

$$\frac{\partial u}{\partial z} = -\frac{g}{fT}\left[\left(\frac{\partial T}{\partial y}\right)_z + \frac{\partial T}{\partial z}\left(\frac{\partial z}{\partial y}\right)_p\right]$$

From Fig. 12.9 we see that the above expression in brackets is nothing more than the horizontal temperature gradient measured along the isobaric surface. (Compare with page 188.) Thus:

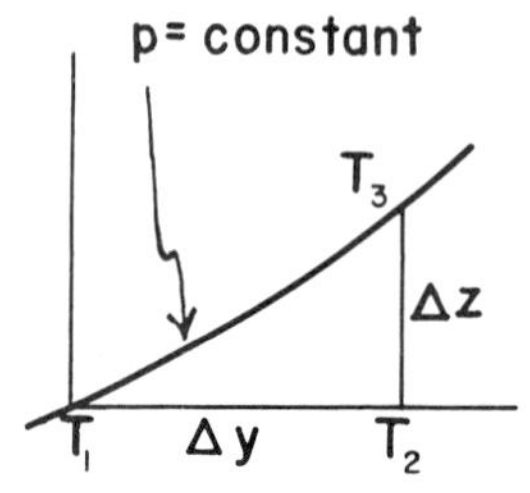

FIG. 12.9. Cross-section in the y,z plane

$$\frac{\partial u}{\partial z} = -\frac{g}{fT}\left(\frac{\partial T}{\partial y}\right)_p$$

and (12.13)

$$\frac{\partial v}{\partial z} = \frac{g}{fT}\left(\frac{\partial T}{\partial x}\right)_p$$

Equations (12.13) are exact equations for the shear of the geostrophic wind with height since they involve no assumptions about the vertical-temperature-gradient term. They should be compared with Eq. (12.12).

This exactness of a simple form of the thermal wind equations on an isobaric surface, as opposed to the approximate nature of the analogously simple form on a constant-level surface, is one of the chief advantages of isobaric analysis over constant-level analysis in practical meteorology. This point is made most clear by the following considerations: The hydrostatic equation may be written:

$$\frac{1}{p}\frac{\partial p}{\partial z} = -\frac{g}{RT}$$

If we integrate between two isobaric surfaces, p_1 and p_2, at elevations z_1 and z_2 we get

$$\ln\frac{p_1}{p_2} = \frac{g}{R}\int_{z_1}^{z_2}\frac{dz}{T}$$

If we extract a mean reciprocal temperature, $1/\overline{T}$, we may write

$$z_2 - z_1 = \left[\frac{R}{g} \ln \frac{p_1}{p_2}\right] \overline{T}$$

If we consider two particular isobaric surfaces, the quantity in brackets is constant. This means the lines of constant value of the difference in height of two pressure surfaces are also mean isotherms for the layer, and the thermal wind shear in the layer is parallel to such lines. The difference lines may be drawn quickly by superposing the two charts in question and subtracting graphically as in Fig. 12.10. This is done

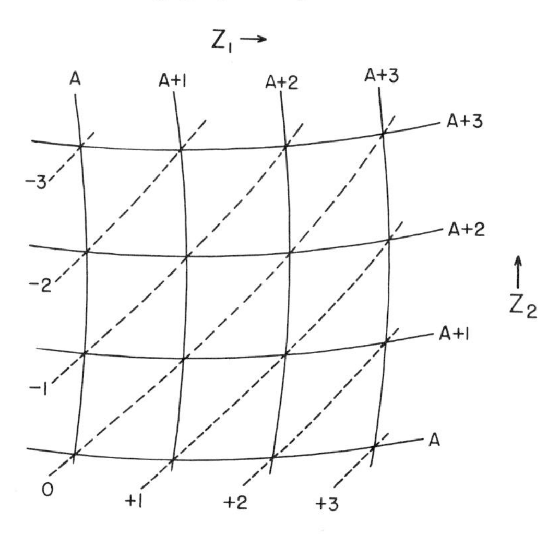

FIG. 12.10. Graphical subtraction of two fields, Z_1 and Z_2

by drawing the lines which pass through appropriate intersections of the two superposed fields. This procedure is an important way of promoting consistency of synoptic analysis at successive levels. It is possible in an exact form only when constant-pressure charts are analyzed rather than constant-level charts, and this is the foremost reason for that widespread practice in current meteorology.

All the relationships derived in this section have counterparts in the case of curved flow. That is, there are thermal wind equations for gradient flow similar to those for geostrophic flow.[8] In the gradient thermal wind equations, the vertical wind shear need not be parallel to the mean isotherms but may deviate somewhat because of the effects of curvature of flow and its variation with height. Without going into

[8] G. E. FORSYTHE, A Generalization of the Thermal Wind Equation to Arbitrary Horizontal Flow. *Bull. Am. Meteor. Soc.*, 26, 1945, pp. 371-5.

detail, it is possible to see that if the wind shear is approximately parallel to the mean isotherms with cold air on the left, then the shear vector points cyclonically around cold centers, and anticyclonically around warm centers. Thus the geostrophic cyclonic flow around a low will increase in strength with height if the low has a cold core, but decrease in strength with height if the low has a warm core. On the other hand, the geostrophic anticyclonic flow around a high will decrease in strength if the high has a cold core, but increase in strength with height if the high has a warm core. Furthermore, if the circulation and temperature centers do not coincide, the cyclones will shift with height towards the cold air, while the anticyclones will shift with height towards the warm air. These rules are of fundamental importance in synoptic analysis and are based firmly on the properties of the thermal wind.

PROBLEMS

1. Calculate the geostrophic wind speed in knots for (a) a pressure gradient of 3 mb/100 km, density $= 1.20 \times 10^{-3}$ g cm^{-3}, at latitude 45°; (b) a contour gradient on the 700 mb surface of 200 geopotential feet in 3 degrees of latitude at latitude 60°. Show your computations.

2. Show that the geostrophic wind equation applied to an isobaric surface at 45° may be written $c = 100/\Delta n$ to a sufficient degree of approximation if c is expressed in knots and Δn is the separation in degrees of latitude between contours drawn for every 200 ft.

3. Calculate the gradient wind speed in knots for case (a) of Problem 1 when the radius of curvature is 10° of latitude. Do this for all four mathematically possible cases. Show your computations.

4. Show that the maximum possible normal (nonanomalous) anticyclonic wind speed around a high is twice the geostrophic wind speed corresponding to the same pressure gradient.

5. The equations of geostrophic flow on a surface of constant temperature may be written $fu = -(\partial\psi/\partial y)_T$; $fv = (\partial\psi/\partial x)_T$. Derive an expression for ψ.

6. A radiosonde observation at a station at 43° N indicated nearly isothermal conditions up to 3 km with $t = 0°$ C. A simultaneous pibal gave the following results:

hgt (m)	*wind*	
500	180°	10.00 m/s
1500	210°	11.53
2500	180°	10.00

(a) Calculate the horizontal mean temperature gradients in the layers 500-1500 m and 1500-2500 m. (in °C/100 km).

(b) Calculate the rate of advective mean temperature change in each of these layers (in °C/hour).

(c) Determine the average lapse rate between 1000 m and 2000 m which would result if these advective temperature changes persisted for twelve hours.

(d) What assumptions did you have to make in order to carry out these calculations?

7. Calculate the numerical relationship between the atmospheric thickness (difference of height in feet) from the 1000 mb surface to the 500 mb surface, and the mean virtual temperature of the layer from 1000 to 500 mb (in °C). If thickness lines are drawn at intervals of 400 geopotential feet, to what mean temperature interval does this correspond?

CHAPTER 13

KINEMATICS OF FLUID FLOW

13.1 Kinematics versus Dynamics

Until now we have been studying certain topics in the dynamics of the atmosphere—the forces governing the motion of air. We now turn to the *kinematics* of the atmosphere—the motion of air alone, without reference to the forces at work. It should be stated immediately that kinematics is potentially less powerful a tool than dynamics for the study of fluids. Although complete solutions to specific flow problems cannot be obtained from kinematics alone, a number of important results may yet be derived in this way, and a number of concepts arise that find application in dynamics.

13.2 Resolution of a Linear Velocity Field

For simplicity we shall restrict our attention to two-dimensional, horizontal motion in a Cartesian coordinate system. The velocity components, at a given instant, may be expressed in Taylor's series about the origin:

$$u = u_0 + \left(\frac{\partial u}{\partial x}\right)_0 x + \left(\frac{\partial u}{\partial y}\right)_0 y + \tfrac{1}{2}\left(\frac{\partial^2 u}{\partial x^2}\right)_0 x^2 + \tfrac{1}{2}\left(\frac{\partial^2 u}{\partial y^2}\right)_0 y^2 + \left(\frac{\partial^2 u}{\partial x \partial y}\right)_0 xy + \ldots$$

$$v = v_0 + \left(\frac{\partial v}{\partial x}\right)_0 x + \left(\frac{\partial v}{\partial y}\right)_0 y + \tfrac{1}{2}\left(\frac{\partial^2 v}{\partial x^2}\right)_0 x^2 + \tfrac{1}{2}\left(\frac{\partial^2 v}{\partial y^2}\right)_0 y^2 + \left(\frac{\partial^2 v}{\partial x \partial y}\right)_0 xy + \ldots$$

where a subscript zero means the quantity is to be evaluated at the origin. Such subscripted quantities are, of course, constants. If we further restrict attention to the immediate vicinity of the origin, x and y are then small and it becomes possible to approximate u and v by linear expressions:

$$\begin{aligned} u &\approx u_0 + \left(\frac{\partial u}{\partial x}\right)_0 x + \left(\frac{\partial u}{\partial y}\right)_0 y \\ v &\approx v_0 + \left(\frac{\partial v}{\partial x}\right)_0 x + \left(\frac{\partial v}{\partial y}\right)_0 y \end{aligned} \tag{13.1}$$

The significance of the several velocity derivatives in Eq. (13.1) may be approached by inquiring if there are any of these derivatives or their combinations which are invariant during a rotation of the coordinate axes. Any such invariant combinations may then represent fundamental properties of such linear motions, and as such will be proper subjects for further study.

Let the initial coordinate system be x, y, and let it be rotated through a fixed arbitrary angle θ into an x', y' system, as in Fig. 13.1. It is clear that

$$\begin{aligned} x' &= x \cos\theta + y \sin\theta \\ y' &= -x \sin\theta + y \cos\theta \end{aligned}$$

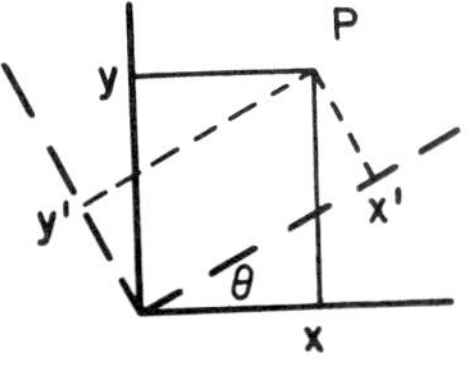

FIG. 13.1. The coordinates of a point P in two Cartesian systems differing by the angle θ

Thus, by differentiation with respect to time

$$\begin{aligned} u' &= u \cos\theta + v \sin\theta \\ v' &= -u \sin\theta + v \cos\theta \end{aligned}$$

or upon solution for u and v:

$$\begin{aligned} u &= u' \cos\theta - v' \sin\theta \\ v &= u' \sin\theta + v' \cos\theta \end{aligned}$$

We shall now form the four first derivatives contained in Eq. (13.1) remembering that for any function, F,

$$\begin{aligned} \frac{\partial F}{\partial x} &= \frac{\partial x'}{\partial x}\frac{\partial F}{\partial x'} + \frac{\partial y'}{\partial x}\frac{\partial F}{\partial y'} = \cos\theta \frac{\partial F}{\partial x'} - \sin\theta \frac{\partial F}{\partial y'} \\ \frac{\partial F}{\partial y} &= \frac{\partial x'}{\partial y}\frac{\partial F}{\partial x'} + \frac{\partial y'}{\partial y}\frac{\partial F}{\partial y'} = \sin\theta \frac{\partial F}{\partial x'} + \cos\theta \frac{\partial F}{\partial y'} \end{aligned}$$

The results are:

$$\frac{\partial u}{\partial x} = \cos\theta\left[\frac{\partial u'}{\partial x'}\cos\theta - \frac{\partial v'}{\partial x'}\sin\theta\right] - \sin\theta\left[\frac{\partial u'}{\partial y'}\cos\theta - \frac{\partial v'}{\partial y'}\sin\theta\right] \tag{13.2}$$

$$\frac{\partial u}{\partial y} = \sin\theta\left[\frac{\partial u'}{\partial x'}\cos\theta - \frac{\partial v'}{\partial x'}\sin\theta\right] + \cos\theta\left[\frac{\partial u'}{\partial y'}\cos\theta - \frac{\partial v'}{\partial y'}\sin\theta\right] \tag{13.3}$$

$$\frac{\partial v}{\partial x} = \cos\theta\left[\frac{\partial u'}{\partial x'}\sin\theta + \frac{\partial v'}{\partial x'}\cos\theta\right] - \sin\theta\left[\frac{\partial u'}{\partial y'}\sin\theta + \frac{\partial v'}{\partial y'}\cos\theta\right] \tag{13.4}$$

$$\frac{\partial v}{\partial y} = \sin\theta\left[\frac{\partial u'}{\partial x'}\sin\theta + \frac{\partial v'}{\partial x'}\cos\theta\right] + \cos\theta\left[\frac{\partial u'}{\partial y'}\sin\theta + \frac{\partial v'}{\partial y'}\cos\theta\right] \tag{13.5}$$

If we subtract Eq. (13.3) from Eq. (13.4), we get

$$\frac{\partial v}{\partial x} - \frac{\partial u}{\partial y} = \frac{\partial v'}{\partial x'} - \frac{\partial u'}{\partial y'} \tag{13.6}$$

If we add Eq. (13.2) and Eq. (13.5) we get

$$\frac{\partial u}{\partial x} + \frac{\partial v}{\partial y} = \frac{\partial u'}{\partial x'} + \frac{\partial v'}{\partial y'} \tag{13.7}$$

These, then, are two combinations of the first derivatives of the velocity components which are invariant under a rotation of the coordinate axes. They are referred to as the *vorticity* (Eq. 13.6) and *divergence* (Eq. 13.7), respectively. Other combinations of interest are

$$\begin{aligned} \frac{\partial v}{\partial x} + \frac{\partial u}{\partial y} &= \sin 2\theta \left[\frac{\partial u'}{\partial x'} - \frac{\partial v'}{\partial y'}\right] + \cos 2\theta \left[\frac{\partial v'}{\partial x'} + \frac{\partial u'}{\partial y'}\right] \\ \frac{\partial u}{\partial x} - \frac{\partial v}{\partial y} &= \cos 2\theta \left[\frac{\partial u'}{\partial x'} - \frac{\partial v'}{\partial y'}\right] - \sin 2\theta \left[\frac{\partial v'}{\partial x'} + \frac{\partial u'}{\partial y'}\right] \end{aligned} \tag{13.8}$$

These combinations are not themselves invariant during a rotation of the axes but the sum of their squares is invariant:

$$\left(\frac{\partial v}{\partial x} + \frac{\partial u}{\partial y}\right)^2 + \left(\frac{\partial u}{\partial x} - \frac{\partial v}{\partial y}\right)^2 = \left(\frac{\partial v'}{\partial x'} + \frac{\partial u'}{\partial y'}\right)^2 + \left(\frac{\partial u'}{\partial x'} - \frac{\partial v'}{\partial y'}\right)^2 \tag{13.9}$$

The combinations of derivatives whose squares appear in Eq. (13.9) are each referred to as *deformations.*

Before investigating the properties of these presumably fundamental quantities, the vorticity, divergence, and two kinds of deformations, we shall first show that a linear velocity field is composed of a linear combination of these quantities, together with a constant velocity called the translation. Their combination to form a linear velocity field is, of course, the chief reason these quantities are important. We shall take advantage of the invariance of vorticity and divergence during a rotation of the axes to place the axes at such an angle that one of the deformations becomes zero. Thus, from the first of Eq. (13.8) it may be seen that $\partial v/\partial x + \partial u/\partial y$ may be made to vanish, whatever the values of the primed derivatives, by choosing the appropriate value of θ. In this we have adopted the view that either the primed or the unprimed system may be considered the initial one from which we rotate through θ, as suits our convenience. In this case the primed system will be considered as given and the unprimed as the one that results from the rotation. Because of the simplicity so introduced let us rotate to these axes, called the *principal axes*, wherever they may be in a given flow pattern. One may verify from Eq. (13.1) that

$$u \approx u_0 - \tfrac{1}{2}\left(\frac{\partial v}{\partial x} - \frac{\partial u}{\partial y}\right)_0 y + \tfrac{1}{2}\left(\frac{\partial u}{\partial x} + \frac{\partial v}{\partial y}\right)_0 x + \tfrac{1}{2}\left(\frac{\partial u}{\partial x} - \frac{\partial v}{\partial y}\right)_0 x + \underline{\tfrac{1}{2}\left(\frac{\partial v}{\partial x} + \frac{\partial u}{\partial y}\right)_0 y}$$

$$(13.10)$$

$$v \approx v_0 + \tfrac{1}{2}\left(\frac{\partial v}{\partial x} - \frac{\partial u}{\partial y}\right)_0 x + \tfrac{1}{2}\left(\frac{\partial u}{\partial x} + \frac{\partial v}{\partial y}\right)_0 y - \tfrac{1}{2}\left(\frac{\partial u}{\partial x} - \frac{\partial v}{\partial y}\right)_0 y + \underline{\tfrac{1}{2}\left(\frac{\partial v}{\partial x} + \frac{\partial u}{\partial y}\right)_0 x}$$

Here the last terms (underlined) are zero if we rotate to the principal axes. The constants u_0 and v_0 are the parts of the velocity called pure *translation*. The remaining terms represent motions which are purely vortical, divergent, and deformative, respectively. We shall take advantage of the fact that a velocity may be looked upon, approximately, as the sum of four such pure types of motion to study the individual types of flow rather than all the complex combinations of them that exist.

13.3 Streamlines, Trajectories, and Streak Lines

In order to study the flow patterns associated with each of these four basic fields of motion we shall introduce a number of ways of representing flow patterns. A *streamline* is a line to which, at any given

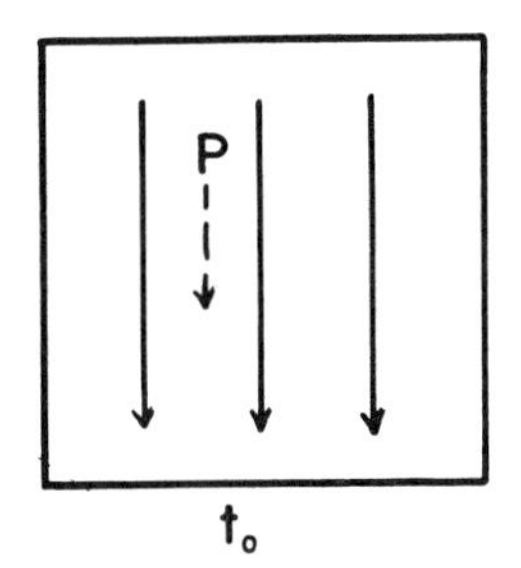

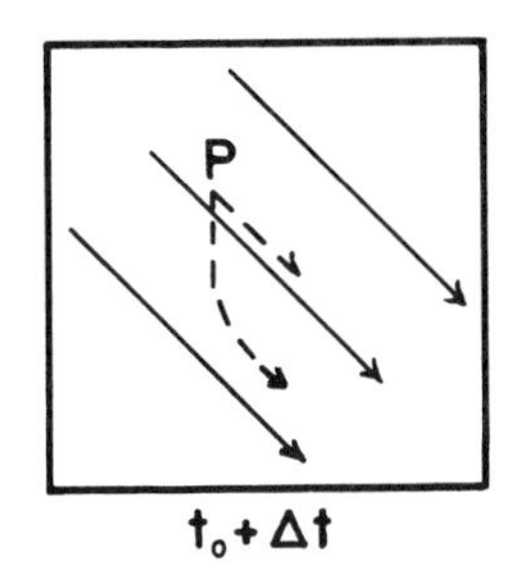

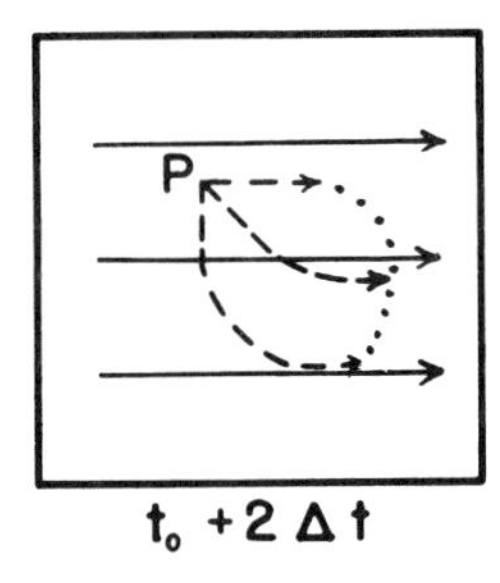

FIG. 13.2. Streamlines (solid), trajectories (dashed), and streak line (dotted) in a pattern of straight parallel streamlines which turn at a constant rate

instant, are tangent all the velocity vectors of the points through which the line passes; thus the flow is along the streamlines at any given moment. A *trajectory* is a line along which a given fluid particle has moved; for example, the path traced by a radioactive cloud. A *streak line* is a line connecting all the particles that have passed a given geometrical point; thus a plume of smoke from a chimney is a streak line. These three means of describing a fluid motion are not necessarily identical, as may be seen from Fig. 13.2. The three coincide only in the case of steady motion.

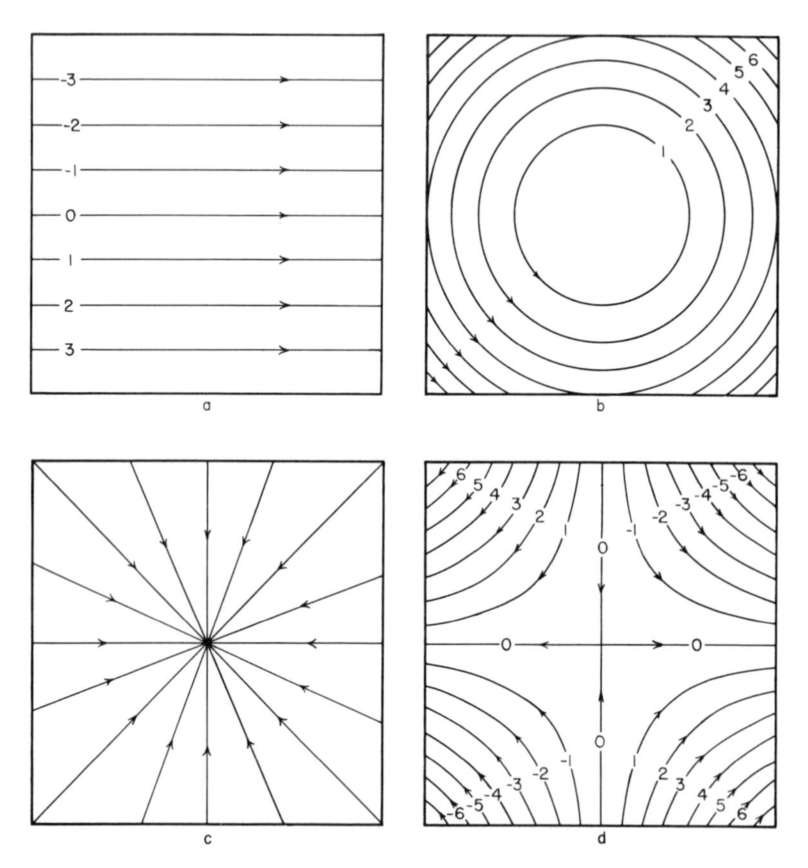

FIG. 13.3. Streamlines of pure constant—
(a) translation, (b) vorticity, (c) divergence, and (d) deformation

We shall devote most of our attention to the streamlines. The requirement that such a line be parallel to the flow means, if dx and dy are components of an increment of distance along a streamline, that $dy/dx = v/u$. This statement is the differential equation of the streamlines. If u and v are given as functions of x and y it is possible by means of this relationship to calculate the equation for the set of curves making up the streamlines. We shall use as examples the four linear velocity fields of Eq. (13.10).

Case 1—pure translation: If no other motion than that of translation is present, the differential equation of the streamlines is $dy/dx = v_0/u_0$. Since the velocities are constant this may be integrated immediately to yield $y = (v_0/u_0)x + K$, where K is a constant of integration. This is the equation of a family of straight lines all of the same slope, v_0/u_0. Each line has a different intercept on the y axis, depending upon the numerical value assigned to K, as shown is Fig. 13.3*a*. This case is called pure translation because a fluid moving in this fashion will merely transport fluid particles in a uniform fashion.

Case 2—pure vorticity: If no other motion is present than that associated with the second terms on the right of Eq. (13.10), the equation of the streamlines is $dy/dx = -x/y$. Upon integration we get $x^2 + y^2 = K$. This is the equation of a family of circles of varying radii, depending upon the values assigned to K, as shown in Fig. 13.3*b*. Thus a purely vortical motion is one in which a fluid element exhibits no other motion than a rotation about some axis. Because of this such a flow is also referred to as a pure *rotation*.

A further insight into the nature of vorticity, $\partial v/\partial x - \partial u/\partial y$, may be gained by considering the case in which both terms make a positive contribution. The velocity shears will then be as shown in Fig. 13.4 where, for pure rotation, $u_0 = v_0 = 0$ so there is no motion at the origin. In this case each term acts to turn a line of fluid cyclonically about the z axis. Thus cyclonic motion is associated with positive vorticity, anticyclonic motion with negative vorticity.[1] We are discussing only rotation about a vertical axis. Vorticity may, of course, exist around an arbitrarily oriented axis.

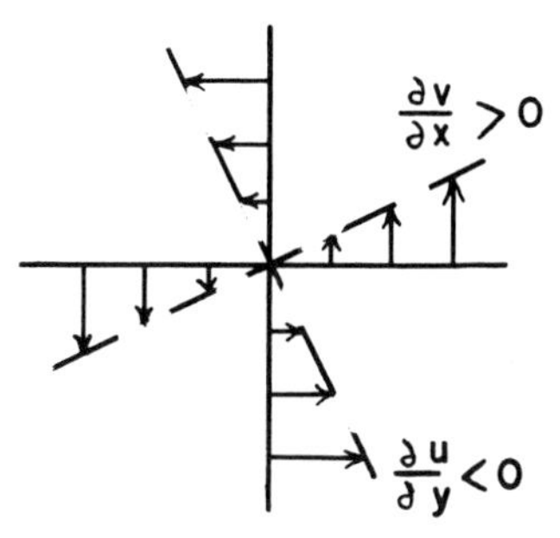

FIG. 13.4. Rotation owing to shear

Case 3—pure divergence: If no other motion is present than that represented by the third terms on the right of Eq. (13.10), the equation of the streamlines is $dy/dx = y/x$. Integration yields $y = Kx$, which is the equation of a family of straight lines passing through the origin

[1] The example of Fig. 13.4 illustrates the importance of shear in vorticity, while that of Fig. 13.3*b* illustrates the importance of curvature of the streamlines. In natural coordinates this dual nature is most evident since the expression for vorticity is $c/R - \partial c/\partial n$, where R is the radius of curvature. These are the expressions for a flat coordinate system. When the curved latitude and longitude grid is used the vorticity (rotation about the vertical) is $\partial v/\partial x - \partial u/\partial y + u \tan \phi/a$, where a is the radius of the earth.

with varying slopes, depending upon the values assigned to K. The family is shown in Fig. 13.3c. When the flow is directed outward it is obvious why this is called pure divergence. When the flow is directed inward it is called convergence.

A further understanding of the nature of divergence, $\partial u/\partial x + \partial v/\partial y$, is obtained by considering the case in which both terms make a positive contribution and the speed is zero at the origin. The velocity derivatives

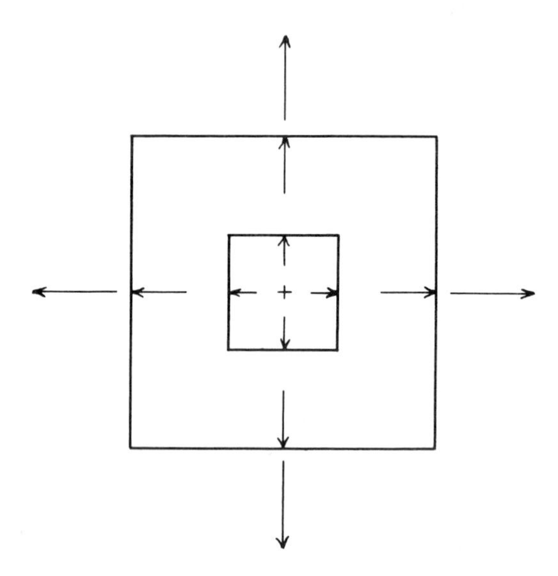

FIG. 13.5. Divergence owing to gradients of speed along the streamlines

will then be as in Fig. 13.5. A fluid curve will be expanded without being translated, rotated, or changed in shape, hence the name divergence.[2]

Case 4—pure deformation: If no other motion is present than that represented by the fourth terms on the right of Eq. (13.10), the equation of the streamlines is $dy/dx = -y/x$. Integration yields $xy = K$, which is the equation of a family of hyperbolas with the x and y axes as asymptotes. The specific value assigned to K determines which member

[2] The divergence discussed here is two-dimensional. In three dimensions it is $\partial u/\partial x + \partial v/\partial y + \partial w/\partial z$. The portion of horizontal divergence which is due to longitudinal stretching is clearly exemplified by this expression and Fig. 13.5. When expressed in natural coordinates the horizontal divergence is $\partial c/\partial s + c\partial\psi/\partial n$, where ψ is the angle a streamline makes with the x axis. The second term shows the additional contribution due to spreading of streamlines. These are expressions for a flat coordinate system. When the curved latitude and longitude grid is used, the three-dimensional divergence is $\partial u/\partial x + \partial v/\partial y + \partial w/\partial z - (v \tan\phi)/a + 2\,w/a$.

of the family one is dealing with. The streamlines of a pure deformation are shown in Fig. 13.3*d*. When the deformation investigated is that represented by the fifth terms on the right of Eq. (13.10), rectangular hyperbolas again result, but the asymptotes are rotated 45°. This is the term, of course, that may be eliminated by rotation to the principal axes. It may be seen from Fig. 13.3*d* that a fluid curve (such as a square centered on the origin) will be altered in shape by this flow pattern (a square will become a rectangle), but no net translation, rotation, or divergence will occur. Hence the name deformation.[3]

These results indicate that in the immediate vicinity of a point in a fluid the motion may be looked upon as a sum of, at most, four different pure kinds of simple motions. Much of our modern attitude towards complex flow patterns is based upon a realization that they may be resolved into these simpler constituents. Thus, from this point of view, one should seek on a weather map indications of the spatial distribution of the basic flows.

13.4 The Stream Function

A few examples of combinations of the basic velocity fields will be studied in order to gain some appreciation of the disguises they may adopt when in combination. We shall consider the special case when the two-dimensional divergence is zero: $\partial u/\partial x + \partial v/\partial y = 0$. This relationship is satisfied by a function $\psi(x, y)$ provided

$$u = -\,\partial\psi/\partial y; \qquad v = \partial\psi/\partial x \tag{13.11}$$

as may be verified by substitution. The function ψ is called the ***stream function***. The similarity of Eq. (13.11) to the geostrophic wind equations should be noted. When such a function exists the equation of the streamlines is

$$\frac{\partial\psi}{\partial x}\,dx + \frac{\partial\psi}{\partial y}\,dy = 0$$

or

$$d\psi = 0$$

where $d\psi$ represents an increment in ψ along a streamline at a given moment. It is clear, therefore, that the equation of a streamline in

[3] Deformation, $\partial u/\partial x - \partial v/\partial y$, results from a change in the sign of one term of the divergence. When a curved coordinate system is used this deformation becomes $\partial u/\partial x - \partial v/\partial y - (v \tan \phi)/a$.

nondivergent flow is ψ = constant. Thus each streamline may be labelled with its value of the stream function and then the streamlines not only indicate the direction of flow but, by their spacing, also determine the magnitude.

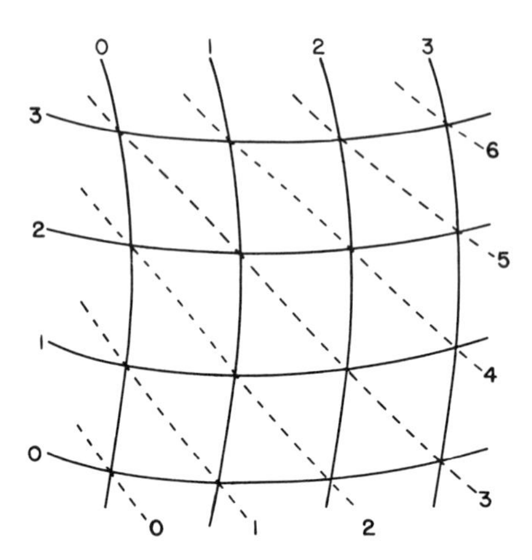

FIG. 13.6. Graphical addition of two fields

These properties of the stream function permit rapid addition of velocity fields by graphical methods. Figure 13.6 shows how the stream

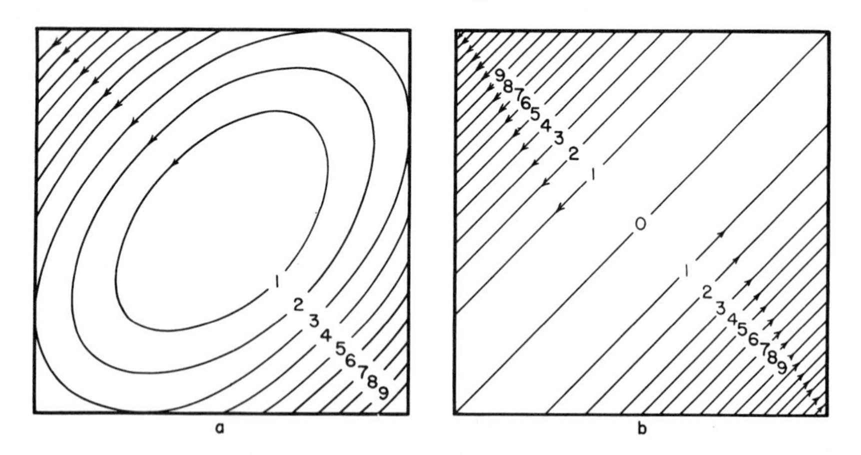

FIG. 13.7. Streamlines of (a) rotation plus a "half-strength" deformation; (b) rotation plus a "full-strength" deformation

functions of two flow patterns may be added graphically to yield immediately the streamlines of the combined flow. In order to utilize this technique the streamlines of Fig. 13.3*a*, *b*, and *d* have been labelled with appropriate values of stream function. These may now be added graphically to obtain the streamlines of combined fields.

The sum of pure translation with either pure rotation or pure deformation is not portrayed here, since it merely results in a shift in the position of the center of the rotation or the deformation without

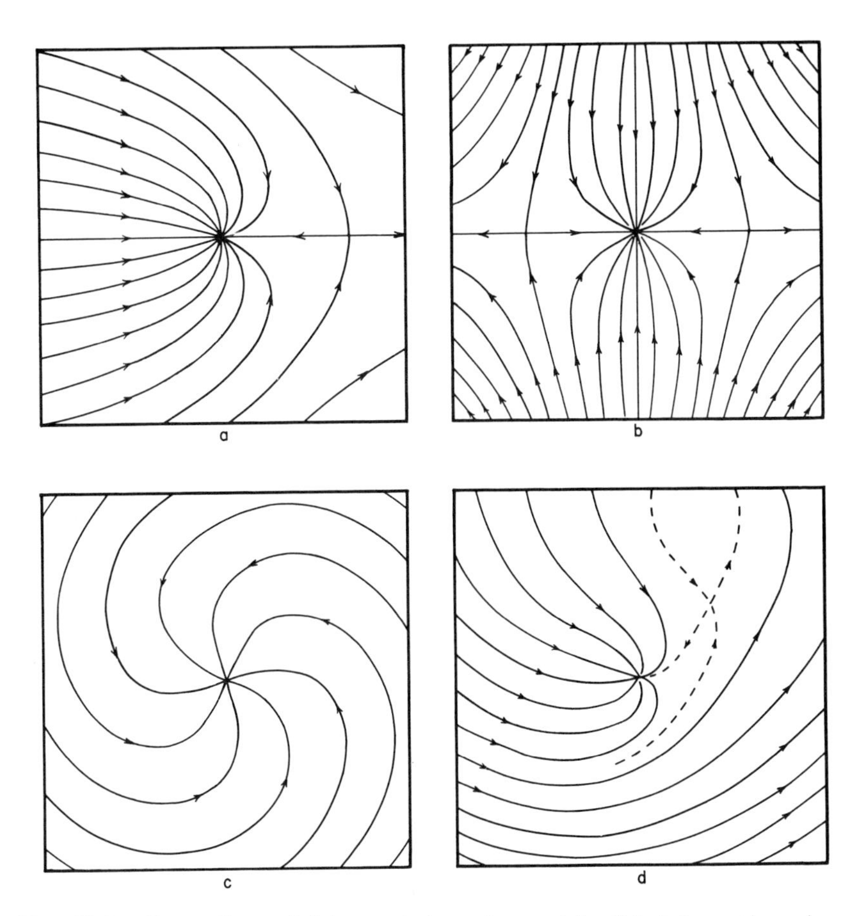

FIG. 13.8. Streamlines of (a) translation plus a sink, (b) deformation plus a sink, (c) rotation plus a sink, (d) translation plus rotation plus a sink

otherwise affecting the pattern. The result of adding the rotation of Fig. 13.3*b* to a deformation is shown in Fig. 13.7*a*. Here the deformation is that of Fig. 13.3*d* with all the speeds cut in half. The result is a series of elliptic streamlines. When this same rotation is added to the "full strength" deformation of Fig. 13.3*d* the result is as shown in Fig. 13.7*b*, where the streamlines are straight with cyclonic shear. It is easy to verify graphically in both these figures that a line of fluid particles centered on the origin will be both rotated and deformed.

The combinations of the four basic, linear fields of motion do not, of course, comprise all the important and interesting combinations. Figure 13.8*a* shows the streamlines of a combination of translation from left to right with a nonlinear kind of "convergent" flow called a *sink*.[4] Figure 13.8*b* shows a combination of this sink with a deformation. Note in both figures the existence of at least one point where streamlines cross each other at right angles. Here the resultant speed is zero, since at such points the velocity of one of the component fields is exactly equal and opposite to the velocity of the other component field. Since in a sink the speed increases towards the center, in both Fig. 13.8*a* and 13.8*b* the flow pattern resembles a sink near the center, but more nearly resembles the other added field near the periphery.

Figure 13.8*c* gives the streamlines of a sink combined with a pure rotation. In such a motion the fluid spirals in towards the center and exhibits the properties of both convergence and rotation. Finally, in Fig. 13.8*d* we have the streamlines resulting from combination of a sink, a pure cyclonic rotation, and a translation from left to right. This resembles the kind of flow pattern which might occur in low levels when a cyclone with bad weather is imbedded in a basic westerly or easterly current.

13.5 Circulation and its Relationship to Vorticity

There are other means of expressing the properties of the basic flow patterns than the combinations of partial derivatives which have been set down above. We now consider another measure of the vorticity of a fluid. The *circulation*, C, around a given closed curve in a fluid is the integral around the curve of the components of the velocities along the curve:

$$C = \oint V \cos \alpha \, ds$$

[4] A sink is a flow pattern in which the motion is directed radially inward and in which the speed is inversely proportional to the distance from the center. In a sink the divergence is zero everywhere except at the origin, where fluid is removed at a fixed rate. The flow resulting from extraction of fluid by a narrow pipe in otherwise resting water closely resembles a sink. It is introduced here because the streamlines of a sink look like those of a pure convergence but nevertheless may be assigned values of the stream function to facilitate graphical addition of fields.

where, as in Fig. 13.9, V represents the magnitude of the total velocity, α the angle between the velocity vector and the curve, ds an increment of distance along the curve, and the symbol $\oint$ means that the integral is to be taken in a closed path along the curve in a counterclockwise

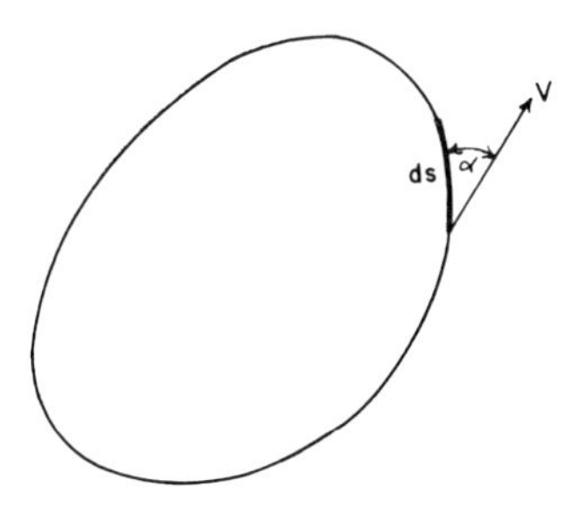

FIG. 13.9. The relationship of the velocity of a fluid particle and an increment of distance along a curve

direction. Thus C is a measure of the extent to which the fluid exhibits rotary motion and is positive for cyclonic circulation. Circulation has dimensions $[L^2T^{-1}]$.

If dx, dy, dz are components of ds, then

$$C = \oint(u\,dx + v\,dy + w\,dz)$$

or, in two dimensions

$$C = \oint(u\,dx + v\,dy)$$

In order to establish the relationship of the circulation to vorticity, consider the circuit formed by an infinitesimal rectangle as in Fig. 13.10.

FIG. 13.10. Velocity components around an infinitesimal rectangle in the x, y plane

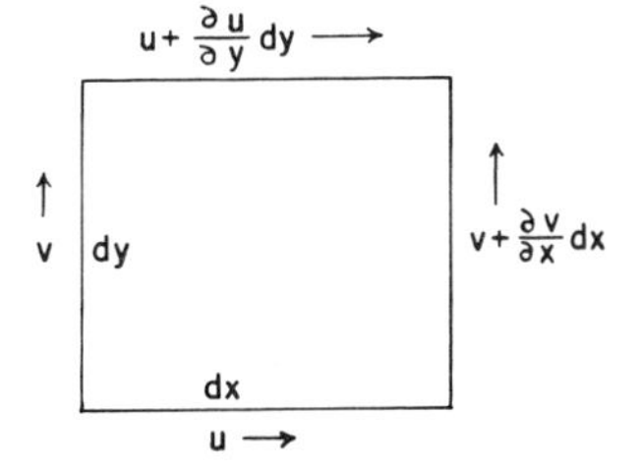

The contribution to circulation around this figure is

$$dC_{dxdy} = udx + \left(v + \frac{\partial v}{\partial x}dx\right)dy - \left(u + \frac{\partial u}{\partial y}dy\right)dx - vdy$$

or

$$dC_{dxdy} = \left(\frac{\partial v}{\partial x} - \frac{\partial u}{\partial y}\right)dx\,dy$$

Thus, for a small element of area the vorticity of the fluid in the area is equal to the circulation around the perimeter per unit area.

Obviously the area contained within and around any closed curve may be divided into a number of small rectangles (Fig. 13.11). One can always select a set of contiguous rectangles whose outer limit will nearly coincide with the given curved circuit, as shown in Fig. 13.11. As the size of the rectangles is made smaller, the outer edges of the

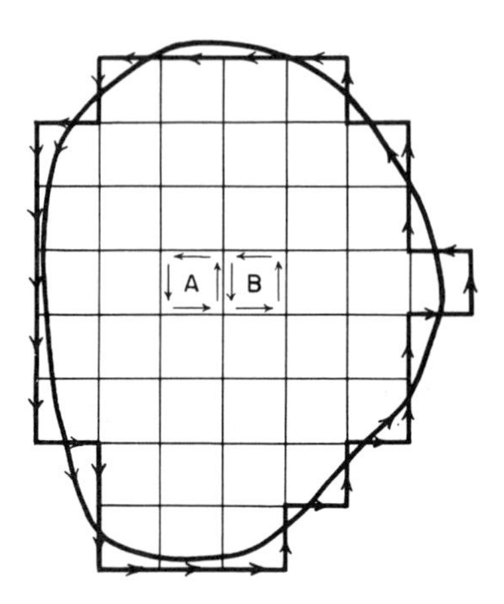

FIG. 13.11. The approximation of a closed curve by the perimeter of a set of small rectangles

rectangular array will coincide more closely with the curve until in the limit they become identical. If one determines the contribution to the total circulation from each small rectangle and sums up over all rectangles the sum will approach the desired circulation around the given curve. This may be seen by considering the two rectangles A and B. When one calculates the circulation for each of these two in the counterclockwise direction (mathematically positive sense) the side common to the two will contribute the same increment to both A and B, but with opposite sign, because we integrate along this side in opposite directions in the two cases. From the foregoing it is clear that all the inside segments cancel each other in the summation and only the heavy outer straight lines can make a contribution. Since this set of outer straight lines approaches the curve as the rectangles become smaller we have as a limit

$$\oint (u\,dx + v\,dy) = \iint \left(\frac{\partial v}{\partial x} - \frac{\partial u}{\partial y}\right) dxdy \tag{13.12}$$

where the left side is the circulation around the given curve and the right side is the integral of vorticity over the area enclosed by the curve. This is the two-dimensional form of *Stokes' theorem*, which relates circulation to vorticity for figures of finite extent. Circulation is an areal measure of the rotational tendency of a fluid and vorticity is a point measure of that same tendency. A fluid which has zero vorticity everywhere has, by Eq. (13.12), zero circulation around all possible closed paths and is said to be *irrotational.* The use of this technical term does not mean that the flow cannot be curved; it is possible for curved flow to be irrotational if it is accompanied by a shear in velocity of opposite sign.

As an example of the two ways of calculating circulation afforded by Stokes' theorem, consider the simple case where $u = u_0 + ay$, $v = 0$ (where u_0 and a are constants) and it is desired to get C around the rectangular path of Fig. 13.12. By line integration starting at the point 0,0,

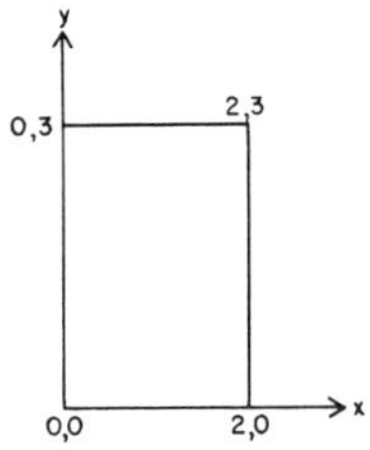

$$C = \oint (u\,dx + v\,dy) = \int_0^2 u_0 dx + 0 + \int_2^0 (u_0 + 3a)\,dx + 0$$

or

$$C = 2u_0 - 2u_0 - 6a = -6a$$

By the method of area integration,

$$C = \iint \left(\frac{\partial v}{\partial x} - \frac{\partial u}{\partial y}\right) dxdy = \iint -a\,dxdy$$

or

$$C = -a \iint dxdy = -6a$$

FIG. 13.12. Example of the calculation of circulation

which is the same answer, as it should be.

We can now quantitatively relate these two measures of rotation to the angular velocity of rotation of a small fluid element. Consider a finite closed curve of fluid in the x, y plane (Fig. 13.13). Let O be the instantaneous center of rotation of a fluid particle as it moves through the distance Vdt sweeping out the angle $d\theta$ at O. The radial lines from O to the initial and final positions of the particle will intersect the given curve and define an increment of distance dS along the curve. Let us define a quantity ζ as twice the angular velocity around O as the center. Then the contribution to circulation from this element of arc is $dC = r\,\frac{1}{2}\,\zeta \cos \alpha\, dS$. But $\cos \alpha = rd\theta/dS$, therefore, $dC = r^2\,\frac{1}{2}\,\zeta d\theta$. When we integrate over the entire curve we get $C = \oint r^2\,\frac{1}{2}\,\zeta d\theta$.

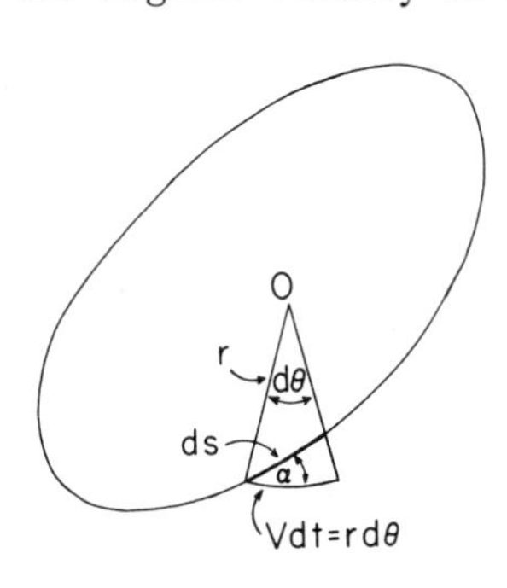

FIG. 13.13. Calculation of the relationship among circulation, vorticity, and angular velocity

If we now let the curve be a circle and permit it to shrink to an infinitesimal radius, we may consider the angular velocity at the central point, $\frac{1}{2}\,\zeta$, to be a constant independent of θ. We then get $C = \pi r^2 \zeta$.

Thus ζ is equal to the vorticity which we now find to be twice the angular velocity at the point under consideration.

These kinematic results concerning vorticity, circulation, and angular velocity will be useful later on when we consider the dynamics of these quantities. As an immediate application we may now look upon the Coriolis parameter, $f = 2\Omega \sin \phi$, which is twice the angular velocity of the earth's rotation about the vertical at any latitude, as the vorticity about the vertical due to the earth's rotation.[5]

13.6 The Equation of Continuity

An important restriction upon the velocity of a fluid may be obtained by kinematic methods applied to the law of conservation of mass. We shall find that the result of this application will involve one of the basic velocity fields we have studied.

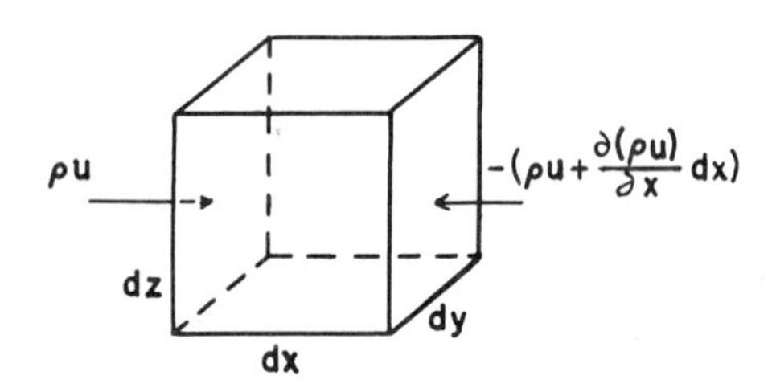

FIG. 13.14. Flow of mass into an infinitesimal parallelopiped

Consider an infinitesimal box of fluid, fixed in space, whose sides are dx, dy, dz, as in Fig. 13.14. We shall compute the rate at which mass enters the box as a result of fluid motion and, since mass must be conserved, we shall equate this to the rate of change with time of the mass contained within the cube.

In an interval of time dt the amount of mass transported into the

[5] The reader is now in a position to understand the footnote on page 181. Since f is the vorticity about the vertical due to the earth's rotation and

$$\zeta = \frac{\partial v_\theta}{\partial r} + \frac{V_\theta}{r}$$

is the vorticity of air flow relative to the earth, their sum is the absolute vorticity, or twice the angular velocity of the air as seen by an inertial observer. For the absolute vorticity to be zero the relative vorticity must exactly cancel f. This is so special a state that it rarely occurs.

box by flow in the x direction through the left-hand vertical face will be $(\rho u) dt\, dy\, dz$. The gain of mass in the same time by flow through the right-hand vertical face will be

$$-\left[\rho u + \frac{\partial(\rho u)}{\partial x} dx\right] dt\, dy\, dz$$

where the minus sign denotes the fact that a positive value of u means a loss in this case, and we have taken account of a possible variation in the quantity ρu along the x axis. The net gain of mass due to motion along the x axis will then be the algebraic sum of these two or

$$-\frac{\partial(\rho u)}{\partial x} dt\, dx\, dy\, dz$$

Similarly the net accretions of mass due to motion along the y and z axes will be, respectively:

$$-\frac{\partial(\rho v)}{\partial y} dt\, dx\, dy\, dz \qquad \text{and} \qquad -\frac{\partial(\rho w)}{\partial z} dt\, dx\, dy\, dz$$

The increase in mass inside the box during this time is $(\partial\rho/\partial t)\, dt\, dx\, dy\, dz$. Here the local derivative is used because the box is fixed in location. When we set this equal to the sum of all the contributions to an increase in mass we get

$$\frac{\partial\rho}{\partial t} = -\left[\frac{\partial(\rho u)}{\partial x} + \frac{\partial(\rho v)}{\partial y} + \frac{\partial(\rho w)}{\partial z}\right] \tag{13.13}$$

This is one form of the *equation of continuity* for mass.

There are a number of other useful forms of the continuity equation. If we expand the partial derivatives in (13.13) and rearrange we get

$$-\frac{1}{\rho}\frac{d\rho}{dt} = \frac{\partial u}{\partial x} + \frac{\partial v}{\partial y} + \frac{\partial w}{\partial z} \tag{13.14}$$

Equation (13.14) requires that the individual rate of change of density be proportional to the three-dimensional *velocity divergence*, while Eq. (13.13) requires that the local rate of change of density be proportional to what is called the three-dimensional *mass divergence*. That this should be so is physically understandable, since divergence will act to remove fluid from a given fixed volume.

It is frequently convenient to consider only the horizontal divergence, $\partial u/\partial x + \partial v/\partial y$. It was realized earlier that this was related to increases or decreases of the area encompassed by a curve of fluid particles. The exact quantitative relationship may be obtained by considering the infinitesimal area in Fig. 13.15 which at the initial moment has area $A = dx\ dy$. After a small finite time, Δt, the fluid particles bounding the area will have moved and the new area will be

$$\left(dx + \frac{\partial u}{\partial x} dx\, \Delta t\right)\left(dy + \frac{\partial v}{\partial y} dy\, \Delta t\right)$$

or

$$dx\ dy\left(1 + \frac{\partial u}{\partial x}\Delta t\right)\left(1 + \frac{\partial v}{\partial y}\Delta t\right) = dx\, dy\left[1 + \left(\frac{\partial u}{\partial x} + \frac{\partial v}{\partial y}\right)\Delta t + \frac{\partial u}{\partial x}\frac{\partial v}{\partial y}(\Delta t)^2\right]$$

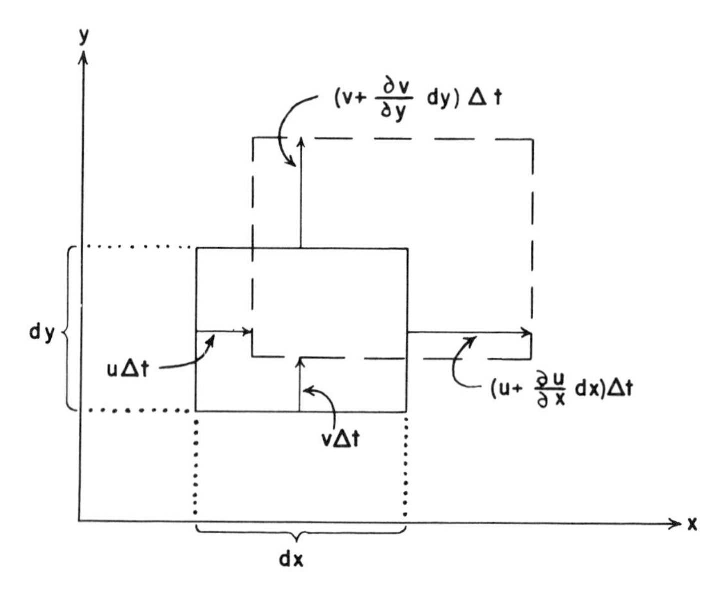

FIG. 13.15. Horizontal growth of an infinitesimal rectangle of moving fluid

The change in area is

$$\Delta A = dx\, dy\left[\left(\frac{\partial u}{\partial x} + \frac{\partial v}{\partial y}\right)\Delta t + \frac{\partial u}{\partial x}\frac{\partial v}{\partial y}(\Delta t)^2\right]$$

and, since $dx\ dy = A$, the rate of change of area is

$$\frac{\Delta A}{\Delta t} = A\left[\frac{\partial u}{\partial x} + \frac{\partial v}{\partial y} + \frac{\partial u}{\partial x}\frac{\partial v}{\partial y}\Delta t\right]$$

In the limit as Δt approaches zero we obtain

$$\frac{1}{A}\frac{dA}{dt} = \frac{\partial u}{\partial x} + \frac{\partial v}{\partial y} \tag{13.15}$$

Thus the horizontal divergence at a point is equal to the fractional rate of change of the area enclosed by a small chain of fluid particles surrounding the point.

A fluid in which the individual rate of change of density with time, $d\rho/dt$, differs from zero is said to be *compressible*. A fluid in which an individual parcel experiences no change of density with time is said to be *incompressible*. Although the atmosphere is actually compressible, there are many phenomena in which the compressibility plays no significant role. For these phenomena, and these alone, it is valid to neglect individual variations of density with time.

For an incompressible fluid, it is clear that conservation of mass requires the volume of a given fluid element to be constant. If we consider cylindrical fluid columns of depth D, the constant volume is DA. Thus,

$$\frac{1}{A}\frac{dA}{dt} = -\frac{1}{D}\frac{dD}{dt}$$

From Eq. (13.15), the equation of continuity becomes

$$-\frac{1}{D}\frac{dD}{dt} = \frac{\partial u}{\partial x} + \frac{\partial v}{\partial y} \tag{13.16}$$

Equations (13.15) and (13.16) may be used to eliminate the horizontal divergence wherever it appears in a problem. Equation (13.16) shows that the horizontal divergence of an incompressible fluid need not be zero even though the three-dimensional divergence must vanish. The horizontal divergence causes a simultaneous change in the depth and base area of a fluid column, and may be estimated from the rate at which these quantities change with time. Applications are often made of each of the various forms of the equation of continuity which have been derived.

13.7 The Complete Set of Equations Governing the Atmosphere

We are now in a position to set down a complete set of equations describing the behavior of our atmosphere. They consist, in part, of the equations of motion:

$$\frac{\partial u}{\partial t} + u\frac{\partial u}{\partial x} + v\frac{\partial u}{\partial y} + w\frac{\partial u}{\partial z} = 2\Omega\,(v\sin\phi - w\cos\phi) - \frac{1}{\rho}\frac{\partial p}{\partial x} + \frac{1}{m}F_x$$

$$\frac{\partial v}{\partial t} + u\frac{\partial v}{\partial x} + v\frac{\partial v}{\partial y} + w\frac{\partial v}{\partial z} = -2\Omega\,u\sin\phi - \frac{1}{\rho}\frac{\partial p}{\partial y} + \frac{1}{m}F_y$$

$$\frac{\partial w}{\partial t} + u\frac{\partial w}{\partial x} + v\frac{\partial w}{\partial y} + w\frac{\partial w}{\partial z} = 2\Omega\,u\cos\phi - \frac{1}{\rho}\frac{\partial p}{\partial z} - g + \frac{1}{m}F_z$$

and the equation of continuity:

$$\frac{\partial \rho}{\partial t} + u\frac{\partial \rho}{\partial x} + v\frac{\partial \rho}{\partial y} + w\frac{\partial \rho}{\partial z} = -\rho\left(\frac{\partial u}{\partial x} + \frac{\partial v}{\partial y} + \frac{\partial w}{\partial z}\right)$$

It will be noted that these are four equations in five dependent variables: u, v, w, ρ, p. A fifth independent relationship is needed in order to make this a theoretically adequate set of equations. This fifth expression is usually derived from some physical restriction on the system based upon the First Law of Thermodynamics:

$$dh = c_p dT - \frac{1}{\rho} dp$$

When $dh = 0$ the motion is adiabatic and the density of each individual parcel is determined solely by the pressure. Thus Poisson's equation is the fifth, physical equation. Other simple physical equations are those for an isothermal process, p/ρ = constant, and for an incompressible process, ρ = constant. Any such equation relating no more than two of the variables of state is called a *piezotropic equation*, and one such is necessary to complete the set of governing equations. In any theoretical investigation based upon these equations the specific piezotropic relationship to be chosen from those available depends upon the nature of the physical phenomena which one wishes to treat. The usual course is to select as realistic a piezotropic equation as possible, without causing the problem to become mathematically intractable.

One should distinguish carefully between the conditions of piezotropy and barotropy. While both consist of a relationship between two of the variables of state, the former relates them for an individual particle from time to time, whereas the latter relates them from point to point in the fluid at a given instant. As an example, consider an atmosphere which has the same temperature everywhere at a given moment, but in which individual parcels move adiabatically. Since the atmosphere is isothermal the barotropic relationship is $p = k\rho$, but the piezotropic relationship is the adiabatic law, $p = K\rho^{\gamma}$, where γ is the ratio of the specific heat capacities of air. If vertical motions take place the adiabatically moving air will change temperature and the atmosphere will no longer be isothermal. That is, the barotropic relationship will no longer hold and the fluid will become baroclinic even though it still obeys the adiabatic-piezotropic law. It can be seen that an initially barotropic fluid will remain so only if the equations of barotropy and piezotropy happen to be identical. Such a fluid is said to be *autobarotropic*.

The five governing equations given above are prognostic in nature since they involve derivatives with respect to time of the dependent variables. It is theoretically possible to insert observed conditions in the atmosphere at a given moment as initial conditions and to proceed to solve this set for the state of the atmosphere at a future time. Such a mathematical forecast is, of course, one of the chief goals of theoretical meteorology. The basic relationships, however, are nonlinear, partial differential equations of considerable generality. Such equations are so difficult that there is no way of obtaining solutions in terms of elementary functions without gross simplifications which vitiate the results to lesser or greater degree. This suggests solution by numerical methods using high-speed electronic computing machinery. This approach has developed rapidly in the last ten years and will be discussed in a later chapter.

PROBLEMS

1. Derive the equation of the streamlines of horizontal flow if $u = u_0$, $v = v_0 \cos ax$, where u_0, v_0, and a are constants. Sketch a few of the streamlines. Derive formulae for the divergence and vorticity of this motion. Where in the streamline pattern is the vorticity a maximum?

2. Derive the horizontal divergence and vorticity of the geostrophic wind assuming (a) f and ρ are constants; (b) $f = 2\,\Omega \sin(y/a)$ and ρ is constant, where a is the radius of the earth.

3. Show that the lines of constant horizontal wind speed (isotachs) are concentric circles in the case of (a) pure constant divergence, (b) pure constant vorticity, and (c) pure constant deformation.

4. Figure 13.16 is a diagram of circular streamlines in which the velocity

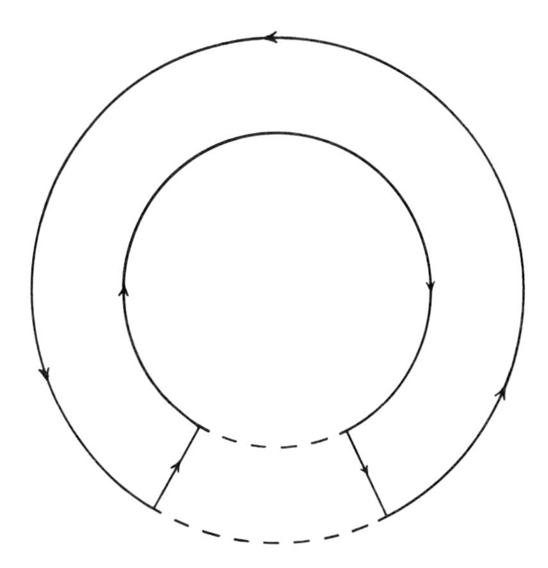

FIG. 13.16. Path of integration

is inversely proportional to distance from the center. Calculate the circulation around the closed path indicated by the heavy curve. The essential characteristic of this curve is that it does not enclose the center. What does your result prove about the vorticity of this motion at all points but the center?

5. Show that the equation of continuity for a compressible fluid may be written as

$$\frac{1}{V}\frac{dV}{dt} = \frac{\partial u}{\partial x} + \frac{\partial v}{\partial y} + \frac{\partial w}{\partial z}$$

where V is the volume of a fluid element.

CHAPTER 14

THE MECHANISM AND INFLUENCE OF PRESSURE CHANGES

14.1 The Tendency Equation

It is a well-known fact that in most latitudes at most times of the year the atmospheric pressure varies appreciably with time. Thus geostrophic or gradient balance is never reached and maintained for any length of time, but constant mutual readjustments of the wind and the changing pressure gradient prevail. What are the physical factors responsible for the variations of pressure with time? To answer this we shall assume the pressure to be simply the weight per unit area of the overlying air. That is, we consider the hydrostatic equation to be valid, and the pressure at some level h to be

$$p = \int_h^\infty \rho g \, dz$$

If we differentiate with respect to time and assume g constant we obtain

$$\frac{\partial p}{\partial t} = g \int_h^\infty \frac{\partial \rho}{\partial t} \, dz$$

where the differentiation operation may be performed under the integral since time and height are independent. We now substitute from the equation of continuity:

$$\frac{\partial p}{\partial t} = -g \int_h^\infty \left[\frac{\partial(\rho u)}{\partial x} + \frac{\partial(\rho v)}{\partial y} + \frac{\partial(\rho w)}{\partial z} \right] dz$$

The last term in the brackets can be integrated, with the result that

$$\frac{\partial p}{\partial t} = -g \int_h^\infty \left[\frac{\partial(\rho u)}{\partial x} + \frac{\partial(\rho v)}{\partial y} \right] dz + g(\rho w)_h \tag{14.1}$$

The upper limit gives zero for the vertical velocity term (provided w remains finite) because ρ tends toward zero as z approaches infinity.

Equation (14.1) is called the *tendency equation*, and it asserts that the pressure tendency at the level h is due to, first, the integrated effect of horizontal mass divergence in the column above and, second, transport of mass vertically through the bottom of the column. Basically this means the pressure can change only if the mass of the column overhead changes, which can occur in either of two ways—mass can be imported or exported through the sides of the column, or it can be imported or exported by means of vertical motion of air through the base of the column. If the base of the column is at the solid surface of the earth, assumed to be level, we may be certain that the vertical velocity is zero, since physically we cannot have any separation between the air and an adjacent material boundary. In this case pressure changes can arise only from mass divergence in the vertical column.

This result immediately suggests a practical use for this prognostic equation. All we need do is determine the mass divergence through an appreciable depth of the atmosphere, sum it up, and thereby obtain the instantaneous rate of change of pressure with time. This value may be suitably extrapolated to obtain a forecast of the pressure field at some future time. This simple procedure, however, is at present incapable of adequate execution. The reasons are twofold. First, measurements of the large-scale mass divergence in middle latitudes show that even in a layer only a few kilometers thick it will yield pressure tendencies of several tens of millibars per three hours. Such magnitudes generally persist throughout the troposphere and low stratosphere. Since observed surface tendencies are usually much smaller than this it is clear that successive layers of the atmosphere must have mass-divergence values of opposite sign so that rapid accumulations or depletions of the total mass in a vertical column cannot occur (see Fig. 14.1). This means that the integral of Eq. (14.1) represents the relatively small difference between at least two large numbers. Second, the accuracy with which we can presently determine mass divergence in the atmosphere is not great. Generally, therefore, the required small difference cannot be measured adequately. These unfortunate properties are not limited to the tendency equation;

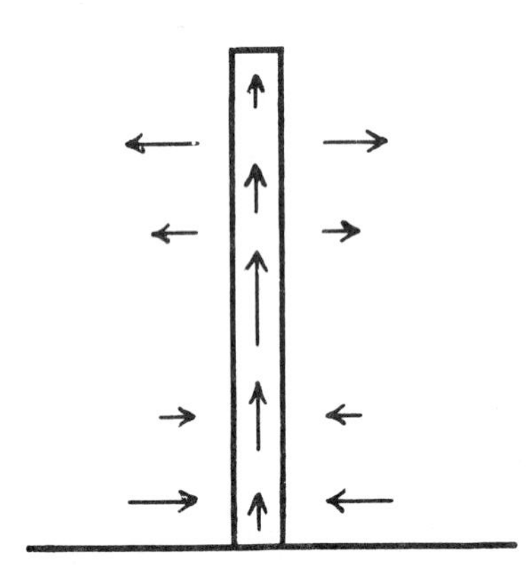

FIG. 14.1. Vertical compensation of divergence in a column

it is not uncommon in meteorological phenomena that the required quantities are small differences between large, imperfectly measured variables.

One further example of the frustrating nature of the tendency equation may be had by expanding the integrand of Eq. (14.1):

$$\frac{\partial p}{\partial t} = -g \int_h^\infty \left(\frac{\partial u}{\partial x} + \frac{\partial v}{\partial y}\right) \rho \, dz - g \int_h^\infty \left(u \frac{\partial \rho}{\partial x} + v \frac{\partial \rho}{\partial y}\right) dz + g(\rho w)_h$$

The first integral involves the horizontal velocity divergence, which is large, vertically compensated, cumbersome to measure, and known with only poor accuracy. The second integral involves the horizontal advection of mass, which is more easily measured, far less compensated vertically, and often small compared with the first integral. The third term may likewise be large and it is difficult to measure. Thus those quantities which we can determine adequately are not ordinarily important, and those which are significant we cannot measure properly.

14.2 The Bjerknes-Holmboe Theory

The discussion given above indicates the futility, with present measurement techniques, of a direct, prognostic application of the tendency equation. It does not preclude, however, the use of this equation in the construction of a theory of the motion of pressure systems. All that is basically required is a way of relating the horizontal distribution of net mass divergence to a pattern of highs and lows. The highs will then move towards regions of rising pressure (mass convergence), and the lows towards regions of falling pressure (mass divergence). We shall discuss such a theory from a heuristic and qualitative point of view, although it has been given a quantitative treatment by its authors. It is commonly known as the *Bjerknes-Holmboe theory*.[1]

It would be fruitless to base such a theory on the assumption of geostrophic flow since we have found (Chap. 13, Problem 2) that the divergence of the geostrophic wind is essentially zero (except for the variations of f and ρ). However, the divergence of the gradient wind is not zero. Let us therefore examine the pattern of divergence associated with certain simple pressure fields if gradient flow obtains. We shall consider (1) sinusoidal isobars extending from west to east, (2) circular,

[1] J. Bjerknes and J. Holmboe, On the Theory of Cyclones. *J. Meteor.*, 1, 1944, p. 1.

concentric isobars, and (3) a superposition of sinusoidal isobars at high levels and circular isobars at low levels.

In gradient flow the motion is along the isobars, but even if the isobars are parallel and equally spaced there may be changes in speed along them because of two effects. The first of these is the curvature effect. As we have already shown, for a given isobar spacing at a given latitude the gradient wind for a low is less than the geostrophic value, while the magnitude of the gradient wind in a high is greater than either of these. Thus, owing to this curvature effect we expect a distribution of wind speeds as shown in Fig. 14.2. If such a pattern existed through

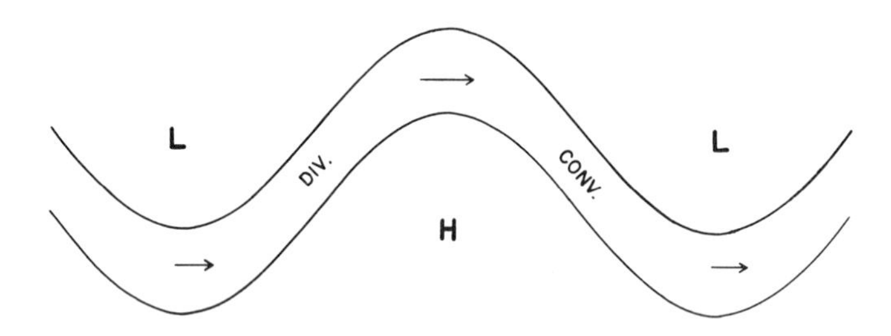

FIG. 14.2. The distribution of divergence owing to the curvature effect in a pattern of sinusoidal isobars

a considerable layer it would lead to falling pressures east of the troughs and rising pressure east of the ridges. We would then expect the pressure systems to move eastward. The second effect is that due to latitude. If all other parameters are kept the same but the latitude is increased, both the geostrophic and gradient wind speeds decrease. This may be seen by writing the gradient wind equation in the form

$$c = \frac{1}{f}\left(\frac{1}{\rho}\frac{\partial p}{\partial r} - \frac{c^2}{r}\right)$$

Usually c^2/r is a small correction term, so it is accurate to say that the speed varies inversely as the latitude. Due to this influence of latitude we expect a distribution of wind speeds as shown in Fig. 14.3. We would then expect the pressure systems to move westward, the curvature and the latitude effects thus working in opposite directions. For low wind speeds, long wave lengths, and given fixed amplitude, one may expect the curvature term to be small; thus the waves will move westward as determined by the latitude effect. For high wind speeds, short wave lengths, and the same fixed amplitude, the curvature term should dominate and the waves should move eastward. It is clear that there is a certain critical wind speed at each wave length for which the waves will be stationary. These qualitative results compare favorably with synoptic experience at upper levels (e.g., 500 mb) where the contour pattern in often quasi-sinusoidal in nature. We frequently see these

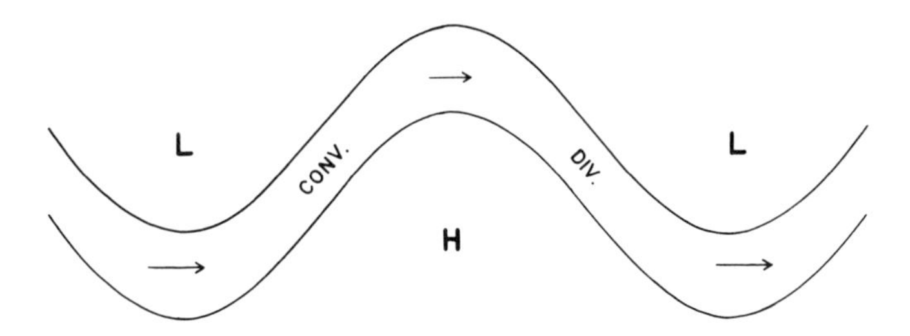

FIG. 14.3. The distribution of divergence owing to the latitude effect in a pattern of sinusoidal isobars

waves of fairly long length moving only slowly eastward. In the Bjerknes-Holmboe theory, this is interpreted as a slight predominance of the curvature effect.

If we now turn to the case of equally spaced, concentric, circular isobars as in Fig. 14.4 we see that the curvature effect will be essentially

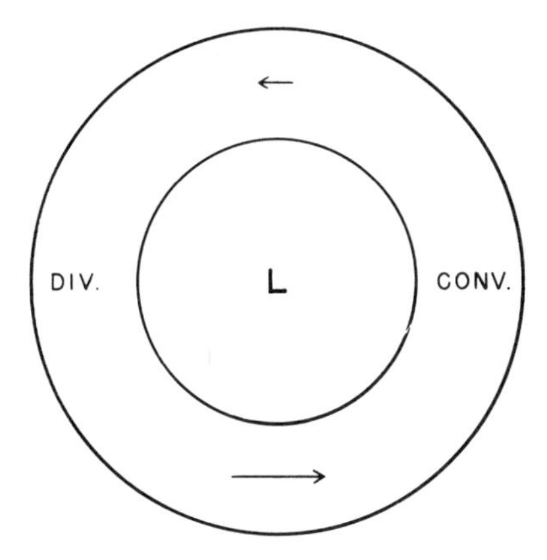

FIG. 14.4. The distribution of divergence in a pattern of circular concentric isobars

the same everywhere, but the latitude effect will produce higher winds on the equatorward side. Thus convergence with pressure rises will occur to the east and divergence with pressure falls to the west of a low. Such systems then will be expected to move westward.

We may now build up a rather realistic schematic system by considering closed isobars in the low levels and sinusoidal isobars aloft, with the proper displacement of the upper trough relative to the center below, as in Fig. 14.5. The upper sinusoidal pattern is assumed to be the usual one observed, in which the curvature effect is somewhat larger than the latitude effect, so divergence appears east of the trough line and convergence ahead of the ridge line. Thus, east of the center of the surface low we find low-level convergence and upper-level divergence with ascending motion. At the center of the surface low we find low-level frictional convergence and upper-level divergence again with ascending motion. To the west of the upper-level trough we find the

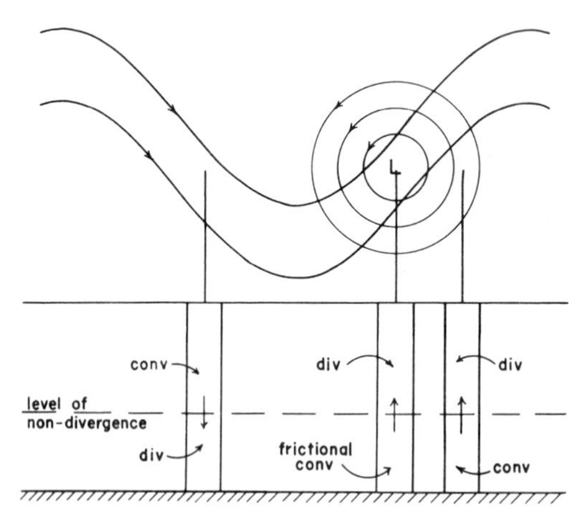

Fig. 14.5. A plan view of upper-level sinusoidal isobars superimposed on lower-level circular isobars and the resulting vertical distribution of divergence

opposite—low-level divergence and upper-level convergence with descending motion. This picture incorporates, then, the vertical compensation which we previously found to be required in the actual atmosphere. Because of this compensation it is not possible to decide immediately whether the system will move eastward or westward; this much may be said qualitatively. If the level of transition from divergence of one sign to divergence of the opposite sign (*level of nondivergence*) is low the high altitude pattern will predominate and the system will move eastward. If the level of nondivergence is high the low altitude pattern will predominate and the system will move westward.

Inspection of Fig. 14.5 reveals a considerable qualitative agreement with synoptic experience (compare the distribution of vertical velocities with the normally observed distribution of clouds and precipitation in middle-latitude storms). Furthermore, the interplay of a few simple physical factors in this model is clearly capable of complexity of result compatible with the observed complexity of behavior of the atmosphere.

14.3 The Isallobaric Wind

Now that we have some knowledge of the properties of the wind under balanced forces and also some knowledge of the factors that cause pressure and wind systems to change with time, let us investigate the effect upon the wind of time variations of the pressure field. We shall approach this matter in a particularly simple fashion. Let the motion to be studied be horizontal and frictionless. The equations of motion in a tangent plane are:

$$\frac{du}{dt} = fv - \frac{1}{\rho}\frac{\partial p}{\partial x}$$

$$\frac{dv}{dt} = -fu - \frac{1}{\rho}\frac{\partial p}{\partial y}$$

We shall consider straight isobars which are oriented east-west. This involves no loss of generality except through the omission of curvature effects. If we now assume that the pressure-gradient force is a linear function of time we have

$$-\frac{1}{\rho}\frac{\partial p}{\partial x} = 0; \qquad -\frac{1}{\rho}\frac{\partial p}{\partial y} = P = P_0 + \left(\frac{\partial P}{\partial t}\right)_0 t$$

Here P represents the north-south pressure-gradient force per unit mass; a quantity with subscript zero is a constant evaluated at the initial moment. The equations of motion become

$$\frac{du}{dt} = fv$$

$$\frac{dv}{dt} = -fu + P$$

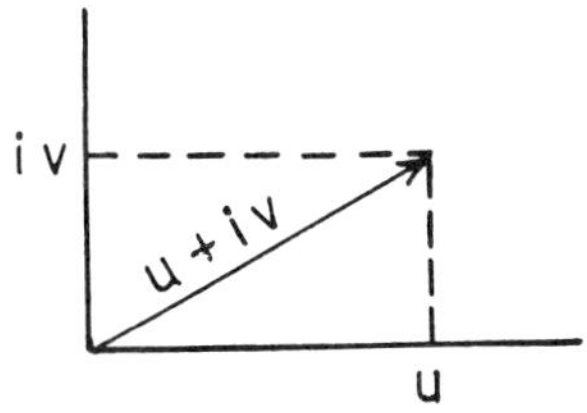

We shall now combine these two equation into one by multiplying the second equation by i (the square root of -1) and adding the two expressions. The result may be written

$$\frac{d}{dt}(u + iv) = -if(u + iv) + iP$$

FIG. 14.6. Two-dimensional velocity vector in the complex plane

It may been seen that a quantity of the form $u + iv$ is analogous to a two-dimensional vector. In the complex plane of Fig. 14.6 we plot all imaginary quantities on the ordinate and all real quantities on the abscissa. Thus $u + iv$ is the directed line, or vector, shown. This kind of compression is quite convenient in two-dimensional problems. We shall denote the complex velocity by V, which will be the dependent variable. After solving the differential equation for V we shall separate the real and imaginary parts in order to obtain final solutions for u

and v. Since we deal with the individual time derivative, we are investigating the behavior of individual parcels of air, which is a problem in particle dynamics.

The equation to be solved, $dV/dt = -\mathrm{i}fV + \mathrm{i}P$, is a first order, linear, ordinary, differential equation with constant coefficients, provided f is assumed to be constant, as we do here. Such an equation may be solved by standard methods:

$$Ve^{\mathrm{i}ft} = \int \mathrm{i}Pe^{\mathrm{i}ft}\,dt + K$$

where K is a constant of integration. If we insert our previous assumption that P is a linear function of time, integrate, and rearrange we get

$$V = \frac{P_0}{f} + \frac{1}{f}\left(\frac{\partial P}{\partial t}\right)_0 t + \frac{\mathrm{i}}{f^2}\left(\frac{\partial P}{\partial t}\right)_0 + Ke^{-\mathrm{i}ft}$$

This solution may be verified by direct substitution into the differential equation. In order to evaluate K we shall adopt the initial condition that at $t = 0$ the wind is geostrophic, i.e., $V_0 = P_0/f$. Then

$$V = \frac{P}{f} + \frac{\mathrm{i}}{f^2}\left(\frac{\partial P}{\partial t}\right)_0 \left[1 - e^{-\mathrm{i}ft}\right]$$

The next step is to equate the real and imaginary parts of V to the real and imaginary parts of the solution on the right. To do this we shall use Euler's expression for an imaginary exponential:

$$e^{\pm \mathrm{i}\theta} = \cos\theta \pm \mathrm{i}\sin\theta$$

With this equality we find, as our final form of the solution:

$$\begin{aligned} u &= -\frac{1}{\rho f}\frac{\partial p}{\partial y} + \frac{1}{\rho f^2}\frac{\partial}{\partial y}\left(\frac{\partial p}{\partial t}\right)\sin ft \\ v &= -\frac{1}{\rho f^2}\frac{\partial}{\partial y}\left(\frac{\partial p}{\partial t}\right) + \frac{1}{\rho f^2}\frac{\partial}{\partial y}\left(\frac{\partial p}{\partial t}\right)\cos ft \end{aligned} \tag{14.2}$$

This solution is for the case of isobars and isallobars (lines of constant pressure tendency) which are oriented east-west. Equations (14.2) indicate that there will be three kinds of contributions to the velocity field. First, as represented by the last term in each equation, we have a periodic time variation with period $2\pi/f$, the period of inertia motion. This is, in a way, a fundamental mode of vibration of horizontally moving particles on the rotating earth. It arises in this case because the initial equilibrium between Coriolis and pressure-gradient forces has been upset by the changing pressure field. We have already seen that the energy of such oscillations is dissipated rapidly by friction and is dispersed quickly into the surrounding fluid. As a practical matter, therefore, this periodic behavior may be neglected. Second, as represented by the first term on the right in the first of

Eq. (14.2), we have a geostrophic component—that is, there is one part of the velocity field which is in geostrophic balance with the pressure gradient. Third, as represented by the first term on the right of the second of Eq. (14.2), we have a component whose magnitude depends upon the gradient of the pressure tendency and which blows perpendicular to the isallobars towards falling pressure. This component is the *isallobaric wind*, first derived by Brunt and Douglas.[2] Note that f^2 is involved so that no change of sign occurs in crossing the equator. This isallobaric wind blows toward falling pressure in both hemispheres.

Physically this is precisely the main effect that one would expect in a field of changing pressure gradient. A part of the wind will always be in balance with the pressure gradient at a given moment, but another part will not be in balance and will therefore blow toward the region of greatest negative pressure anomaly. This is toward the falling pressure.[3] Furthermore, this transport of mass towards the region of most rapidly falling pressure tends to counteract the pressure fall. Thus we have another example of the strong tendency of the atmosphere to restore equilibrium (Le Chatelier's principle).

A number of other theories of the isallobaric wind have been advanced in which special assumptions led to different results than those obtained here and by Brunt and Douglas. However, there is no real need, either observational or logical, for these special assumptions.

Measurements on weather maps indicate that the magnitude of the isallobaric wind is often not negligible. It may reach speeds of 5 m s^{-1} and may be used to account for the bad weather often encountered in the vicinity of strong negative isallobaric centers.

PROBLEMS

1. Calculate the magnitude of the pressure tendencies in mb per three hours at the base of an atmospheric column 3 km thick due to (a) horizontal velocity divergence, (b) horizontal mass advection, and (c) vertical velocities, given the following reasonable properties of the atmosphere:

horizontal velocity divergence $= 3 \times 10^{-5}\ s^{-1}$
horizontal wind speed $= 15\ m\ s^{-1}$

---

[2] D. Brunt and C. K. M. Douglas, On the Modification of the Strophic Balance. *Mem. Roy. Meteor. Soc.*, 3, 1928, p. 22.

[3] The solutions for cases in which the pressure-gradient force depends upon time in certain more complex ways than the linear dependence assumed here have been given by G. E. Forsythe, Exact Particle Trajectories for Non-Viscous Flow in a Plane with a Constant Coriolis Parameter. *J. Meteor.*, 6, 1949, p. 337.

horizontal temperature gradient = 1° C per degree of latitude
angle between wind and isotherms = 30°
average pressure = 700 mb
average temperature = −3° C
vertical wind speed at bottom of column = + 3 cm s^{-1}
vertical wind speed at top of column = + 1 cm s^{-1}

2. Prove that $e^{i\theta} = \cos\theta + i \sin\theta$. (Hint—use series expansion.)

3. If the isallobaric wind is given by

$$u = -\frac{1}{\rho f^2}\frac{\partial}{\partial x}\left(\frac{\partial p}{\partial t}\right); \qquad v = -\frac{1}{\rho f^2}\frac{\partial}{\partial y}\left(\frac{\partial p}{\partial t}\right)$$

derive the horizontal divergence and the vorticity about the vertical of the isallobaric wind, assuming ρ and f to be constant.

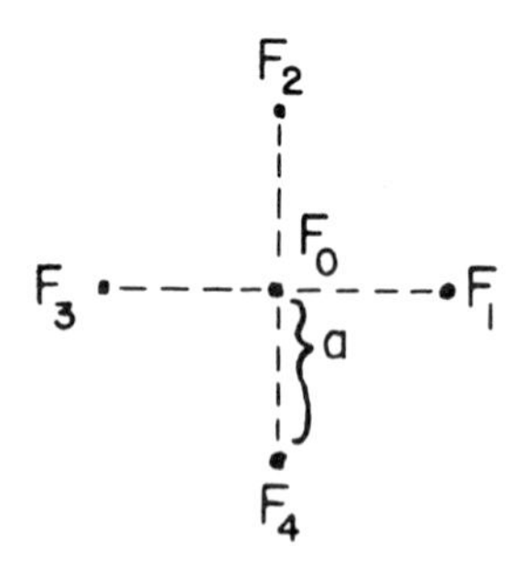

FIG. 14.7. Finite difference grid for evaluation of the LaPlacian of a function F

4. In a number of physical problems a function $F(x, y)$ may appear repeatedly in the form $\partial^2F/\partial x^2 + \partial^2F/\partial y^2$. This is called the two-dimensional LaPlacian of F. To evaluate the LaPlacian from a mapped field of F it is necessary to replace derivatives by ratios of finite differences. If the values of F are read at the five points in the grid of Fig. 14.7, show that

$$\frac{\partial^2F}{\partial x^2} + \frac{\partial^2F}{\partial y^2} \approx \frac{F_1 + F_2 + F_3 + F_4 - 4F_0}{a^2}$$

where a is the constant distance separating points 1, 2, 3, 4 from point 0.

5. Calculate the divergence of the isallobaric wind if at the center of an isallobaric system the tendency is − 4 mb/3 hr, and 2° of latitude away in all directions the tendency is − 3 mb/3 hr. Assume $f = 10^{-4}$ s^{-1}, density $= 10^{-3}$ g cm^{-3}.

6. If the divergence obtained in answer to problem 5 above were to persist through a layer three kilometers thick, calculate the pressure tendency it would produce by this effect alone.

7. Show that the magnitude of the normal gradient wind speed decreases with increasing latitude for a constant pressure-gradient force per unit mass and constant radius of curvature, without further approximation.

CHAPTER 15

SURFACES OF DISCONTINUITY

15.1 Discontinuities

In the real atmosphere there are probably no true spatial discontinuities in any of its properties so long as one observes at a scale greater than microscopic. Nevertheless one frequently can find zones in which some property changes value considerably in a small distance. The larger the scale of view the more nearly do such transition zones approximate discontinuities and the more convenient it is to treat them as such.

We often wish to take a synoptic view of the atmosphere over an area as large as the North American continent. Since the over-all linear dimensions of such an area approximate 3000 miles, any transition zone of the order of 50 miles wide will appear on such a map essentially as a discontinuity. This, of course, is the situation with atmospheric fronts.

We shall introduce certain standard nomenclature concerning discontinuities. If we consider a property which is itself discontinuous, that property is said to possess a *zero order discontinuity*. Density, for example, may be such a property. If however, the density is continuous but its first derivative with respect to distance is discontinuous, then it is said to possess a *first order discontinuity*. In general a discontinuity in the *n*th derivative of a quantity is described as an *nth order discontinuity* of that quantity. A graphical example is given in Fig. 15.1.

Each line of discontinuity in a fluid is to be treated as a boundary at which certain conditions must apply. The first of these is the *dynamic boundary condition*. This requirement is based on the physical impossibility of infinite forces. Thus, if at an internal boundary there were a zero order discontinuity in the pressure we would have to deal with a finite change in pressure through an infinitely small distance. The infinite pressure-gradient force that would result is not permissible, so this dynamic condition requires that pressure must be continuous through an internal boundary, that is, the pressures immediately on either side of a discontinuity must be the same.[1]

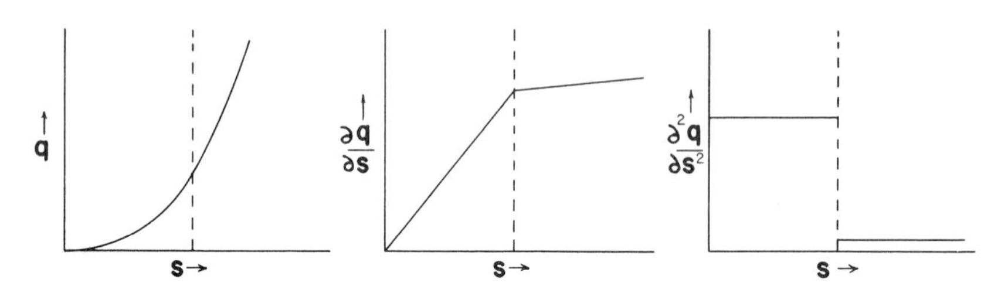

FIG. 15.1. Graphs of a function $q(s)$ and its first and second derivatives when a second order discontinuity in q exists

The second requirement is the *kinematic boundary condition.* This concept is based upon the physical supposition that no pockets of vapor or near vacuum will form during a fluid motion. When applied to a discontinuity this requirement means that the fluid particles immediately on either side of the internal boundary must have the same component of velocity perpendicular to the boundary, for if they did not a cavity (infinite divergence) would develop at the discontinuity. For an ideal or frictionless fluid there is no restriction whatever on the two components of motion tangent to the internal boundary. When viscosity is introduced even these tangential components on either side of the boundary must be equal. This is called the *no-slip boundary condition.*

15.2 Fronts

We shall define an idealized front as a zero order discontinuity in the density. Without loss of generality we may consider the intersection of such a front with the ground as lying east-west with the dense air (denoted by subscript D) to the north, and the light air (denoted by subscript L) to the south. In general the surface of discontinuity between

[1] In recent years analyses on a scale smaller than the synoptic one have revealed rapid changes in pressure over small distances. It is convenient to treat these "pressure jumps" as zero order discontinuities in the pressure despite the requirements of the dynamic boundary condition.

the two air masses may have a certain inclination in a vertical north-south plane. We shall calculate this inclination from the dynamic boundary condition: $p_D - p_L = 0$. Differentiation along the frontal surface yields

$$d(p_D - p_L) = \left[\left(\frac{\partial p}{\partial x}\right)_D - \left(\frac{\partial p}{\partial x}\right)_L\right] dx + \left[\left(\frac{\partial p}{\partial y}\right)_D - \left(\frac{\partial p}{\partial y}\right)_L\right] dy + \left[\left(\frac{\partial p}{\partial z}\right)_D - \left(\frac{\partial p}{\partial z}\right)_L\right] dz = 0$$

where dx, dy, and dz represent components of an increment of distance along the surface of discontinuity. Since the front is parallel to the x axis the dynamic boundary condition also requires that

$$(\partial p/\partial x)_D - (\partial p/\partial x)_L = 0 \tag{15.1}$$

This simply says that as one moves parallel to the x axis the pressures on either side must remain equal. When this is inserted above we have the condition that the pressures shall remain equal as one moves vertically and north-south:

$$\frac{dz}{dy} = -\frac{(\partial p/\partial y)_D - (\partial p/\partial y)_L}{(\partial p/\partial z)_D - (\partial p/\partial z)_L} \tag{15.2}$$

Thus the dynamic boundary condition is now expressed as a requirement that the horizontal and vertical pressure gradients in the two air masses conform to the slope of the front. In the case we are considering the frontal slope is positive, since the light air must overlie the dense air as in Fig. 15.2. From the hydrostatic equation we know that the denominator of Eq. (15.2) is negative; therefore

$$\left(\frac{\partial p}{\partial y}\right)_D > \left(\frac{\partial p}{\partial y}\right)_L \tag{15.3}$$

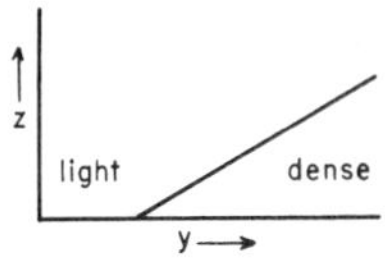

FIG. 15.2. Vertical cross-section through an idealized front

Thus we have the result from Eq. (15.1) that the horizontal pressure variations along a front are equal in the two air masses, but from Eq. (15.3) the horizontal pressure variation perpendicular to the front is greater in the dense air than in the light air.

These results mean that the isobars intersecting a front must change inclination sharply so that a kink is formed, with the smaller angle towards low pressure. This may be seen graphically by setting up various combinations of the conditions expressed in Eqs. (15.1) and (15.3). Examples are given in Fig. 15.3 for the case of pressure increasing to the north in the cold air, and increasing to the south in the warm air. This result may be proved similarly for the cases where pressure increases to the north in both air masses, where it decreases to the north in both air masses, and where the pressure remains constant with latitude in one of the air masses. This exhausts all the possibilities and leads to the rule of weather-map analysis that isobars crossing a front must

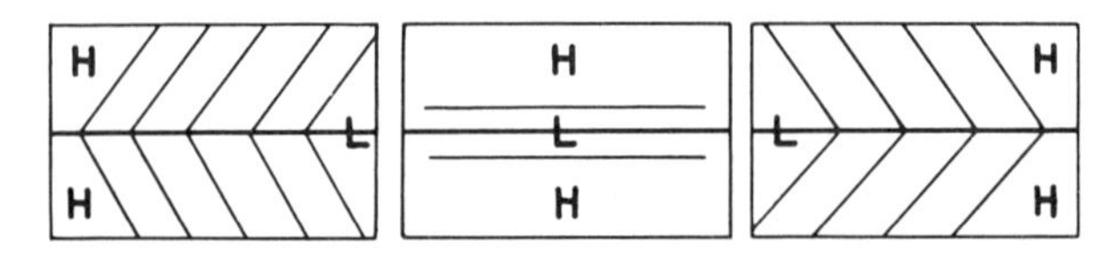

FIG. 15.3. Pressure distribution in the horizontal plane around an east-west front for

$$\left(\frac{\partial p}{\partial y}\right)_D > 0 > \left(\frac{\partial p}{\partial y}\right)_L$$

and for the three cases from left to right of

$$\frac{\partial p}{\partial x} < 0, \quad \frac{\partial p}{\partial x} = 0, \quad \frac{\partial p}{\partial x} > 0$$

kink towards the high pressure. This rule is based upon the properties of discontinuities and the hydrostatic equation. It is applicable without change in either hemisphere.

15.3 Fronts in a Geostrophic Wind Field

When the wind is geostrophic the results of the previous section indicate that the wind must have a cyclonic shear across a front. Since the pressure gradient along a front is the same in the two air masses, the geostrophic wind component perpendicular to the front is essentially the same in the two air masses (except for the difference in density). This is consistent with the kinematic boundary condition. The shear is confined to the component of the wind parallel to the front and is cyclonic.

When the geostrophic and hydrostatic assumptions are made, Eq. (15.2) becomes

$$\frac{dz}{dy} = \frac{f}{g} \frac{(\rho u)_L - (\rho u)_D}{\rho_D - \rho_L}$$

If we substitute for density from the equation of state and remember that the pressures are the same on either side of the front we get

$$\frac{dz}{dy} = \frac{f}{g} \frac{T_D u_L - T_L u_D}{T_L - T_D}$$

If the air is moist one should use the virtual temperature instead of the actual temperature.

The percentage difference in absolute temperature across a front is ordinarily small compared to the percentage difference in wind. Therefore a mean temperature, $\bar{T}$, may be extracted from the numerator:

$$\frac{dz}{dy} \approx \frac{f\bar{T}}{g} \frac{u_L - u_D}{T_L - T_D} \tag{15.4}$$

Equation (15.4) is the equation for the equilibrium slope of a front under geostrophic and hydrostatic conditions first derived by Margules.[2] Note that the greater the temperature contrast the smaller the equilibrium slope. This relationship exists because the minimum potential energy state of such a system occurs when the cold air lies completely under the warm air with zero slope of the discontinuity. The greater the temperature contrast, other things being equal, the closer the approach to this final state. On the other hand, the greater the wind shear across the front the greater the slope. This is because the only factor considered which keeps the cold air from spreading out under the warm air in response to the horizontal pressure gradients (so as to minimize potential energy) is the Coriolis force which balances the pressure gradients. One might say that the Coriolis force keeps the fronts inclined to the horizontal. These ideas do not apply, of course, near the equator.

These results have been for a front oriented east-west. The physical significance and the mathematical form are unaltered, except for sign, if one considers any other orientation of the front.

15.4 Fronts as Zones of Transition

In the actual atmosphere there is not a discontinuity in temperature or density at a front but a rapid transition from one temperature to another as in Fig. 15.4. Here we have two first order discontinuities of temperature instead of one zero order discontinuity.

The slope of such a transition zone is simply the vertical thickness of the zone divided by its horizontal width, $\Delta z_F/\Delta y_F$. If we write for the vertical geostrophic wind shear through the zone $\Delta u/\Delta z_F$ and call the temperature difference across the front $-\Delta T$ the approximate thermal wind equation (12.12) may be written

$$\frac{\Delta z_F}{\Delta y_F} = \frac{f\bar{T}}{g} \frac{\Delta u}{\Delta T}$$

[2] M. Margules, Über Temperaturschichtung in stationär bewegter und ruhender Luft. *Met. Zeit., Hann-Vol.*, 1906, p. 293.

This is Margules' formula for a transition layer rather than for a true discontinuity. We have shown, therefore, that the wind variation through a continuous frontal layer has the same form as the wind variation through an idealized discontinuous front, in the absence of accelerations.

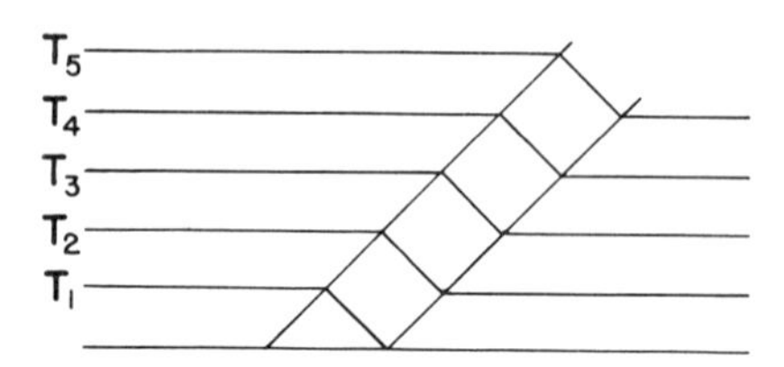

FIG. 15.4. Temperature distribution in a vertical cross-section through a zone of transition

The frontal slopes indicated by Margules' equation range from 1/50 to 1/400 which is the approximate observed range of slope of actual fronts. This is to be contrasted to the much smaller slopes of isobaric surfaces which rarely exceed 1/1000. Although the equilibrium theory gives the right order of magnitude of the slope of a front it is not to be expected that it will predict the slope of a front accurately when it is subject to significant non-geostrophic effects such as friction and unbalanced accelerations. The effects of curvature (gradient flow) have been investigated.[3] The result is that a correction factor should be applied to the right side of Eq. (15.4) of the form

$$\left(1 + \frac{c_L + c_D}{fr}\right)$$

where c_L and c_D are the winds in the two air masses, and r is the radius of curvature. This correction factor does not ordinarily differ greatly from one, but can be appreciable when strong sharply curved flows occur.

15.5 The Tropopause

The boundary between the troposphere and the stratosphere is a sloping surface which separates two regions of markedly different lapse rate. Therefore, it is a suitable idealization of the tropopause to state

[3] H. KOSCHMIEDER, *Dynamische Meteorologie*, Ak. Verlag., Leipzig, 1933, p. 252.

that it is a surface across which there is a discontinuity in lapse rate (first order discontinuity in temperature). From the dynamic boundary condition the pressure must be continuous across the tropopause, just as in the case of a front. Since both pressure and temperature are continuous it follows from the equation of state that density is continuous also, provided we neglect possible discontinuities in water vapor content.

We shall consider the case in which the intersection of the tropopause with a horizontal surface is parallel to the x axis, as in the earlier discussion of fronts. Then, by the same argument which led from the dynamic boundary condition to Eq. (15.2) we obtain for the slope of the tropopause

$$\frac{dz}{dy} = \frac{(\partial p/\partial y)_T - (\partial p/\partial y)_S}{g(\rho_T - \rho_S)} \tag{15.5}$$

where a subscript T refers to the troposphere and a subscript S to the stratosphere. Here the vertical pressure gradients have been replaced by means of the hydrostatic equation. Since ρ is continuous the denominator of Eq. (15.5) is zero and, if the slope is to differ from 90° as it almost always does, then

$$\left(\frac{\partial p}{\partial y}\right)_T = \left(\frac{\partial p}{\partial y}\right)_S$$

From the dynamic boundary condition directly

$$\left(\frac{\partial p}{\partial x}\right)_T = \left(\frac{\partial p}{\partial x}\right)_S$$

and, from the hydrostatic equation and the requirement that ρ is continuous,

$$\left(\frac{\partial p}{\partial z}\right)_T = \left(\frac{\partial p}{\partial z}\right)_S$$

Since all three components are continuous we conclude that the total pressure gradient is continuous across the tropopause.

We can now repeat the derivation of the slope of such a surface starting with the condition that the pressure gradients, instead of the pressure, be continuous. The procedure is entirely parallel to the development of Eq. (15.2) and since the function kept continuous is one order of differentiation higher, we find that there is a second order discontinuity in pressure across the tropopause in place of the first order discontinuity in pressure across a front. That is, the existence of a first order discontinuity in temperature implies that the curvature of the isobars must change abruptly at the tropopause.

More important properties of the tropopause may be demonstrated from the continuity of temperature. Since the argument is exactly the same as in the case of a front, except that T instead of p is the continuous variable, the result is Eq. (15.2) with T in place of p:

$$\frac{dz}{dy} = -\frac{(\partial T/\partial y)_T - (\partial T/\partial y)_S}{(\partial T/\partial z)_T - (\partial T/\partial z)_S} \tag{15.6}$$

Since the lapse rate in the stratosphere is nearly zero and since $(\partial T/\partial z)_T = -\gamma_T$ is not zero, this may be written

$$\frac{dz}{dy} \approx \frac{(\partial T/\partial y)_T - (\partial T/\partial y)_S}{\gamma_T}$$

We shall consider the normal case in which γ_T is positive and the tropopause slopes downward towards the pole. Then

$$\left(\frac{\partial T}{\partial y}\right)_T < \left(\frac{\partial T}{\partial y}\right)_S$$

We shall also require that temperature decrease poleward in the troposphere and increase poleward in the stratosphere as normally observed. A vertical section through the tropopause for these circumstances is shown in Fig. 15.5*a*. A horizontal temperature distribution at the level of the dashed line *AB* consistent with this vertical picture is shown in Figure 15.5*b*.

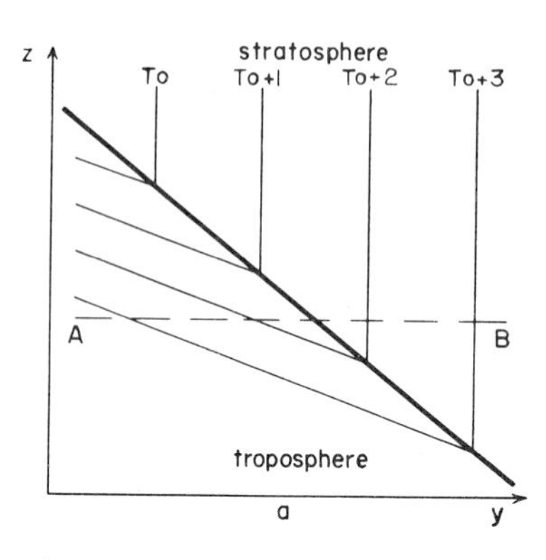

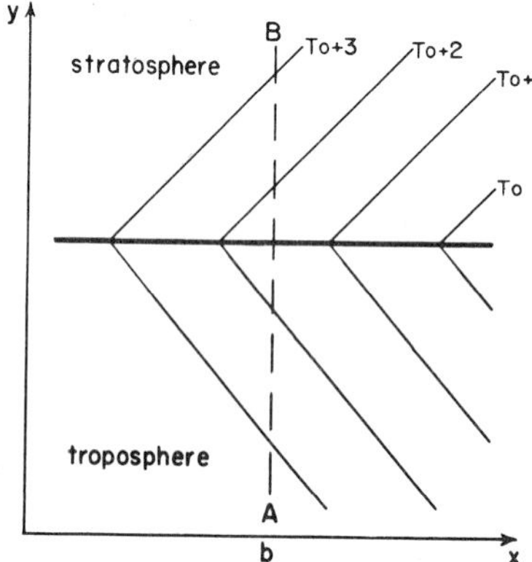

FIG. 15.5. (*a*) Vertical cross section and (*b*) plan view of the temperature distribution around the tropopause (heavy line)

Note that for this normal case the tropopause is a line of relatively low temperature in both vertical and horizontal views. Since $(\partial T/\partial y)_T < 0$ the geostrophic west wind increases up to the tropopause, and since $(\partial T/\partial y)_S > 0$ the geostrophic west wind decreases through the stratosphere. Thus the tropopause is usually a level of maximum wind speed (jet).

PROBLEM

1. The pressure around an east-west front may be distributed horizontally in any of fifteen distinct ways. This is because there are three possibilities for the pressure gradient along the front ($\partial p/\partial x < 0$; $\partial p/\partial x = 0$; $\partial p/\partial x > 0$), each

of which may occur with any of five possibilities for the pressure gradients perpendicular to the front:

$$\left(\frac{\partial p}{\partial y}\right)_D > \left(\frac{\partial p}{\partial y}\right)_L > 0$$

$$\left(\frac{\partial p}{\partial y}\right)_D > \left(\frac{\partial p}{\partial y}\right)_L = 0$$

$$\left(\frac{\partial p}{\partial y}\right)_D > 0 > \left(\frac{\partial p}{\partial y}\right)_L$$

$$\left(\frac{\partial p}{\partial y}\right)_D = 0 > \left(\frac{\partial p}{\partial y}\right)_L$$

$$0 > \left(\frac{\partial p}{\partial y}\right)_D > \left(\frac{\partial p}{\partial y}\right)_L$$

Three of these cases were worked out in Fig. 15.3. Determine the horizontal distribution of pressure and make labelled sketches of the pressure field for each of these twelve remaining cases.

CHAPTER 16

CIRCULATION, VORTICITY, AND DIVERGENCE THEOREMS

16.1 The Circulation Theorem

We have seen that circulation is a parameter which measures the rotational tendency of a sample of fluid. Such a measure is of great importance in meteorology for several reasons. For example, the common occurence of rotational storms invests such a quantity with considerable interest. Furthermore, the rate of change with time of the circulation is important, since it is involved in prognostication. We begin, therefore, by deriving an expression for the individual time rate of change of the circulation.

Let us consider a closed chain of fluid particles. The circulation around such a circuit is

$$C = \oint (udx + vdy + wdz)$$

We shall follow this chain of particles about and determine the change in circulation with time. This means that we shall take the total derivative:

$$\frac{dC}{dt} = \oint\left[\frac{du}{dt}dx + \frac{dv}{dt}dy + \frac{dw}{dt}dz\right] + \oint\left[u\frac{d}{dt}(dx) + v\frac{d}{dt}(dy) + w\frac{d}{dt}(dz)\right] \quad (16.1)$$

Here we have differentiated the increments of distance along the curve, as well as the velocity components, because as the fluid chain moves the distance separating two particles may well change with time. Terms of the sort $d/dt(dx)$ may be evaluated with the aid of Fig. 16.1, in which an infinitesimal line of fluid particles is shown at two moments separated by the time interval Δt. The change in the x component (dx) of the length of the fluid element is

$$\Delta(dx) = \left(\frac{\partial u}{\partial x}dx + \frac{\partial u}{\partial y}dy + \frac{\partial u}{\partial z}dz\right)\Delta t$$

or in the limit as Δt approaches zero

$$\frac{d}{dt}(dx) = du$$

A similar result may be demonstrated for the changes in the y and z components of length of a fluid element. Thus the last integral on the right of Eq. (16.1) becomes

$$\oint (udu + vdv + wdw) = \oint d\left(\frac{u^2 + v^2 + w^2}{2}\right)$$

This integral is equal to zero since it is the closed line integral of an exact differential.

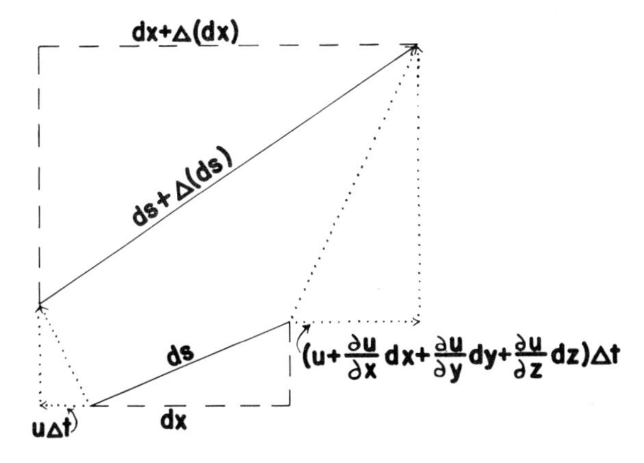

FIG. 16.1. Change in the length of a fluid line element, ds. The solid lines represent the fluid element before and after a time interval, Δt.

We may deal with the remainder of Eq. (16.1) by substituting for the accelerations, du/dt, dv/dt, and dw/dt, from the equations of motion, Eq. (11.10). We get

$$\frac{dC}{dt} = -\oint \frac{1}{\rho}\left(\frac{\partial p}{\partial x}dx + \frac{\partial p}{\partial y}dy + \frac{\partial p}{\partial z}dz\right)$$
$$+ \oint 2\Omega\left[(v \sin\phi - w\cos\phi)\,dx - u\sin\phi\,dy + u\cos\phi\,dz\right]$$
$$-\oint g\,dz + \oint\left(\frac{Fx}{m}dx + \frac{Fy}{m}dy + \frac{Fz}{m}dz\right) \quad (16.2)$$

We shall first simplify these terms mathematically then interpret them physically. The first term on the right of Eq. (16.2) may be written as

$$-\oint \frac{dp}{\rho}$$

Since the second term on the right is a Coriolis term, it is best understood by considering a projection of the closed chain of fluid particles upon the equatorial plane as in Fig. 16.2. Consider an x', y' system in the equatorial plane in which x' is simply the projection of the x axis on the equatorial plane, and y' is the projection of the y, z plane on the equatorial plane, in which y' points toward the earth's axis. Then

$$dx' = dx$$
$$dy' = dy \sin\phi - dz \cos\phi$$

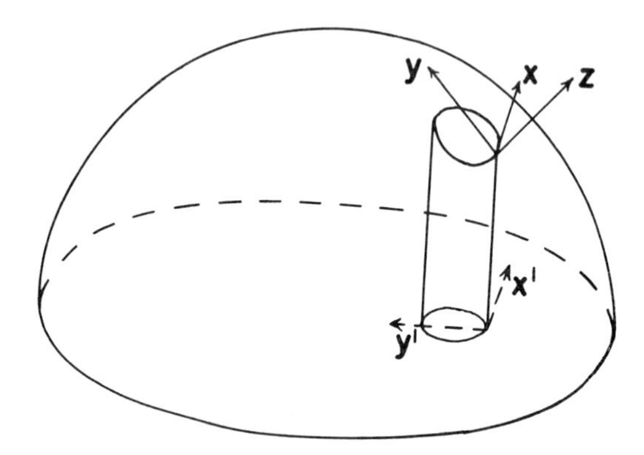

Fig. 16.2. Projection of a closed circuit on the equatorial plane

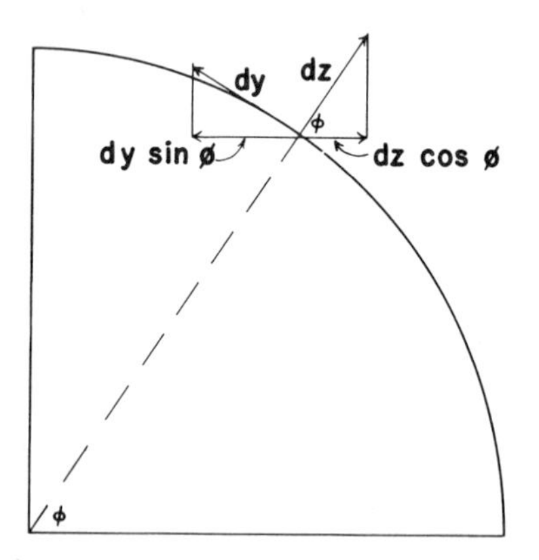

Fig. 16.3. Projection of an element of the y, z plane on an equatorial plane

The second of these formulae may be verified from Figure 16.3. By similar projections

$$u' = u$$
$$v' = v \sin\phi - w \cos\phi$$

The second integral of Eq. (16.2) then becomes $2\Omega \oint (v'dx' - u'dy')$. As one integrates around the circuit in the positive sense (counter-clockwise) it can be seen that $u'\,dy'$ is a contribution to a rate of expansion of the enclosed area, while $v'\,dx'$ is a contribution to a rate of contraction of this area. Thus, if F represents the area enclosed by the projection of the chosen closed circuit on the equatorial plane, the second integral of Eq. (16.2) becomes $-\,2\Omega dF/dt$.

The third integral of Eq. (16.2) is zero because it is the closed line integral of an exact differential. The fourth term of Eq. (16.2) does not vanish in general, and will be retained as an expression of the possible effect of all other forces than pressure-gradient, Coriolis, and gravity forces. Equation (16.2) now becomes

$$\frac{dC}{dt} = -\oint \frac{dp}{\rho} - 2\Omega\frac{dF}{dt} + \oint\left(\frac{Fx}{m}\,dx + \frac{Fy}{m}\,dy + \frac{Fz}{m}\,dz\right) \tag{16.3}$$

This is the *circulation theorem* of V. Bjerknes.[1]

16.2 Physical Interpretation of the Circulation Theorem

The individual derivative of circulation is the fluid analog of angular acceleration in solid dynamics, since it describes the time rate of change of a measure of the fluid's rotation. In the dynamics of solids such an angular acceleration is produced by torques. Thus the terms on the right are analogous to torques of various kinds. We shall examine each term individually to ascertain the physical nature of the torque-like action.

The first term

$$-\oint \frac{dp}{\rho}$$

is not zero, in general. However, if the fluid is barotropic (density a function of pressure alone) this term becomes the closed line integral of an exact differential, and so is zero. This argument is good for any given moment at which the fluid is barotropic. If the barotropy persists in time (autobarotropy) this term never makes a contribution to the generation of circulation.

In the general baroclinic case the density is not determined solely by the pressure; thus isobaric and isosteric surfaces intersect each other. For example, in the vertical cross section of Fig. 16.4 the intersecting

[1] V. BJERKNES, Über die Bildung von Circulationsbewegung, *Videskabs. Skrifter*, Oslo, 1898.

isobars and isosteres form a net of approximate parallelograms. If one attempts to evaluate

$$-\oint \frac{dp}{\rho}$$

around the indicated path one finds no contribution in going from 1 to 2 and from 3 to 4 since along these portions of the path $dp = 0$. On the leg from 2 to 3, dp is negative so the contribution to the integral in question is positive, while on the leg from 4 to 1 dp is positive, so that the contribution to the value of the integral is negative. However, the mean density along 2 to 3 is less than that along 4 to 1. Since density

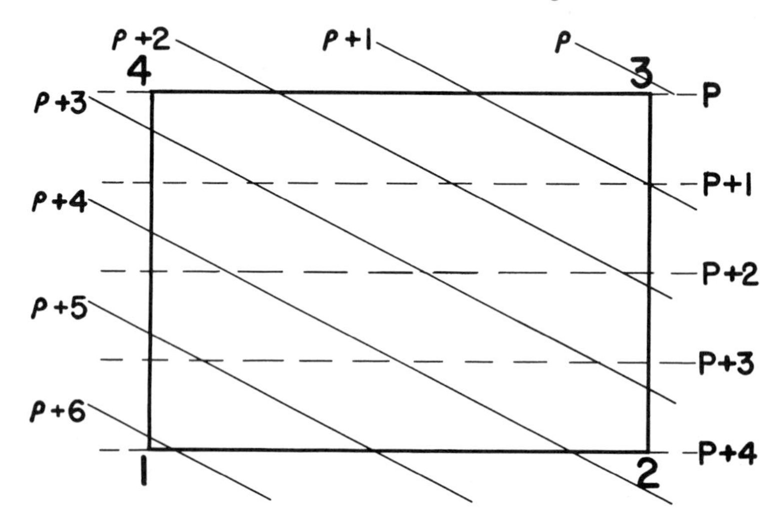

FIG. 16.4. Isobars and isoteres in a baroclinic fluid

appears in the denominator of the integrand this means that the positive contribution (2 to 3) overcomes the negative contribution (4 to 1) and the entire term makes a positive contribution to dC/dt in this example. A positive value of dC/dt means a growth of circulation in the positive (counterclockwise) sense about the curve. Note that this means the fluid circulation will develop in such a direction as to let the dense fluid sink while the light fluid rises. In general, the direction of the circulation which grows from a state of rest as a result of baroclinity turns the isosteres more nearly parallel to the isobars by moving the dense fluid towards high pressure and the light fluid towards low pressure. One can see immediately that this is a process which attempts to convert potential energy of the mass distribution of the fluid into kinetic energy of circulation, thus minimizing the potential energy.

It may be shown that the numerical value of the integral is directly proportional to the number of parallelograms delineated by the intersection of isosteres and isobars. Such parallelograms are called *solenoids* (Greek, "like a tube") and the existence of baroclinity is revealed by the presence of solenoids. Any pair of thermodynamic variables may be used in place of ρ and p to portray baroclinity.

One should note that Eq. (16.3) deals only with contributions to the time rate of change of circulation, not with the circulation itself. Thus the contribution from the solenoid term may be positive yet the circulation may be negative. Any circulation which is actually moving in the same sense as the direction in which the solenoid term impels it is called a *direct solenoidal circulation.* It is one in which potential energy is being converted into kinetic energy, and at an accelerated rate if other effects are negligible. Any circulation that is actually moving in the opposite sense from the direction in which the solenoid term impels it is called an *indirect solenoidal circulation.* It is one in which kinetic energy is being converted into potential energy, but at a decelerated rate if other effects are negligible. Clearly other terms in the circulation theorem must have set up this indirect circulation.

The second term of Eq. (16.3), $-2\Omega dF/dt$, is a Coriolis effect. It is analogous to the contribution to angular acceleration which arises when the radius of rotation of a particle is changed. With increasing radius such a particle will slow down in its rotation, while with decreasing radius it will speed up, as required by the law of conservation of absolute angular momentum. In the fluid case, if the area of the projection on the equatorial plane increases with time, the term in question is negative and the circulation will decrease (comparable to the case in particle mechanics of an increasing radius). If the projected area F shrinks with time, the term in question will be positive and the circulation will increase (comparable to the case in particle mechanics of a decreasing radius).

The Coriolis term may lead to changes in circulation by any of three processes. These are illustrated by dealing with chains of fluid particles which lie initially in a horizontal plane:

(a) The area enclosed may change because of horizontal convergence or divergence, thus producing a change in F, the projected area. Convergence builds up counterclockwise circulation while divergence builds up clockwise circulation. We shall call this the divergence effect.

(b) The area enclosed may experience no horizontal divergence, but may change latitude while remaining horizontal. Clearly this will alter F. If the circuit moves equatorward in the Northern Hemisphere,

F will decrease and counterclockwise circulation will be built up. Conversely, poleward motion will cause clockwise circulation to grow. These results must be reversed for the Southern Hemisphere. This will be referred to as the latitude effect.

(c) The area concerned may experience neither divergence nor changes in latitude, but if a gradient of vertical velocity exists across the circuit the individual particles will be moved out of a horizontal plane and F will change with time. This will be referred to as the tipping effect.

Finally, the term $\oint (F_x dx + F_y dy + F_z dz)$ in Eq. (16.3) represents the torquelike action of any other forces which may be present. Not all force fields will make a contribution here. For example, the field of gravity produces no change in circulation, as we have already seen. The chief additional force which is often important in the atmosphere is that of viscosity.

From the preceding discussion it is clear that the primary source of the growth of circulation in the atmosphere must be the mass distribution as expressed in the solenoid term. The Coriolis and viscous terms may alter the motion profoundly, but the basic source of the energy of the atmosphere lies in the baroclinity produced mostly by differential solar heating and condensation or evaporation of water.

16.3 Selected Applications of the Circulation Theorem

A number of examples of physical phenomena governed by the circulation theorem will now be discussed from a qualitative point of view.

(a) *Land and sea breeze systems*: The frequent occurence at sea coasts of a surface wind from the sea during the day (sea breeze) and from the land during the night (land breeze) may be explained by means of the solenoid term of Eq. (16.3). Figure 16.5 shows a vertical cross section through the atmosphere near a sea coast during the day. Since the land usually becomes warmer than the sea during the day, and since density decreases upward, the surfaces of constant density slope upward from land to sea. Because the isobaric surfaces are usually more nearly horizontal, the two sets of lines intersect to form a solenoid field in the vertical plane. The rate of change of circulation around a vertical circuit half over land and half over water will be negative—that is, clockwise. If the initial circulation is zero or nearly so, the subsequent circulation will be such as to bring cool air from the sea towards the warm land at low levels. At upper levels one would expect a return flow of warm air from the land to the sea. At night the land cools by radiation while the sea surface maintains a more uniform temperature because of the enormous reserve of heat available in the sea. Thus the isosteres will have the opposite slope from those in Fig. 16.5 and a circulation of opposite sense will be developed. This results in a land breeze.

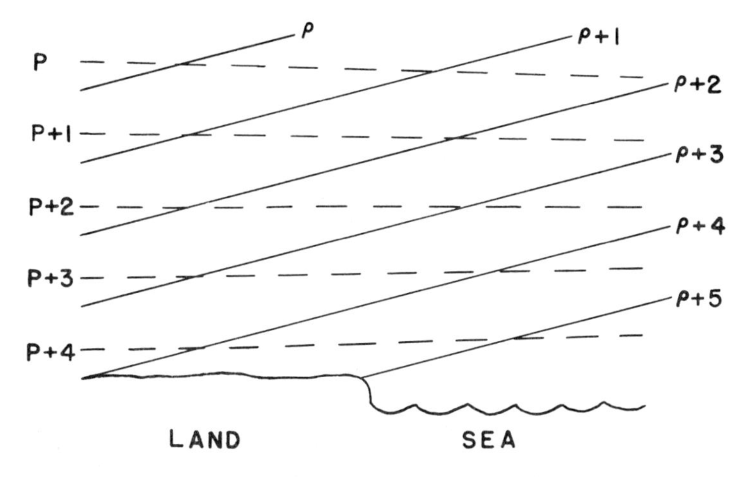

FIG. 16.5. Isobars and isosteres above a sea coast during the day

Land and sea breezes may be much modified by friction and Coriolis effects.[2] For example, the direct solenoidal circulation discussed above would cause progressively increasing wind speeds in the sea breeze during the day. However, the existence of surface friction exerts a braking effect via the last term of Eq. (16.3). Thus a balance is possible between the increase in circulation caused by the mass distribution and the decrease caused by retarding friction, and steady winds become theoretically possible.

If the land-sea breeze system is of large enough scale the Coriolis effect will be important. For example, in Fig. 16.5 an initially vertical chain of air particles parallel to the coast will be moved out of the vertical by the direct solenoidal circulation. The projection of the enclosed area on the equatorial plane will increase with time and this will tend to develop a negative circulation. Thus the upper seaward-bound air will gain a component coming out of the page, and the lower landward-

[2] B. HAURWITZ, Comments on the Sea-Breeze Circulation. *J. Meteor.*, 4, 1947, p. 1.

bound air will gain a component into the page. Note that this is the same result one gets by considering the effect of a right-hand deflecting force on the sea breeze.

(b) *The slope of frontal surfaces*: When there exists a sloping zone of transition from one air mass to another as in Fig. 16.6, the kinetic-energy-producing circulation is that which will cause the dense air to sink under the warm air. That is, the solenoids tend to produce a positive or counterclockwise circulation around the indicated circuit. If however,

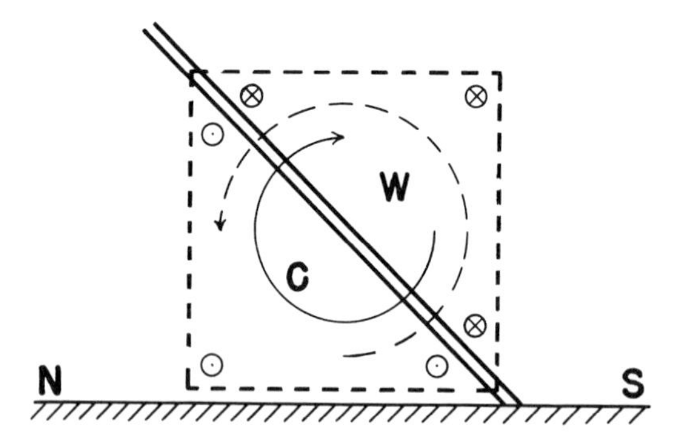

FIG. 16.6. Vertical north-south cross section through a front. The light dashed curve shows the direction in which the solenoids impel the circulation, and the solid curve shows the direction in which the Coriolis term impels the circulation. A circle with a dot indicates wind coming out of the page, while a circle with a cross indicates wind blowing into the page.

there is a cyclonic wind shear across the front, as shown, an originally vertical circuit with zero projection on the equatorial plane becomes inclined to the vertical with a positive area projected on the equatorial plane. Thus the Coriolis term, $-2\Omega dF/dt$, will make a negative contribution to the rate of growth of circulation. The solenoid term and the Coriolis term oppose each other. It is easy to see, therefore, that the length of time required for a given cold air mass to spread out under the warm air is greater in the presence of earth rotation than in its absence. Furthermore, the possibility exists of a balance between the two terms so that dC/dt becomes zero. This is the situation assumed in deriving the Margules equation (15.4) and the thermal wind equations (12.12).

(c) *Convergence, divergence, and latitude effects*: As has been pointed out already, convergence tends to build cyclonic relative circulation and divergence tends to build anticyclonic relative circulation. This effect appears strongly in the normal behavior of cyclones and anticyclones, for low-level cyclones are generally regions of convergence while low-level anticyclones are generally regions of divergence. If one considers the behavior of circulatory systems in a horizontal plane as they move in latitude, the circulation theorem indicates that a cyclone moving poleward should weaken and might eventually develop an anticyclonic circulation. Correspondingly an anticyclone moving equatorward should also weaken and might develop cyclonic circulation. In the main this partial result from the circulation theorem is contrary to experience. It is well known, for example, that cyclones moving poleward frequently intensify, and in a circuit surrounding the storm the circulation increases with time. Presumably this action is due to the domination of the latitude effect by the convergence influence.

16.4 A Vorticity Theorem

As we have seen before, circulation is a measure of fluid rotation while vorticity is a related measure of the same fluid property for an infinitely small area. Frequently it is more convenient to deal with vorticity than with circulation. While it is possible to transform the circulation theorem to a vorticity theorem, it proves easier to begin anew with the equations of motion in order to obtain a dynamical theorem concerning the rate of change of vorticity.

In expanded form the horizontal equations of motion are

$$\frac{\partial u}{\partial t} + u\frac{\partial u}{\partial x} + v\frac{\partial u}{\partial y} + w\frac{\partial u}{\partial z} = fv - \alpha\frac{\partial p}{\partial x} + F_x$$

$$\frac{\partial v}{\partial t} + u\frac{\partial v}{\partial x} + v\frac{\partial v}{\partial y} + w\frac{\partial v}{\partial z} = -fu - \alpha\frac{\partial p}{\partial y} + F_y$$

where the small term $2\Omega w \cos\phi$ has been neglected. Since we are, interested in a theorem involving the vorticity about a vertical axis,

$$\zeta = \frac{\partial v}{\partial x} - \frac{\partial u}{\partial y}$$

we take the partial derivative with respect to x of the second equation and subtract the partial derivative with respect to y of the first equation. After rearrangement the result is

$$\frac{\partial \zeta}{\partial t} + u\frac{\partial \zeta}{\partial x} + v\frac{\partial \zeta}{\partial y} + w\frac{\partial \zeta}{\partial z} + v\frac{\partial f}{\partial y} = -(f+\zeta)D - \left[\frac{\partial w}{\partial x}\frac{\partial v}{\partial z} - \frac{\partial w}{\partial y}\frac{\partial u}{\partial z}\right] - \left[\frac{\partial \alpha}{\partial x}\frac{\partial p}{\partial y} - \frac{\partial \alpha}{\partial y}\frac{\partial p}{\partial x}\right] + \left[\frac{\partial F_y}{\partial x} - \frac{\partial F_x}{\partial y}\right] \tag{16.4}$$

where

$$D = \frac{\partial u}{\partial x} + \frac{\partial v}{\partial y}$$

is the horizontal divergence. Since f depends only upon y, the last term on the left may be written

$$v \frac{\partial f}{\partial y} = \frac{df}{dt}$$

Thus Eq. (16.4) may be written as

$$\frac{d(f + \zeta)}{dt} = -(f + \zeta) D - \left[\frac{\partial w}{\partial x}\frac{\partial v}{\partial z} - \frac{\partial w}{\partial y}\frac{\partial u}{\partial z}\right] - \left[\frac{\partial \alpha}{\partial x}\frac{\partial p}{\partial y} - \frac{\partial \alpha}{\partial y}\frac{\partial p}{\partial x}\right] + \frac{\partial F_y}{\partial x} - \frac{\partial F_x}{\partial y} \tag{16.5}$$

This is the vorticity equation we sought. On the left we have the individual derivative of $f + \zeta$, the vorticity about the vertical of the earth's rotation plus the vorticity about the vertical of motion relative to the earth. This sum is the *absolute vorticity* about the vertical which would be seen by an inertial observer. Equation (16.5) says that the absolute vorticity of a parcel of air can change only through the contributions made by the four terms on the right. We shall now give a physical interpretation of these four terms.

The first term, $-(f + \zeta) D$, is called the "divergence term." It is the fluid analog of the contribution to angular acceleration which arises in rigid dynamics from a change in radius of rotation. Positive divergence is equivalent to an increase in effective radius and, by analogy, should decrease the magnitude of the absolute vorticity. Because of the minus sign this is precisely what the term in question does in this case. On the other hand, convergence will increase the magnitude of the absolute vorticity.

The second term on the right,

$$-\left[\frac{\partial w}{\partial x}\frac{\partial v}{\partial z} - \frac{\partial w}{\partial y}\frac{\partial u}{\partial z}\right]$$

has various names; we shall call it the "tipping term." Consider the first term in brackets when w decreases in the x direction and v increases upward. Because of the minus sign the contribution will be positive; that is, $f + \zeta$ will increase with time. This vertical shear constitutes a vorticity about a horizontal, east-west axis. Its analog in rigid dynamics is the wheel of Fig. 16.7 which has an east-west axle and is rotating counterclockwise as seen from the west. If the west end of the axle is lifted and the east end lowered ($\partial w/\partial x < 0$) the wheel will become more nearly horizontal and a component of cyclonic rotation about a vertical axis will appear. This corresponds to the increase with time of $f + \zeta$ which we found for this case. A corresponding analogy is easily made for the second half of this term. From the analogy it can be seen that this is called the tipping term because it contributes when the field of vertical velocity tips horizontally oriented vorticity into the vertical.

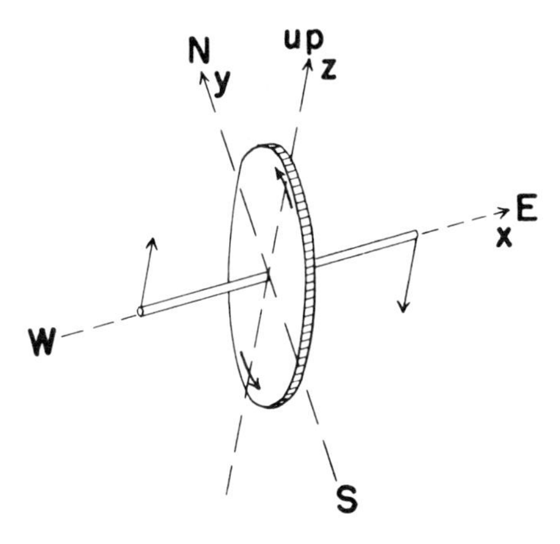

FIG. 16.7. A rotating wheel illustrating the rigid-dynamics analogy of the tipping term of the vorticity theorem

The third term on the right of Eq. (16.5) is the solenoid term of the circulation theorem, Eq. (16.3), applied to an infinitely small horizontal circuit. The circulation solenoid term may be written

$$-\oint \frac{dp}{\rho} = -\oint \left(\alpha \frac{\partial p}{\partial x}\, dx + \alpha \frac{\partial p}{\partial y}\, dy \right)$$

Now Stokes' theorem, Eq. (13.12), was derived for the variables u and v, the components of the horizontal velocity vector. However, examination of that derivation will reveal that the result is valid for the components of any vector, including those of the pressure-gradient force. Thus if we put $\alpha\, \partial p/\partial x$ in place of u, and $\alpha\, \partial p/\partial y$ in place of v we may write

$$-\oint \frac{dp}{\rho} = -\iint \left(\frac{\partial \alpha}{\partial x} \frac{\partial p}{\partial y} - \frac{\partial \alpha}{\partial y} \frac{\partial p}{\partial x} \right) dx dy$$

If we now utilize the restriction to a small area, within which the derivatives assume constant mean values for the area, the contribution to circulation from the solenoid term alone may be written

$$\frac{1}{A} \frac{dC}{dt} = -\left(\frac{\partial \alpha}{\partial x} \frac{\partial p}{\partial y} - \frac{\partial \alpha}{\partial y} \frac{\partial p}{\partial x} \right)$$

where A is the small area concerned. Since $C = \zeta A$, one can see how the third term on the right of Eq. (16.5) contributes to a change of vorticity and is an alternative way of expressing the solenoid term. Solenoids are peculiarly a property of fluids; they have no analog in rigid dynamics.

The last term on the right of Eq. (16.5),

$$\left[\frac{\partial F_y}{\partial x} - \frac{\partial F_x}{\partial y}\right]$$

represents the effect of any other forces which may be present, such as viscosity or turbulence. If these exert a torquelike action upon fluid elements, a change of vorticity will be observed. In most applications the additional forces are ignored.

Calculations from synoptic data indicate that there are times when each of the right-hand terms of the vorticity equation may make appreciable contributions to $d(f+\zeta)/dt$ in large-scale weather systems. However, the divergence term seems to be more often important than the others, but even it may be negligible for appreciable periods of time.

If little divergence exists, vertical motions are not appreciable, the solenoid field is weak, and other forces are neglected, the vorticity theorem reduces to

$$\frac{d}{dt}(f + \zeta) = 0 \tag{16.6}$$

This is a statement of the conservation of absolute vorticity since, under the assumed conditions, an individual parcel of air can experience no change in its absolute vorticity. If air moves poleward under these conditions, then as f increases ζ has to decrease in order to keep their sum constant. If, furthermore, the relative vorticity is expressed mostly as curvature this means that poleward moving air must lose cyclonic curvature and ultimately must curve anticyclonically. The converse is true for air moving equatorward. It is possible to compute the trajectory of an air parcel under conservation of absolute vertical vorticity given its initial speed, direction, latitude, and curvature. Such trajectories are called *constant absolute vorticity trajectories*.[3] Some typical examples are shown in Fig. 16.8.

Such trajectories as these have been much used as a convenient way of applying dynamical reasoning, based upon the vorticity theorem, to weather forecasting. It should be recognized, however, that this can

[3] See, for example, D. FULTZ, *Upper-Air Trajectories and Weather Forecasting*. Misc. Report No. 19, Dept. of Meteorology, Univ. of Chicago, Univ. of Chicago Press, 1945.

only represent a point of departure, since a number of significant factors had to be neglected in order to arrive at Eq. (16.6).

In order to exhibit the effect of the divergence term we shall return to the more complete vorticity equation (16.5). Let us neglect the tipping term, the solenoid term, and other forces, and assume the atmosphere to be incompressible. Then the horizontal divergence may be replaced by $-1/D(dD/dt)$ via Eq. (13.16), where D is the depth of an air column. The vorticity theorem is then

$$\frac{d}{dt}\left(\frac{f+\zeta}{D}\right)=0 \tag{16.7}$$

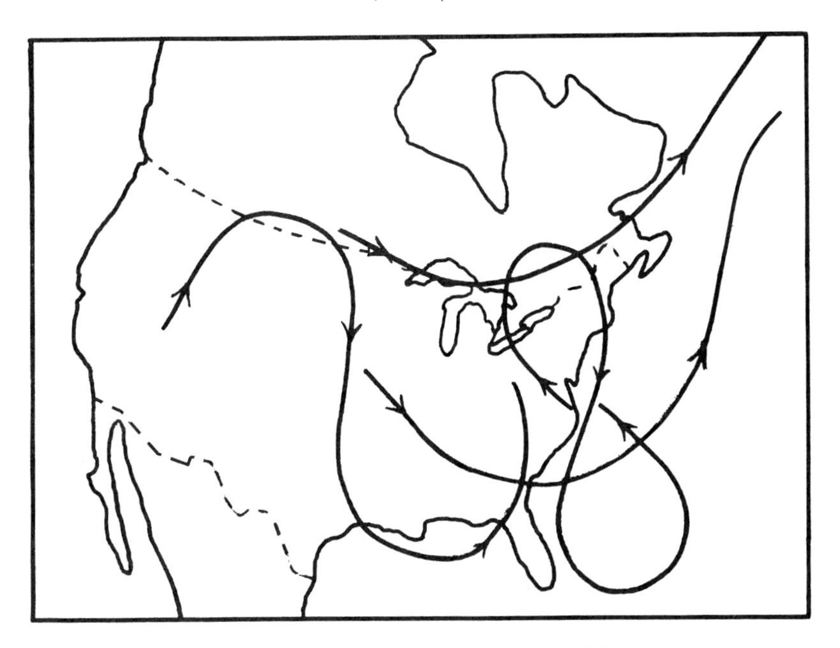

FIG. 16.8. Examples of constant absolute vorticity trajectories

When divergence is included in this way, it is not merely the absolute vorticity which is conserved, but the ratio of this vorticity to the depth of the air column being considered.

As an application of this form of the vorticity theorem we shall consider flow over a mountain range. We shall assume a straight, uniform, westerly current of zero relative vorticity impinging upon a north-south range. In order to simplify the discussion we also assume that any vorticity which may appear will do so as curvature rather than as shear. Furthermore, we assume that the vertical perturbation of the air as it moves over the mountain decreases in magnitude with elevation.

Thus there will be some elevation above which no perceptible effect of the mountain can be found.[4] The behavior of such a zonal current as governed by Eq. (16.7) is shown in Fig. 16.9.

Until the air reaches the mountain there are no changes in depth or latitude to cause it to curve. When it begins to ascend the windward side of the barrier the depth begins to decrease, causing anticyclonic curvature to appear. The depth continues to decrease until the top of the barrier has been reached. At this point the current has its maximum anticyclonic curvature because there it has minimum depth.

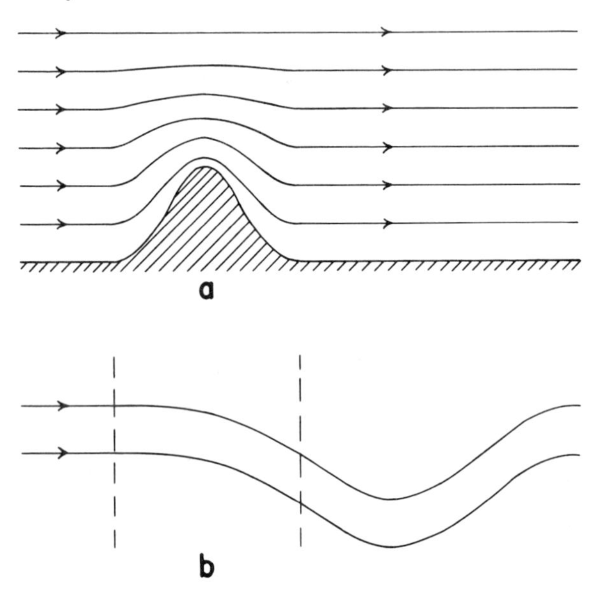

FIG. 16.9. (a) Vertical cross section and (b) plan view of streamlines of an initially zonal current crossing a mountain range in the northern hemisphere

During the descent on the leeward side, the depth gradually increases again while the vorticity returns toward zero. If the divergence effect alone is considered, the current should reach the leeward plain with the same relative vorticity it had to begin with, namely zero. It would thus be a straight, uniform current, but the mountain's influence would have turned it from a westerly to a northwesterly current. However, the current is travelling towards decreasing latitude during the descent. On that account it arrives at the foot of the barrier with some cyclonic curvature. Once past the mountain the air is controlled solely by the

[4] This assumption is known to be far too simple. However, it serves to elucidate one of the essential features of the application of the vorticity theorem to flow over topographical barriers.

latitude effect, since depth no longer changes. The current will then undulate back and forth in a sinusoidal fashion as described in the discussion of constant absolute vorticity trajectories.

This qualitative discussion represents a simplified theoretical explanation of two observed phenomena. First, it predicts that a region of cyclonic curvature should appear on the leeward side of the barrier. This corresponds to the commonly observed "lee-of-the-mountain trough." Such troughs of low pressure are found frequently when well-defined atmospheric currents flow over mountain ranges.

The second observed phenomenon whose origin is suggested by these considerations is the sequence of quasi-stationary troughs and ridges to be found in the upper-level westerlies of middle latitudes. There is reason to believe that these may be the sinusoidal undulations induced in a zonal current by the mechanism described above.

16.5 The Theory of Long Waves in the Westerlies

The equations of motion are a very rich matrix from which a wide range of physical phenomena can be derived. The vorticity theorem shares this complexity since it is derived from the equations of motion. Among the phenomena permitted by the equations are wave motions. We shall now investigate one particular kind of wave motion which is dependent upon the rotation of the earth for its existence.

If we make the additional assumptions of zero horizontal divergence, purely horizontal flow, autobarotropy, and negligible additional forces, the vorticity theorem is $d(f + \zeta)/dt = 0$ or

$$\frac{\partial \zeta}{\partial t} + u\frac{\partial \zeta}{\partial x} + v\frac{\partial \zeta}{\partial y} + v\frac{\partial f}{\partial y} = 0$$

We shall define $\beta \equiv \partial f/\partial y = 2\Omega \cos \phi/a$. If we now assume that we are dealing with a broad westerly current of infinite lateral extent which is undulating north and south in a wavelike fashion, then we may take the dependent variable to be independent of y. This means that we are dealing with a system of waves in which the streamlines at any latitude are parallel to those at any other latitude.[5] Thus $\zeta = \partial v/\partial x$ and we have

$$\frac{\partial^2 v}{\partial x \partial t} + u\frac{\partial^2 v}{\partial x^2} + \beta v = 0$$

[5] We have implicitly neglected the geometric effects of the curvature of the earth by using the tangent-plane equations of motion.

We shall seek as a solution, waves which move in the x direction with a constant speed c, without change of shape. Thus if one follows a point moving in the east-west direction with the speed c, no changes will be observed in any of the variables. That is, the operator $\mathfrak{D}/\mathfrak{D}t = \partial/\partial t + c(\partial/\partial x) = 0$, where $\mathfrak{D}/\mathfrak{D}t$ means an individual derivative following a point moving with speed c. The operation $\mathfrak{D}/\mathfrak{D}t$ is, of course, different from the operation d/dt, since the latter represents an individual derivative following a particle of air. With this assumption the vorticity theorem becomes

$$(u - c)\frac{\partial^2 v}{\partial x^2} + \beta v = 0 \tag{16.8}$$

This equation is nonlinear, since it involves a product of dependent variables, and is therefore difficult to solve. We shall now apply the *perturbation method* in order to reduce it to a linear differential equation. This consists of assuming the horizontal flow to be made up of a large constant basic current, U, flowing from west to east, upon which are superimposed small variable perturbations, u' and v', which are functions of x and t. Thus

$$u(x, t) = U + u'(x, t)$$
$$v(x, t) = v'(x, t)$$

All magnitudes will be referred to U as a standard through a logarithmic scale. Thus U is considered to be a quantity of zero order of magnitude, while u' and v', which are one step smaller on the scale, are of the first order of magnitude. If, as is common, one adopts a logarithmic scale to the base ten then u' and v' will be about 1/10 as large as U. Equation (16.8) now becomes

$$(U - c)\frac{\partial^2 v'}{\partial x^2} + \beta v' + u'\frac{\partial^2 v'}{\partial x^2} = 0$$

The first two terms are first order in size but the last term consists of the product of two first-order quantities and so is second order in magnitude (1/100 of a zero-order term when the base ten is adopted). Within this degree of precision it is possible to neglect the second-order term thus reducing the equation to a linear one.[6] We may now write

$$\frac{\partial^2 v'}{\partial x^2} + \frac{\beta}{U - c} v' = 0 \tag{16.9}$$

Since x is the only independent variable this equation corresponds to an ordinary second-order differential equation with constant coefficients.

[6] It is assumed implicitly that differentiation does not alter the order of magnitude of a term. Clearly if v' were to vary rapidly with x, the derivatives of v' might be larger on the relative scale of magnitudes than v' itself. Thus there is a restriction here to slowly varying functions of x, that is, to long wave lengths.

The solution for v' is known to be sinusoidal in x. To show this let us try a solution of the form $v' = v_0 \sin 2\pi x/L$, where v_0 is a constant amplitude and L is the constant wave length. Substitution into Eq. (16.9) yields

$$U - c = \frac{\beta L^2}{4\pi^2} \tag{16.10}$$

Equation (16.10) is a relationship that must exist between the wave speed and wave length of sinusoidal waves if Eq. (16.9) is to be satisfied. Such an expression is one of the main results of a theoretical investigation of wave phenomena and is called a *frequency equation.*

Equation (16.10) holds the promise of considerable practical forecasting value since it gives the instantaneous speed of propagation of the waves in terms of quantities, U and L, which may be measured on a map. This result has had a considerable effect upon both synoptic and theoretical thinking since it was introduced in 1939 by Rossby.[7]

It may be seen that when c is positive (the waves progress from west to east), $U - c$ is smaller than U and the wave lengths are relatively short. When c is negative (the waves retrogress from east to west) $U - c$ is larger than U and the wave lengths are relatively long. The agreement with the results of the Bjerknes-Holmboe theory should be noted.

The size of the waves described by this theory is indicated in Table 16.1, in which the value of $U - c$ is given as a function of the number of waves which could fit around the circumference of the earth at latitude 45°. Since in most cases in the atmosphere $U - c$ ranges between 10 and 50 m s^{-1}, the number of these waves in middle latitudes ranges from 3 to 6. Thus the distance between successive waves is several thousand kilometers.

TABLE 16.1

NUMBER OF WAVES AROUND THE EARTH AT LATITUDE 45° AS A FUNCTION OF U - c

No. of waves	2	3	4	5	6	7
$U - c$ (m s^{-1})	82.0	36.5	20.5	13.1	9.1	6.7

It is important to recognize that Rossby waves are only one of a whole hierarchy of atmospheric waves which are possible. For example, the shortest and most rapidly moving waves are those due to sound.

[7] C.-G. ROSSBY, Relation between Variations in the Intensity of the Zonal Circulation of the Atmosphere and the Displacements of the Semi-Permanent Centers of Action. *J. Marine Res.*, 2, 1939, p. 38.

In these, compressibility is all important and large values of divergence are found. As we go up the scale we find long gravitational waves which owe their existence to the effect of gravity on a disturbed fluid surface, inertia waves due to a similar effect of the earth's rotation, shearing waves (billow clouds) due to a shear of wind across a discontinuity, cyclone waves due to a combination of several of the factors mentioned, and Rossby waves which owe their existence to the variation with latitude of the Coriolis parameter.[8] This scale of wave motions is summarized in Table 16.2 in which the waves are characterized by the order of magnitude of the frequency of passage of such waves past a given point.

TABLE 16.2

CHARACTERISTIC FREQUENCIES OF VARIOUS TYPES OF ATMOSPHERIC WAVES

Type of wave	*Frequency*	
	Waves per second	*Waves per day*
Sound	10^{3}	10^{8}
Long gravitational	10^{-1}	10^{4}
Inertia, and shearing	10^{-2}	10^{3}
Cyclone	10^{-5}	10^{0}
Rossby	10^{-6}	10^{-1}

As the frequency decreases the time available for the Coriolis force to operate increases and the accelerations decrease, generally speaking. Thus the likelihood of geostrophic balance increases. The most important frequencies for phenomena on the usual scale of a weather map are those smaller than 10^{-4} s^{-1} (10 waves per day)—that is, cyclone and Rossby waves for which the geostrophic assumption is fairly accurate.

16.6 A Divergence Theorem

A dynamical equation for the individual rate of change of horizontal divergence may be obtained in a fashion parallel to the derivation of

[8] It is possible to generalize the simple treatment given here to include large amplitudes, north-south variations, and curvature of the earth. See, for example, S. NEAMTAN, The Motion of Harmonic Waves in the Atmosphere, *J. Meteor.*, 3, 1946, pp. 53-56; R. A. CRAIG, A Solution of the Nonlinear Vorticity Equation for Atmospheric Motion, *J. Meteor.*, 2, 1945, pp. 175-178.

the vertical vorticity equation. The equations of horizontal motion in expanded form are:

$$\frac{\partial u}{\partial t} + u\frac{\partial u}{\partial x} + v\frac{\partial u}{\partial y} + w\frac{\partial u}{\partial z} = fv - \alpha\frac{\partial p}{\partial x} + F_x$$

$$\frac{\partial v}{\partial t} + u\frac{\partial v}{\partial x} + v\frac{\partial v}{\partial y} + w\frac{\partial v}{\partial z} = -fu - \alpha\frac{\partial p}{\partial y} + F_y$$

Since we are looking for a theorem involving the horizontal divergence, $D = \partial u/\partial x + \partial v/\partial y$, we take the partial derivative with respect to x of the first equation and add the partial derivative with respect to y of the second equation. After rearrangement the result is

$$\begin{aligned}\frac{\partial D}{\partial t} + u\frac{\partial D}{\partial x} + v\frac{\partial D}{\partial y} + w\frac{\partial D}{\partial z} + u\frac{\partial f}{\partial y} = [f\zeta - D^2] - \left[\frac{\partial w}{\partial x}\frac{\partial u}{\partial z} + \frac{\partial w}{\partial y}\frac{\partial v}{\partial z}\right] \\ - \left[\frac{\partial \alpha}{\partial x}\frac{\partial p}{\partial x} + \frac{\partial \alpha}{\partial y}\frac{\partial p}{\partial y}\right] + \left[\frac{\partial F_x}{\partial x} + \frac{\partial F_y}{\partial y}\right] \\ - \alpha\left[\frac{\partial^2 p}{\partial x^2} + \frac{\partial^2 p}{\partial y^2}\right] + 2\left[\frac{\partial u}{\partial x}\frac{\partial v}{\partial y} - \frac{\partial u}{\partial y}\frac{\partial v}{\partial x}\right]\end{aligned} \qquad (16.11)$$

This result cannot be described by analogy to rigid dynamics as was possible for the vorticity theorem because divergence is peculiarly a property of fluids and is absent in rigid bodies. Nevertheless, there are certain formal mathematical similarities between Eq. (16.4) and (16.11). All the terms on the left sides and the first four terms on the right sides of the two equations are parallel in form, but with different physical interpretations. The last two terms on the right of Eq. (16.11) have no analog in Eq. (16.4) and arise because we are dealing with the rates of change of quite different quantities.

The divergence equation (16.11) is rather lengthy and complex. One recent use which has been made of it is the following. Suppose we wish to investigate essentially horizontal motion in which there is negligible divergence. If, furthermore the term involving products of derivatives of α and p is small, and no additional forces need be considered, the divergence equation reduces to

$$\alpha\left[\frac{\partial^2 p}{\partial x^2} + \frac{\partial^2 p}{\partial y^2}\right] - f\zeta + \beta u - 2\left[\frac{\partial u}{\partial x}\frac{\partial v}{\partial y} - \frac{\partial u}{\partial y}\frac{\partial v}{\partial x}\right] = 0 \qquad (16.12)$$

This is a differential relationship between horizontal wind and pressure which is, in general, different from the geostrophic and gradient wind equations. It is called the *balance equation*. Because the velocity components as well as the pressure appear in differentiated form it is much more difficult to solve than the two previous wind-pressure relations we have obtained. Consequently the balance equation is used only when high-speed electronic computing machines are employed.

PROBLEMS

1. Consider a westerly jet stream as in Fig. 16.10 in which there is initially no circulation and no rate of change of circulation with time around the rectangular circuit shown. Suppose the jet increases in speed but the solenoid field remains unchanged. Deduce from the circulation theorem the direction of the vertical motions to be expected on either side of the jet below 12 km. Describe qualitatively the trajectory of an air particle in the vicinity of the jet.

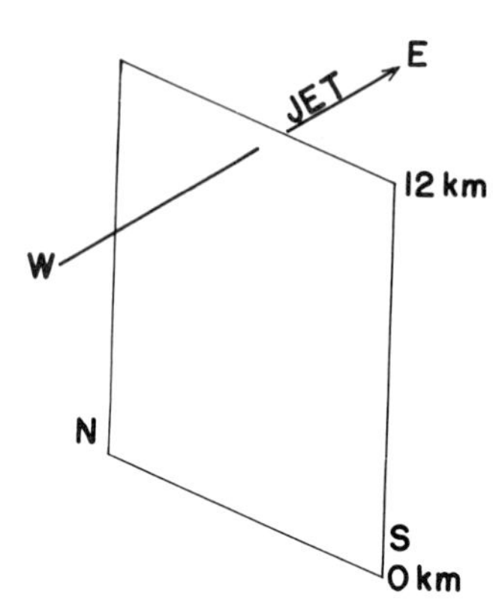

FIG. 16.10. Vertical, north-south, closed curve in the vicinity of a jet stream

2. Express the solenoid term of the vorticity equation in terms of the gradients of (a) temperature and pressure, and (b) temperature and potential temperature.

3. Suppose in the theory of long waves in the westerlies we had assumed a solution of the form

$$v' = v_0 \sin \frac{2\pi}{L}(x - Bt)$$

where B is a constant. Show that this is equivalent to assuming waves which move eastward with constant speed, c, without change of shape if $B = c$.

4. Suppose the wind is purely horizontal and characterized by pure constant vorticity. Show that the frictionless steady-state equations of motion (gradient wind equations) are then identical with the balance equation, provided gradients of α are neglected.

CHAPTER 17

THE FUNDAMENTAL EQUATIONS USING PRESSURE AS AN INDEPENDENT COORDINATE

17.1 Substitution of Pressure for Height

As we have seen in our discussion of the geostrophic and thermal-wind relationships, there are many practical advantages in measuring various quantities and their derivatives along surfaces of constant pressure rather than along level surfaces. Since the hydrostatic approximation is one of the most reliable in meteorology, and since it relates pressure increments to height increments, it would seem possible to express all of dynamics in terms of pressure as the vertical coordinate, rather than height. This procedure produces a significant simplification in some of the relationships and is well worth while in solving many problems.

We shall adopt a Cartesian coordinate system in which x and y are the two horizontal axes and in which position on the vertical axis is specified by pressure rather than by distance along the vertical. This means that partial derivatives with respect to x and y will be taken holding p constant in the fashion we have already seen, and partial derivatives with respect to p will be taken holding x and y constant. Thus the pressure coordinate is not normal to the isobaric surfaces but is exactly vertical.

Along such a vertical z depends upon only one independent coordinate, p. Therefore a derivative with respect to z may be written

$$\frac{\partial}{\partial z} = \frac{\partial p}{\partial z} \frac{\partial}{\partial p}$$

If we now make the hydrostatic assumption the transformation equation becomes

$$\frac{\partial}{\partial z} = -\rho g \frac{\partial}{\partial p} \tag{17.1}$$

17.2 Horizontal Derivatives and Time Derivatives

We shall now investigate how the remaining derivatives in a system where z is the vertical coordinate transform when p is the vertical coordinate. We have already achieved this for x and y (page 188). The results were

$$\left(\frac{\partial}{\partial x}\right)_z = \left(\frac{\partial}{\partial x}\right)_p + \rho g \left(\frac{\partial z}{\partial x}\right)_p \frac{\partial}{\partial p} \tag{17.2}$$

$$\left(\frac{\partial}{\partial y}\right)_z = \left(\frac{\partial}{\partial y}\right)_p + \rho g \left(\frac{\partial z}{\partial y}\right)_p \frac{\partial}{\partial p} \tag{17.3}$$

Partial derivatives with respect to time can be evaluated by exactly the same process with the result:

$$\left(\frac{\partial}{\partial t}\right)_z = \left(\frac{\partial}{\partial t}\right)_p + \rho g \left(\frac{\partial z}{\partial t}\right)_p \frac{\partial}{\partial p} \tag{17.4}$$

Note that the last term of Eq. (17.4) contains, as was to be expected, the rate at which the height of an isobaric surface rises or falls with time.

Equations (17.1) to (17.4) contain all that we need to transform the basic equations of meteorology to an x, y, p, t system of independent coordinates (hereafter called a p-system). Let us apply them to an individual derivative with respect to time. In an x, y, z, t system (hereafter called a z-system),

$$\frac{d}{dt} = \frac{\partial}{\partial t} + u\frac{\partial}{\partial x} + v\frac{\partial}{\partial y} + w\frac{\partial}{\partial z} \tag{17.5}$$

Substitution of the transformation equations makes this

$$\frac{d}{dt} = \left(\frac{\partial}{\partial t}\right)_p + u\left(\frac{\partial}{\partial x}\right)_p + v\left(\frac{\partial}{\partial y}\right)_p - \rho g w\frac{\partial}{\partial p} + \rho g\left[\left(\frac{\partial z}{\partial t}\right)_p + u\left(\frac{\partial z}{\partial x}\right)_p + v\left(\frac{\partial z}{\partial y}\right)_p\right]\frac{\partial}{\partial p} \tag{17.6}$$

If we apply this operator to p itself and define[1] $\omega \equiv dp/dt$ we get

$$\omega \equiv \frac{dp}{dt} = -\rho g w + \rho g\left[\left(\frac{\partial z}{\partial t}\right)_p + u\left(\frac{\partial z}{\partial x}\right)_p + v\left(\frac{\partial z}{\partial y}\right)_p\right]$$

When we substitute from this for $\rho g w$ in Eq. (17.6) we get

$$\frac{d}{dt} = \left(\frac{\partial}{\partial t}\right)_p + u\left(\frac{\partial}{\partial x}\right)_p + v\left(\frac{\partial}{\partial y}\right)_p + \omega\frac{\partial}{\partial p} \tag{17.7}$$

Thus the individual derivative in a p-system takes the same form as in a z-system except that $\omega\,\partial/\partial p$ replaces $w\,\partial/\partial z$. The variable ω is the equivalent of the vertical velocity. When measured in microbars per second it is numerically approximately equal to w in cm per second in the

[1] This ω is *not* the angular velocity of the earth.

lower atmosphere but has the opposite sign. The reason the expansion of an individual derivative in the p-system has the same form as in a z-system is that we are still dealing with Cartesian coordinates. We only measure one of them differently. As a matter of fact, Eq. (17.7) could have been written from first principles exactly as we did originally for Eq. (17.5).

17.3 The Equations of Motion

We have just seen how an individual derivative looks in a p-system, and we have already determined how to express the horizontal pressure-gradient force in terms of variables along a constant-pressure surface (Chap. 12). Therefore the horizontal equations of nonviscous motion become

$$\begin{aligned} \left(\frac{\partial u}{\partial t}\right)_p + u\left(\frac{\partial u}{\partial x}\right)_p + v\left(\frac{\partial u}{\partial y}\right)_p + \omega\frac{\partial u}{\partial p} &= fv - g\left(\frac{\partial z}{\partial x}\right)_p \\ \left(\frac{\partial v}{\partial t}\right)_p + u\left(\frac{\partial v}{\partial x}\right)_p + v\left(\frac{\partial v}{\partial y}\right)_p + \omega\frac{\partial v}{\partial p} &= -fu - g\left(\frac{\partial z}{\partial y}\right)_p \end{aligned} \tag{17.8}$$

where the term $2\Omega w \cos\phi$ has been omitted from the first of Eq. (17.8) on the grounds that it is relatively small. The equation of motion in the vertical direction is simply the hydrostatic equation, which has been used in the transformations which lead to Eq. (17.8). These equations are simpler than their counterparts for the z-system because the density does not appear explicitly.

17.4 The Equation of Continuity

As previously derived for the z-system, the law of conservation of mass is

$$-\frac{1}{\rho}\frac{d\rho}{dt} = \frac{\partial u}{\partial x} + \frac{\partial v}{\partial y} + \frac{\partial w}{\partial z}$$

When we apply expressions (17.1) through (17.3) and (17.7), we get

$$\begin{aligned} &-\frac{1}{\rho}\left[\left(\frac{\partial \rho}{\partial t}\right)_p + u\left(\frac{\partial \rho}{\partial x}\right)_p + v\left(\frac{\partial \rho}{\partial y}\right)_p + \omega\frac{\partial \rho}{\partial p}\right] = \\ &\left(\frac{\partial u}{\partial x}\right)_p + \left(\frac{\partial v}{\partial y}\right)_p + \frac{\partial \omega}{\partial p} + \rho g\left[\left(\frac{\partial z}{\partial x}\right)_p\frac{\partial u}{\partial p} + \left(\frac{\partial z}{\partial y}\right)_p\frac{\partial v}{\partial p}\right] \\ &-\frac{\omega}{\rho}\frac{\partial \rho}{\partial p} - \rho g\frac{\partial}{\partial p}\left[\left(\frac{\partial z}{\partial t}\right)_p + u\left(\frac{\partial z}{\partial x}\right)_p + v\left(\frac{\partial z}{\partial y}\right)_p\right] \end{aligned}$$

or

$$\left(\frac{\partial u}{\partial x}\right)_p + \left(\frac{\partial v}{\partial y}\right)_p + \frac{\partial \omega}{\partial p} = -\frac{1}{\rho}\left[\left(\frac{\partial \rho}{\partial t}\right)_p + u\left(\frac{\partial \rho}{\partial x}\right)_p + v\left(\frac{\partial \rho}{\partial y}\right)_p\right]$$
$$- \rho g\left[\left(\frac{\partial z}{\partial x}\right)_p \frac{\partial u}{\partial p} + \left(\frac{\partial z}{\partial y}\right)_p \frac{\partial v}{\partial p}\right] + \rho g\left[\left(\frac{\partial}{\partial t}\right)_p + u\left(\frac{\partial}{\partial x}\right)_p + v\left(\frac{\partial}{\partial y}\right)_p\right]\frac{\partial z}{\partial p}$$
$$+ \rho g\left[\left(\frac{\partial z}{\partial x}\right)_p \frac{\partial u}{\partial p} + \left(\frac{\partial z}{\partial y}\right)_p \frac{\partial v}{\partial p}\right]$$

The last term on the right cancels with the second term on the right. If we remember that $\partial z/\partial p = -1/(\rho g)$ we see that the third term on the right cancels with the first term on the right. Thus the continuity equation becomes exactly

$$\left(\frac{\partial u}{\partial x}\right)_p + \left(\frac{\partial v}{\partial y}\right)_p + \frac{\partial \omega}{\partial p} = 0 \tag{17.9}$$

This is a much simpler form than its counterpart in a z-system because of the absence of the term $-1/\rho \, d\rho/dt$. This is physically due to the fact that the mass of air per unit area between two isobaric surfaces is constant for hydrostatic equilibrium. This simplicity of the continuity equation is one of the chief advantages of a p-system in meteorological theory.

17.5 The Vorticity and Divergence Equations

We may now follow the same procedure as in Chapter 16 to derive a vorticity theorem. If we differentiate the first of Eq. (17.8) with respect to y and differentiate the second with respect to x (in both cases holding p constant), and then subtract the first result from the second we get, after rearranging

$$\begin{aligned}\left(\frac{\partial \zeta}{\partial t}\right)_p + u\left(\frac{\partial \zeta}{\partial x}\right)_p + v\left(\frac{\partial \zeta}{\partial y}\right)_p + \omega\frac{\partial \zeta}{\partial p} &= -(f+\zeta)\left(\frac{\partial u}{\partial x} + \frac{\partial v}{\partial y}\right)_p - \beta v \\ &\quad + \left(\frac{\partial \omega}{\partial y}\right)_p \frac{\partial u}{\partial p} - \left(\frac{\partial \omega}{\partial x}\right)_p \frac{\partial v}{\partial p}\end{aligned} \tag{17.10}$$

This is the vorticity equation in a p-system. Each of the terms is a counterpart to a term in the corresponding equation in the z-system and has a similar physical interpretation. Equation (17.10) however, does not contain a counterpart to the solenoid term. This is a direct result of the fact that the equations of motion in a p-system do not contain the density as a coefficient, so that no solenoid term arises from the cross-differentiation. The effects of baroclinity are still contained in Eq. (17.10) but not in an explicit form. As a result of shifting to a p-system, certain changes have occurred in the meaning of various terms and operations and some of these changes are related to the baroclinity. For example, the vertical component of vorticity, ζ, in Eq. (17.10) is not the same as in the z-system. As can be verified easily

$$\left(\frac{\partial v}{\partial x}\right)_z - \left(\frac{\partial u}{\partial y}\right)_z = \left(\frac{\partial v}{\partial x}\right)_p - \left(\frac{\partial u}{\partial y}\right)_p + \rho g \left[\left(\frac{\partial z}{\partial x}\right)_p \frac{\partial v}{\partial p} - \left(\frac{\partial z}{\partial y}\right)_p \frac{\partial u}{\partial p}\right] \qquad (17.11)$$

Similarly in the case of horizontal divergence

$$\left(\frac{\partial u}{\partial x}\right)_z + \left(\frac{\partial v}{\partial y}\right)_z = \left(\frac{\partial u}{\partial x}\right)_p + \left(\frac{\partial v}{\partial y}\right)_p + \rho g \left[\left(\frac{\partial z}{\partial x}\right)_p \frac{\partial u}{\partial p} + \left(\frac{\partial z}{\partial y}\right)_p \frac{\partial v}{\partial p}\right] \qquad (17.12)$$

The last terms on the right of Eq. (17.11) and (17.12) are ordinarily relatively small correction terms since the slopes of isobaric surfaces are usually small. It is such correction terms which account for the missing solenoid term in the vorticity equation.

The horizontal divergence equation can be derived likewise by differentiating the first of Eq. (17.8) with respect to x and the second with respect to y (in both cases holding p constant) and adding. After rearranging we get

$$\left(\frac{\partial D}{\partial t}\right)_p + u\left(\frac{\partial D}{\partial x}\right)_p + v\left(\frac{\partial D}{\partial y}\right)_p + \omega\frac{\partial D}{\partial p} = -D^2 + 2\left(\frac{\partial u}{\partial x}\frac{\partial v}{\partial y} - \frac{\partial v}{\partial x}\frac{\partial u}{\partial y}\right)_p$$
$$- \left(\frac{\partial \omega}{\partial x}\right)_p \frac{\partial u}{\partial p} - \left(\frac{\partial \omega}{\partial y}\right)_p \frac{\partial v}{\partial p} - \beta u + f\zeta - g\left[\frac{\partial^2 z}{\partial x^2} + \frac{\partial^2 z}{\partial y^2}\right]_p \qquad (17.13)$$

where

$$D = \left(\frac{\partial u}{\partial x}\right)_p + \left(\frac{\partial v}{\partial y}\right)_p$$

is the horizontal divergence in a p-system. Each of the terms in Eq. (17.13) corresponds to a similar term in the divergence equation for the z-system except, again, for the absence of a term involving derivatives of the density.

17.6 Geostrophic and Thermal-Wind Approximations

When one neglects the individual accelerations in the equations of motion, Eq. (17.8), the result is the geostrophic wind equations in a p-system:

$$fv = g\left(\frac{\partial z}{\partial x}\right)_p; \qquad fu = -g\left(\frac{\partial z}{\partial y}\right)_p$$

These equations were derived in an earlier chapter.

We also found in Chapter 12 that the thermal-wind equations could be written

$$\frac{\partial u}{\partial z} = -\frac{g}{fT}\left(\frac{\partial T}{\partial y}\right)_p; \qquad \frac{\partial v}{\partial z} = \frac{g}{fT}\left(\frac{\partial T}{\partial x}\right)_p$$

These are not completely in a p-system since derivatives with respect to height appear. When we apply transformation Eq. (17.1) we get

$$p\frac{\partial u}{\partial p} = \frac{R}{f}\left(\frac{\partial T}{\partial y}\right)_p; \qquad p\frac{\partial v}{\partial p} = -\frac{R}{f}\left(\frac{\partial T}{\partial x}\right)_p$$

or

$$\frac{\partial u}{\partial \ln p} = \frac{R}{f}\left(\frac{\partial T}{\partial y}\right)_p; \qquad \frac{\partial v}{\partial \ln p} = -\frac{R}{f}\left(\frac{\partial T}{\partial x}\right)_p$$

CHAPTER 18

VISCOSITY AND TURBULENCE

18.1 The Fundamental Law of Viscosity

Consider a layer of fluid confined between two horizontal rigid plates as in Fig. 18.1. If the bottom plate is fixed in position and the upper plate is set in motion parallel to itself with constant speed V, two facts become apparent. First, it will be necessary to exert a certain force F upon the plate to keep it moving with a constant speed. This force will be proportional to the area of the plate and to the speed V. It will be inversely proportional to the distance z separating the two plates. Thus $F = \mu A\ V/z$, where μ is a constant of proportionality. Second, the fluid between the two plates will move along the direction of motion of the upper plate with speed V next to the upper plate, and zero speed next to the lower plate. In the steady state the variation of speed is linear, so that F is also proportional to the derivative of V with respect to z. We may then write

$$\tau = \mu \frac{\partial V}{\partial z} \tag{18.1}$$

where τ is the stress or applied force per unit area, F/A.

Equation (18.1) was anticipated by Newton and is the basis of our ideas on molecular viscosity. The coefficient μ is the *dynamic coefficient of viscosity*. It has dimensions $[ML^{-1}T^{-1}]$. The higher the value of this coefficient the more viscous is the fluid. An ideal fluid is one in which $\mu = 0$. This condition is an abstraction, however, that does not exist in reality, since all real fluids possess some viscosity.

The existence of viscosity is attributable to the fact that real fluids are not continuous media but are made up of discrete molecules. In Fig. 18.1 we may consider successive layers of fluid Δz thick, within each of which a certain average horizontal speed of translation of the molecules exists, varying from layer to layer. Because of the random motion of the discrete molecules, an exchange of fluid particles must take place between one layer and the next. The exchanged particles carry with them the lateral momentum of their original layers, causing the average momentum of each layer to change as a result. But from Newton's second law of motion, such a change of momentum with

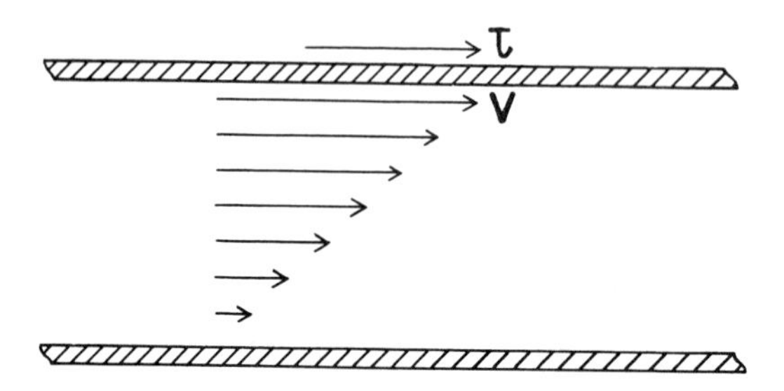

FIG. 18.1. Fluid shear between two parallel plates

time corresponds to a force. Thus contiguous layers of fluid can exert tangential forces upon each other and stress can be transmitted through a fluid. An additional mechanism for the exertion of such stresses exists because of the attraction of molecules of a fluid for each other. Neighboring molecules may therefore be dragged along by these loose binding forces.

So far we have discussed only laminar or molecular viscosity. If in our experiment with the sliding plates (Fig. 18.1) the speed of the upper plate is gradually increased, at a certain speed the smooth, gliding, laminar motion suddenly gives way to an irregular, chaotic motion with visibly large eddies transferring fluid in the z direction as well as parallel to the plates. This state of turbulence will occur when a certain critical value of the Reynolds number, $Vz\rho/\mu$, is exceeded. When this happens the force required to keep the upper plate in motion suddenly increases to many times the previous value. When turbulent flow prevails, momentum is transferred across the direction of average flow mostly by bodily movement of large masses of fluid (eddies), and transport by molecular processes is no longer of direct importance. Since large-scale eddy exchange of momentum is a much more efficient process than small-scale molecular exchange of momentum, the stress required increases greatly. For one-dimensional flow a similar equation may be written for the required stress:

$$\tau = \mu_e \frac{\partial \bar{V}}{\partial z} \tag{18.2}$$

where μ_e is the *coefficient of eddy viscosity* and $\bar{V}$ is the velocity averaged over a period of time. Equation (18.2) is not as valid a relationship as Eq. (18.1), because μ_e is not a constant that is characteristic of the fluid at a given temperature and pressure, as is μ. The eddy viscosity coefficient also depends upon the state of motion.

In turbulent flow, eddies of all sizes can be observed, the largest eddies being limited by the dimensions of the fluid system. Thus a continuous hierarchy exists from the largest down to the smallest eddies, with molecular diffusion occupying the bottom of the scale. Kinetic energy is transferred between eddies of various sizes in a way which has been described clearly in a verse of L. F. Richardson:

Big whirls have little whirls that feed upon their velocity
And little whirls have lesser whirls and so on to viscosity.

18.2 The Equations of Motion Including Viscosity

We shall now discuss the form of the viscous terms in the equations of motion. For the sake of brevity and simplicity the demonstration will not be complete, but will be intuitive and heuristic.

Consider a cube of fluid as in Fig. 18.2. In general there are nine kinds of stresses that must be considered, arising from the three fundamental kinds of cube faces, across any of which a stress may be exerted due to motion in any of the three coordinate directions. We shall denote by the first subscript to τ the kind of face acted upon, and by the second subscript the direction of the motion producing the stress. The face acted upon will be indicated by the coordinate direction which is perpendicular to the face. Thus τ_{yx} means a force per unit area on a face normal to the y axis due to motion in the x direction. The nine possible stresses may be arranged as follows:

$$\begin{Bmatrix} \tau_{xx} & \tau_{xy} & \tau_{xz} \\ \tau_{yx} & \tau_{yy} & \tau_{yz} \\ \tau_{zx} & \tau_{zy} & \tau_{zz} \end{Bmatrix}$$

The above array is called the stress tensor. We shall assert without proof that this array is symmetrical about a diagonal from upper left to lower right.[1] That is, the three stresses to the left of the diagonal are equal to the corresponding three stresses to the right of the diagonal:

$$\tau_{yx} = \tau_{xy}; \qquad \tau_{zx} = \tau_{xz}; \qquad \tau_{zy} = \tau_{yz}$$

[1] See H. LAMB, *Hydrodynamics*, 6th ed., Dover, N.Y., 1945, pp. 571-72.

Thus there are only six independent stresses to deal with. The three elements along the diagonal are perpendicular forces per unit area, the remaining three are tangential stresses.

To calculate the net viscous force on an element of fluid, consider in Fig. 18.2 the stresses on the faces perpendicular to the z axis due

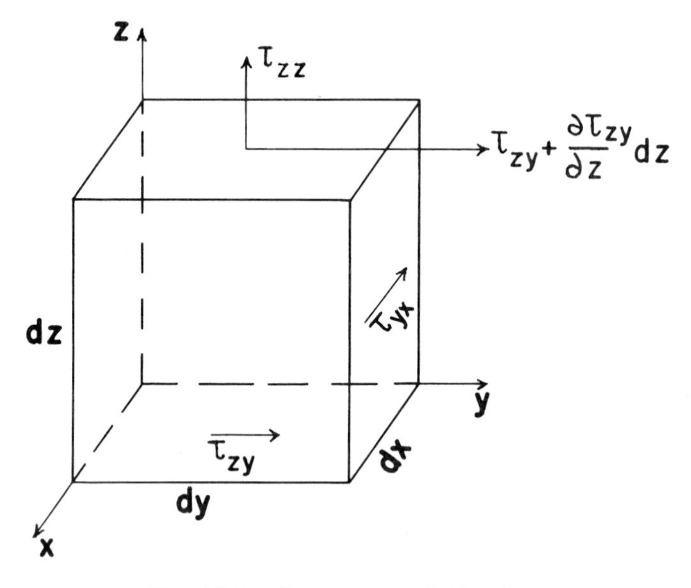

FIG. 18.2. Stresses on a fluid cube

to motion along y. The stress exerted across the bottom z-face by the fluid element on the fluid below is τ_{zy}. If the cube is infinitesimally small, the stress exerted across the upper z-face by the fluid above on the cube is $\tau_{zy} + (\partial\tau_{zy}/\partial z)dz$. The total force on the cube is the difference between these two stresses:

$$\left(\tau_{zy} + \frac{\partial\tau_{zy}}{\partial z}\, dz\right) dxdy - \tau_{zy}\, dxdy = \frac{\partial\tau_{zy}}{\partial z}\, dxdydz$$

Thus the force per unit mass due to this one kind of stress is

$$\frac{1}{\rho}\frac{\partial\tau_{zy}}{\partial z} = \frac{1}{\rho}\frac{\partial}{\partial z}\left(\mu\frac{\partial v}{\partial z}\right)$$

This may be simplified further if μ is constant.

The remaining stresses may be dealt with similarly with the result, when they are properly combined, that an additional term must be added to the appropriate equation of motion. Thus the y-equation of motion becomes

$$\frac{dv}{dt} = -fu - \frac{1}{\rho}\frac{\partial p}{\partial y} + \frac{1}{\rho}\left(\frac{\partial\tau_{xy}}{\partial x} + \frac{\partial\tau_{yy}}{\partial y} + \frac{\partial\tau_{zy}}{\partial z}\right) \tag{18.3}$$

If one inserts the law, Eq. (18.1), and assumes μ to be constant, the additional terms in the case of an incompressible fluid are:

$$\text{in the } x \text{ direction;}\quad \frac{\mu}{\rho}\left(\frac{\partial^2 u}{\partial x^2}+\frac{\partial^2 u}{\partial y^2}+\frac{\partial^2 u}{\partial z^2}\right)$$

$$\text{in the } y \text{ direction;}\quad \frac{\mu}{\rho}\left(\frac{\partial^2 v}{\partial x^2}+\frac{\partial^2 v}{\partial y^2}+\frac{\partial^2 v}{\partial z^2}\right)$$

$$\text{in the } z \text{ direction;}\quad \frac{\mu}{\rho}\left(\frac{\partial^2 w}{\partial x^2}+\frac{\partial^2 w}{\partial y^2}+\frac{\partial^2 w}{\partial z^2}\right)$$

These are the terms arising from viscosity which must be added to each of the equations of motion. They represent the effects of molecular diffusion of momentum. The combination $\mu/\rho = \nu$ is called the *kinematic coefficient of viscosity* and has dimensions $[L^2T^{-1}]$.

18.3 The Equations of Mean Motion in Turbulent Flow

In turbulent motion the velocity components at any given point fluctuate widely as successive random eddies pass by. Nevertheless, if one observes events at a point in such a fluid for a period of time it usually becomes clear that there is a general trend to the motion which underlies the fluctuations. The velocity averaged over a certain period of time will then be a steadier, more slowly changing function than the velocity itself. Thus suggests that one should deal with a mean velocity, $\bar{u}$, defined by

$$\bar{u} = \frac{1}{\Delta t}\int_{t-\Delta t/2}^{t+\Delta t/2} u\,dt \tag{18.4}$$

The time, Δt, is to be chosen just long enough to smooth out the random eddies and reveal the mean underlying flow. We may then write

$$u = \bar{u} + u',$$

where u is the actual instantaneous velocity, $\bar{u}$ is the mean velocity, and u' is the instantaneous departure from the mean (not necessarily small). For a given interval, Δt, it is obvious that the mean of the departures is zero, $\bar{u}' = 0$.

The equations of motion are valid for instantaneous motion. Let us see what they are like for mean motion. The y-equation of motion may be written

$$\rho\frac{\partial v}{\partial t} + \rho u\frac{\partial v}{\partial x} + \rho v\frac{\partial v}{\partial y} + \rho w\frac{\partial v}{\partial z} = -\rho f u - \frac{\partial p}{\partial y} \tag{18.5}$$

where molecular viscosity is neglected. The equation of continuity for an incompressible fluid may be written

$$\rho v \frac{\partial u}{\partial x} + \rho v \frac{\partial v}{\partial y} + \rho v \frac{\partial w}{\partial z} = 0 \tag{18.6}$$

Addition of Eq. (18.5) and (18.6) gives

$$\frac{\partial(\rho v)}{\partial t} + \frac{\partial(\rho v u)}{\partial x} + \frac{\partial(\rho v v)}{\partial y} + \frac{\partial(\rho v w)}{\partial z} = -\rho f u - \frac{\partial p}{\partial y}$$

We shall now form the mean of each term in accordance with the operation defined in Eq. (18.4). Note that the following kind of equality is true by virtue of the vanishing of the mean of a departure:

$$\overline{v\, v} = \overline{(\bar{v} + v')(\bar{v} + v')} = \bar{v}\,\bar{v} + \overline{v'\, v'}$$

where terms like $\overline{v'\, v'}$ do not vanish if there is a non-zero correlation between the two departures whose product is averaged. We then get

$$\frac{\partial(\rho\bar{v})}{\partial t} + \frac{\partial(\rho\bar{v}\bar{u})}{\partial x} + \frac{\partial(\rho\bar{v}\bar{v})}{\partial y} + \frac{\partial(\rho\bar{v}\bar{w})}{\partial z} = -\rho f\bar{u} - \frac{\partial\bar{p}}{\partial y} - \frac{\partial(\rho\overline{v'u'})}{\partial x} - \frac{\partial(\rho\overline{v'v'})}{\partial y} - \frac{\partial(\rho\overline{v'w'})}{\partial z} \tag{18.7}$$

The averaged equation of continuity when multiplied by $\bar{v}$ is

$$\bar{v}\frac{\partial\bar{u}}{\partial x} + \bar{v}\frac{\partial\bar{v}}{\partial y} + \bar{v}\frac{\partial\bar{w}}{\partial z} = 0 \tag{18.8}$$

Subtraction of Eq. (18.8) from Eq. (18.7) gives

$$\frac{\partial\bar{v}}{\partial t} + \bar{u}\frac{\partial\bar{v}}{\partial x} + \bar{v}\frac{\partial\bar{v}}{\partial y} + \bar{w}\frac{\partial\bar{v}}{\partial z} = -f\bar{u} - \frac{1}{\rho}\frac{\partial\bar{p}}{\partial y} - \frac{1}{\rho}\left[\frac{\partial(\rho\overline{v'u'})}{\partial x} + \frac{\partial(\rho\overline{v'v'})}{\partial y} + \frac{\partial(\rho\overline{v'w'})}{\partial z}\right] \tag{18.9}$$

This is an equation of mean motion for a turbulent incompressible fluid. It is completely parallel to Eq. (18.3) in which three terms for viscous stress appeared on the right. Here three eddy-stress terms appear on the right instead. Strictly speaking, Eq. (18.9) should have both viscous- and eddy-stress terms, but the eddy terms are generally so much larger that viscous terms may be neglected. One should note that the eddy stresses arise out of averaging the nonlinear terms of the equations of motion.

It is often assumed that the eddy stresses are proportional to the gradient of the mean wind, i.e.,

$$-\rho\overline{v'w'} = \mu_{ev}\frac{\partial\bar{v}}{\partial z} \tag{18.10}$$

In this way Eq. (18.9) and the other two analogous equations of motion may be put into a form that contains only mean quantities and their derivatives. While such a result has great utility it is not always well grounded. The quantity μ_{ey} is often called an "Austausch" coefficient or exchange coefficient, as well as eddy viscosity.

18.4 Modeling and Dynamic Similitude

In many important cases of fluid flow the phenomena to be studied are on such a large scale that it is advisable to construct smaller models for laboratory experiments. Examples are the general circulation of the atmosphere, tidal flow of water in a harbor, and the diffusion of contaminants emanating from a smoke stack. In order for the model to give results that will be applicable to the prototype, a certain geometric resemblance must exist between them and the ratios of the important forces must be equal in the two cases—in other words, there must be combined geometric and dynamic similitude between the model and the prototype.

In order to determine the important ratios we shall transform the basic equations of motion to nondimensional form. That is, both the dependent and independent variables will be transformed to new variables which have no dimensions. When this is done, certain coefficients of the various terms will appear, and when these coefficients are the same in model and prototype the dynamical equations will be identical for both. We shall deal with the x-equation of motion including viscous stress owing to vertical variation of wind:

$$\frac{\partial u}{\partial t} + u\frac{\partial u}{\partial x} + v\frac{\partial u}{\partial y} + w\frac{\partial u}{\partial z} = fv - \frac{1}{\rho}\frac{\partial p}{\partial x} + \nu\frac{\partial^2 u}{\partial z^2} \tag{18.11}$$

and the hydrostatic equation

$$0 = -g - \frac{1}{\rho}\frac{\partial p}{\partial z} \tag{18.12}$$

Since the form of the y-equation of motion is the same as that of the x-equation, the results will be the same and we may give the argument for only one of the two equations.

The method for rendering these equations nondimensional is to select certain constant reference quantities which are characteristic of the physical system to be studied, and to use them to relate the dimensional variables to new nondimensional ones. The characteristic reference quantities we shall choose are Ω, a characteristic frequency (for example, the earth's angular velocity); U, a representative wind speed; a, a characteristic length (for example, the radius of the earth); and $\neg$, a representative density of the fluid medium. Obviously the choice of these parameters is partly arbitrary; those chosen here have been suggested with the intention that large-scale meteorological systems will be investigated. New nondimensional variables, denoted by an asterisk, are produced by the following transformations which utilize the selected reference quantities.

$$t^* = \Omega t,\ c^* = c/U,$$
$$x^* = x/a,\ y^* = y/a,\ z^* = z/a,$$
$$\alpha^* = \neg\,\alpha,\ p^* = p/\neg U^2$$

Here c and c^* represent any of the three components of velocity. When the dimensional variables in Eq. (18.11) and (18.12) are replaced by the above nondimensional ones we get

$$\Omega U \frac{\partial u^*}{\partial t^*} + \frac{U^2}{a}\left(u^*\frac{\partial u^*}{\partial x^*} + v^*\frac{\partial u^*}{\partial y^*} + w^*\frac{\partial u^*}{\partial z^*}\right) = \Omega U\, 2\sin\phi\, v^* - \frac{U^2}{a}\alpha^*\frac{\partial p^*}{\partial x^*} + \frac{\nu U}{a^2}\frac{\partial^2 u^*}{\partial z^{*2}}$$

and

$$0 = -g - \frac{U^2}{a}\alpha^*\frac{\partial p^*}{\partial z^*}$$

If we divide the first equation by ΩU and multiply the second equation by a/U^2 we get

$$\frac{\partial u^*}{\partial t^*} + \left[\frac{U}{a\Omega}\right]\left(u^*\frac{\partial u^*}{\partial x^*} + v^*\frac{\partial u^*}{\partial y^*} + w^*\frac{\partial u^*}{\partial z^*}\right) = 2\sin\phi\, v^* - \left[\frac{U}{a\Omega}\right]\times \alpha^*\frac{\partial p^*}{\partial x^*} + \frac{[U/a\Omega]}{[Ua/\nu]}\frac{\partial^2 u^*}{\partial z^{*2}}$$

and

$$0 = -\frac{1}{[U^2/ga]} - \alpha^*\frac{\partial p^*}{\partial z^*}$$

These nondimensional equations are parallel in form to the original dimensional relationships except for the appearance of three non-dimensional ratios enclosed in brackets. These are:

the Reynolds number, $Re = \dfrac{Ua}{\nu}$

the Froude number, $F = \dfrac{U^2}{ga}$

the Rossby number, $Ro = \dfrac{U}{a\Omega}$

If the model and prototype have equal values of these three numbers and geometric similitude of the physical boundaries the flow patterns and dynamics of the two systems will be identical. Because it is frequently impossible to satisfy all of these criteria simultaneously, the model is only partially similar to the real phenomenon.

These three numbers have significance which extends beyond their use in modeling. To demonstrate this we shall write these numbers in the following form

$$Re = \frac{U^2/a}{\nu U/a^2}, \qquad F = \frac{U^2/a}{g}, \qquad Ro = \frac{U^2/a}{\Omega U}$$

where each quantity now has a numerator equal to U^2/a. Now consider the advective terms in the equations of motion, represented by $u\,\partial u/\partial x$. Such terms represent the effects of inertia in the fluid since they express the transport of momentum by the motion (inertia) of the fluid. Hence they are referred to as *inertial forces* (per unit mass). When expressed nondimensionally,

$$u\frac{\partial u}{\partial x} = \left[\frac{U^2}{a}\right] u^* \frac{\partial u^*}{\partial x^*}$$

Therefore, inertial forces are proportional to U^2/a and the numerators of *Re*, *F*, and *Ro* represent the magnitude of the inertial forces. In the same fashion one can show quite simply that the denominators represent the magnitudes of the viscous (or eddy) forces, gravitational forces, and Coriolis forces respectively.

Since the Reynolds number is the ratio of the inertial force to the viscous force, its numerical value in any specific case is a measure of the relative importance of these forces. From this it is possible to deduce certain broad characteristics of fluid motion for small and large values of *Re*. Physically we know that turbulence is a result of inertial effects, because high speeds and high gradients of speed are conducive to turbulent mixing. On the other hand, viscous forces inhibit turbulence, since the effect of viscosity is to smooth out irregularity of flow. As an example, it is more difficult to produce turbulent flow in a viscous syrup than in water. Thus we would expect that when the Reynolds number is small, and the inertial forces thus do not dominate the viscous forces, the flow should be laminar. When the Reynolds number is large, and the inertial forces are dominant, the flow should be turbulent. This is precisely the result found by Osborne Reynolds in a classical series of experiments. One must exceed a certain critical value of Reynolds' number in order to produce turbulent flow.

Froude's number is the ratio of inertial force to gravitational force, and must have similar values in model and prototype whenever gravity is an important influence. As an example of its application in modeling, suppose we wish to construct a model of a large-scale flow in which the dimensions are reduced by a factor of 100. Since gravity cannot easily be reduced it is necessary that the speeds be reduced ten times in order to keep F the same.

The Rossby number is the ratio of inertial force to Coriolis force. Very large values of this parameter mean that the Coriolis effect may be neglected. This will occur if the rotation rate is slow (small Ω), or if the dimensions of the system are not large (small a). This is why Coriolis forces are unimportant in the dynamics of everyday, small-scale phenomena despite the rotation of the earth. If the Rossby number is small then rotation is an important influence. This will be the case when the characteristic size of a system is large, as in many meteorological and oceanographic phenomena.

18.5 The Mixing-Length Theory

In molecular viscosity, momentum is exchanged between adjacent layers of fluid by the random component of motion of the molecules. An important concept is then the average distance a molecule can move before it collides with another molecule, called the mean free path. Prandtl[2] has treated the case of turbulent momentum exchange by setting up a reasonable analogy between the motion of molecules and the motion of macroscopic elements of turbulent fluid. Consider a case in which the flow is turbulent but the *mean* motion is in the y-direction with speed increasing upward ($\partial\bar{v}/\partial z > 0$). Prandtl assumed that when a parcel of fluid is moved from its original level by turbulence it carries with it the mean momentum in the y-direction of that level and, after having moved vertically a certain distance l', will produce a turbulent fluctuation v' which will be absorbed abruptly by the fluid at that height. Thus an exchange of momentum between layers can occur by eddy transport of momentum just as it is by molecular transport which ends in absorption of momentum by collision of molecules.

If a mass of fluid originates at the level $z + \Delta z$ with mean speed $\bar{v}_{z+\Delta z}$ and moves downward to the level z where the mean speed is $\bar{v}_z$,

[2] L. PRANDTL, Bericht über Untersuchungen zur ausgebildeten Turbulenz. *Zeitschr. angew. Math. Mech.*, 5, 1925, pp. 136-139.

the perturbation produced will be $v' = \bar{v}_{z+\Delta z} - \bar{v}_z$. But if Δz is small, $v' = \Delta z\, \partial\bar{v}/\partial z$. Now the sign of Δz depends upon whether the fluid mass originates above or below z and the displacement may be written as $l' = -\Delta z$, since fluid originating above z (Δz positive) is displaced downward (l' negative). Thus $v' = -l'\, \partial\bar{v}/\partial z$ and the eddy stress is

$$\tau_{zy} = -\rho\, \overline{v'w'} = \rho\, \overline{w'l'}\, \frac{\partial \bar{v}}{\partial z} \tag{18.13}$$

Mass continuity suggests that horizontal and vertical eddy speeds are equal. That is, $|w'| = |v'|$, for when a fluid mass is transported to another level an equivalent mass should, on the average, move away horizontally with the same speed to make room. Thus $w' = l'\, |\partial\bar{v}/\partial z|$, where the absolute magnitude sign is used because w' and l' must have the same sign, and

$$\tau_{zy} = \rho\, \overline{l'^2} \left|\frac{\partial \bar{v}}{\partial z}\right| \frac{\partial \bar{v}}{\partial z}$$

We may define a root-mean-square mixing length, $l_y \equiv \sqrt{\overline{l'^2}}$ so that

$$\tau_{zy} = \rho\, l_y^2 \left|\frac{\partial \bar{v}}{\partial z}\right| \frac{\partial \bar{v}}{\partial z}$$

This means that the eddy exchange coefficient of Eq. (18.10) can be expressed in terms of the mean mixing length:

$$\mu_{ey} = \rho\, l_y^2 \left|\frac{\partial \bar{v}}{\partial z}\right|$$

A similar result can be derived when the mean flow is in the x-direction with the result that

$$\tau_{zx} = \rho\, l_x^2 \left|\frac{\partial \bar{u}}{\partial z}\right| \frac{\partial \bar{u}}{\partial z} \qquad \text{and} \qquad \mu_{ex} = \rho\, l_x^2 \left|\frac{\partial \bar{u}}{\partial z}\right| \tag{18.14}$$

where we have retained the possibility of having mean mixing lengths and eddy viscosities which are different in the various coordinate directions.

The mean mixing length for a turbulent eddy is a rough analog of the mean free path of a molecule in laminar diffusion. Although these results are based on a number of artificially simple assumptions, the Prandtl theory of the mixing length is a useful first approximation which we shall use to elucidate certain significant turbulent processes. We can see immediately that the eddy viscosities are theoretically expected to be more complex than the analogous molecular viscosities because the eddy coefficients are functions of the mean state of motion, while the molecular coefficients are independent of the state of motion.

Taylor has offered an alternative to Prandtl's mixing-length theory in which vorticity rather than momentum is the property transferred by turbulence. The results are similar to but not identical with Prandtl's. Taylor and others[3] have also developed a statistical theory of turbulence in which the mixing is treated as a continuous rather than a discontinuous process.

18.6 Vertical Structure of the Wind in the Lowest Turbulent Layer

With the preceding ideas in mind we shall now investigate two important cases of turbulence near the earth's surface. In both cases we shall assume equilibrium of forces. That is, the accelerations will be taken to be small, and a balance among pressure gradient, Coriolis, and turbulent forces will be assumed. In both cases we shall seek expressions for the vertical variation of the mean horizontal wind.

The horizontal turbulent stresses are due primarily to vertical variations of the mean wind and vertical mixing by turbulence. Therefore we shall neglect stresses which are due to eddies lying in horizontal planes (that is, τ_{xy}, τ_{yx}) and consider only stresses arising from eddies lying in vertical planes (τ_{zx}, τ_{zy}). This is possible because the vertical shear of the mean wind is ordinarily much greater than its horizontal shear. In the very lowest atmospheric layers (approximately the first 50 m) the horizontal wind stresses are observed to be nearly constant with height. Since the stresses are constant the wind direction is also constant, for if there are appreciable variations in wind direction then the component stresses will have to vary also. Thus the first case to be considered in this section is the one in which stress and wind direction are constant through the lowest few hundred feet of the atmosphere.

Without loss of generality we may orient the x-axis parallel to the eddy stress in the lowest layer. Then, from the mixing-length theory

$$\frac{\partial \bar{u}}{\partial z} = \frac{1}{l_x}\sqrt{\frac{\tau_{zx}}{\rho}} \tag{18.15}$$

where, since ρ and τ_{zx} are essentially constant in so thin a layer, the only quantity on the right which can vary with height is l_x. The constant quantity $u_* = \sqrt{\tau_{zx}/\rho}$ has the dimensions of a speed and is called the *friction velocity*.

We cannot integrate Eq. (18.15) to obtain the wind profile in the

[3] See for example, H. L. DRYDEN, A Review of the Statistical Theory of Turbulence. *Quart. of Appl. Math.*, 1, 1943, pp. 7-12.

surface layer until we specify how l_x varies with height. Very near the earth's surface, mixing cannot take place over large distances because the solid boundary interferes. As we consider points higher in the surface layer the turbulence is less and less inhibited by the solid boundary and it is reasonable to suppose that the eddies can be larger than they are at lower levels. The most simple assumption which incorporates this physical idea is that $l_x = kz$, where k is a positive constant. That is, the mixing length is zero at the earth's surface and increases linearly with height. Then $\partial \bar{u}/\partial z = u_*/kz$ and integration yields $\bar{u} = (u_*/k) \ln z + C$, where C is a constant of integration. Over a very smooth surface $\bar{u}$ will become zero exceedingly close to the surface, but when the underlying ground is rough the wind will vanish before $z = 0$, the mean level of the rough surface. Therefore, we shall apply the boundary condition that $\bar{u} = 0$ at $z = z_0$, where z_0 depends upon the roughness of the surface. This gives the *logarithmic wind law*:

$$\bar{u} = \frac{u_*}{k} \ln \frac{z}{z_0} \tag{18.16}$$

where k is a pure number called the *von Karman constant* and z_0 is the *roughness parameter*. Von Karman's constant is nondimensional and has been found to be approximately 0.38 in a wide variety of experiments. The roughness parameter is proportional to (and smaller than) the mean height of the roughness elements. Table 18.1 gives some approximate values of z_0 for various types of underlying surface. Note that Eq. (18.16) does not apply below $z = z_0$, so it is not required that the wind reverse direction below z_0.

TABLE 18.1

VALUES OF THE ROUGHNESS PARAMETER

Type of surface	z_0, (cm)
smooth snow	0.5
fallow field	2.1
low grass	3.2
high grass	3.9
wheat field	4.5

The simple logarithmic wind law fits the observations reasonably well wherever the temperature lapse rate in the lowest layer is neutrally stable. Then mechanical turbulence predominates and is neither augmented by thermally induced turbulence (unstable case) nor surpressed by thermal stratification (stable case). Figure 18.3 gives the observed mean wind profiles taken on several days at O'Neill, Nebraska[4] at 4:35 p.m., when the lapse rates between 2 and 16 m were most nearly neutral. These profiles are plotted on a logarithmic scale of height and are remarkably straight. This, of course, is required if Eq. (18.16) governs the motion.

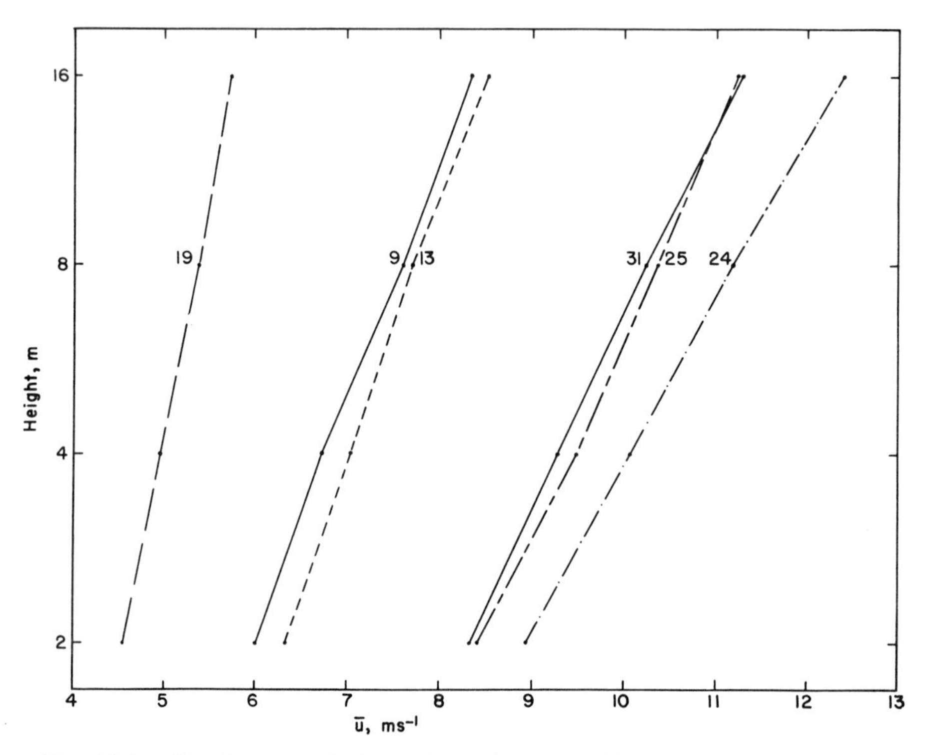

FIG. 18.3. Hourly mean wind speed as a function of height (on a logarithmic scale) at O'Neill, Nebraska. The mean is centered at 1635 CST on the indicated dates in August 1953.

When the lapse rate is not neutral, the logarithmic wind law is not a good description of the wind profile. Various empirical laws have been suggested which incorporate the stability; one of these is the *power law* in which

$$\bar{u} = \bar{u}_1 \left(\frac{z}{z_1}\right)^m$$

where $0 \leq m \leq 1$ is a nondimensional exponent which depends upon the thermal stability. This exponent has been found empirically to

[4] H. H. LETTAU and B. DAVIDSON, ed., *Exploring the Atmosphere's First Mile*, Pergammon, N.Y., 1957.

decrease with increasing lapse rate, with a value about 1/7 for the neutral case.

Recently Deacon[5] has suggested that

$$\frac{\partial \bar{u}}{\partial z} = \frac{u_*}{kz_0}\left(\frac{z}{z_0}\right)^{-\beta}$$

where β decreases with increasing values of the parameter

$$Ri = \frac{\theta_5 - \theta_{0.2}}{\bar{u}_1^{\,2}}$$

Here the numerator measures the mean lapse rate of potential temperature from 0.2 to 5 m, and the denominator, the square of the mean wind speed at 1 m, measures the kinetic energy of the mean motion. Clearly when $\beta = 1$ this case reduced to the logarithmic wind law; this is verified by observation. Integration from $\bar{u} = 0$ at $z = z_0$ to $\bar{u}$ at z yields

$$\bar{u} = \frac{u_*}{k(1-\beta)}\left[\left(\frac{z}{z_0}\right)^{1-\beta} - 1\right] \qquad (18.17)$$

Equation (18.17) is called the *Deacon profile.*

18.7 Vertical Structure of the Wind above the Lowest Turbulent Layer

Above 50 to 100 m it is no longer realistic to assume, as we did in the previous section, that the turbulent stress is independent of height. Observations indicate that there the stress decreases with height until a level is reached where the stress is negligible and the wind blows primarily in response to the pressure-gradient and Coriolis forces. That is, approximate geostrophic balance is reached, outside of the tropics. This height, called the *gradient level,* is found near 1000 m in most cases. Thus the layer marked by decreasing stress is at least ten times thicker than the layer marked by constant stress. The second case to be considered will be this upper turbulent layer.

As before, we shall assume (1) horizontal mean motion; (2) horizontal mean wind shears small compared to the vertical mean wind shears; and (3) a balance among Coriolis, pressure-gradient, and eddy-viscosity forces at every level.

[5] E. L. DEACON, Vertical Diffusion in the Lowest Layers of the Atmosphere. *Quart. J. R. Meteor. Soc.*, 75, 1949, p. 89.

Under these conditions the horizontal equations of motion may be written

$$0 = -\frac{\partial p}{\partial x} + \rho f v + \frac{\partial \tau_{zx}}{\partial z}$$
$$0 = -\frac{\partial p}{\partial y} - \rho f u + \frac{\partial \tau_{zy}}{\partial z} \qquad (18.18)$$

Here we have omitted all eddy forces except those due to vertical shear, in accord with assumption (2) above. This is a quite realistic step. The variables in Eq. (18.18) are all mean quantities but the bars have been omitted for convenience. If we combine these two equations into one complex expression by multiplying the second of Eq. (18.18) by i, the square root of minus one, and adding, we get:

$$0 = -\left(\frac{\partial p}{\partial x} + \mathrm{i}\frac{\partial p}{\partial y}\right) - \rho f \mathrm{i}(u + \mathrm{i}v) + \frac{\partial}{\partial z}(\tau_{zx} + \mathrm{i}\tau_{zy})$$

We shall now assume (4) the existence of an eddy-exchange coefficient, ρK, which is independent of height. This is a quite different assumption than the previous one of constant stress. Since we will find in the result that the shears decrease upward to the gradient level, this assumption means the stress will decrease upward and vanish at the gradient level. In this case

$$\tau_{zx} + \mathrm{i}\tau_{zy} = \rho K\left(\frac{\partial u}{\partial z} + \mathrm{i}\frac{\partial v}{\partial z}\right)$$

and the complex equation of motion becomes

$$0 = -\left(\frac{\partial p}{\partial x} + \mathrm{i}\frac{\partial p}{\partial y}\right) - \rho f \mathrm{i}(u + \mathrm{i}v) + \rho K\frac{\partial^2}{\partial z^2}(u + \mathrm{i}v)$$

We shall also assume that (5) in the shallow layer to be dealt with, pressure gradient and density do not vary appreciably with height and may be taken as constant.

If, in conjunction with this last assumption, we orient the x-axis parallel to the surface isobars with a positive geostrophic wind, u_g, we lose no generality and find that $\partial p/\partial x = 0$, $\partial p/\partial y = -\rho f u_g$, and we get

$$\frac{d^2}{dz^2}(u + \mathrm{i}v - u_g) - \frac{\mathrm{i}f}{K}(u + \mathrm{i}v - u_g) = 0 \qquad (18.19)$$

where partial differentiation with respect to z has been replaced by ordinary differentiation since z is the only independent variable. Equation (18.18) is a second-order, homogeneous, linear, differential equation with constant coefficients. Together with the boundary conditions, it determines the vertical distribution of the dependent variable, $u + \mathrm{i}v - u_g$. Since the upper turbulent layer we are dealing with is so much larger than the lower turbulent layer of the previous section, we shall ignore the surface layer and assume that the wind vanishes exactly at the earth's surface, and approaches the geostrophic value as elevation increases. The two boundary conditions needed for such a second-order equation are then:

$$\text{at } z = 0 \text{ let } u + iv = 0$$
$$\text{at } z = \infty \text{ let } u + iv = u_g$$

The solution to Eq. (18.19) is a standard one. It may be written

$$u + iv - u_g = A\, e^{\sqrt{\frac{if}{K}}\, z} + B\, e^{-\sqrt{\frac{if}{K}}\, z} \tag{18.20}$$

where A and B are two arbitrary constants to be determined. That Eq. (18.20) does satisfy Eq. (18.19) may be verified by direct substitution.

It is known that $\sqrt{i} = (1 + i)/\sqrt{2}$, which may be verified by squaring both sides. Therefore

$$u + iv - u_g = A\, e^{a(1 + i)z} + B e^{-a(1 + i)z}$$

where $a = \sqrt{f/2K}$. Application of the second boundary condition gives $A = 0$. Application of the first boundary condition yields $B = -u_g$. Therefore we may write the solution as

$$u + iv = u_g\, [1 - e^{-az}\, e^{-iaz}]$$

Now we apply Euler's relationship: $e^{-iaz} = \cos az - i \sin az$ (see problem 2, p. 228). When we separately equate the real and imaginary parts we get the final form of the solution:

$$\begin{aligned} u &= u_g\, (1 - e^{-az} \cos az) \\ v &= u_g\, e^{-az} \sin az \end{aligned} \tag{18.21}$$

To interpret this result we shall consider how the wind changes with elevation. The ratio of v to u is the tangent of the angle the wind makes with the eastward direction (and with the isobars). At $z = 0$, $u = v = 0$ and this ratio is indeterminate. If we evaluate this indeterminate quantity by differentiating numerator and denominator separately with respect to z and examine the ratio of the derivatives (L'Hôpital's rule) we find that v/u approaches $+1$ as z approaches zero. Thus the wind direction at or close to $z = 0$ makes an angle of 45° with the isobars and points toward low pressure. The speed is zero precisely at $z = 0$, but increases with increasing elevation up to a certain point. Because of the trigonometric functions present in Eq. (18.21) the wind direction turns continuously with elevation. At $z = \pi/a = \pi\sqrt{2K/f}$ the v component will vanish for the first time and the wind will become parallel to the isobars. Thus the wind vector turns clockwise with elevation in the Northern Hemisphere. Observations indicate that the wind first becomes parallel to the isobars at about 1 km, therefore $K \approx 5 \times 10^4$ cm² s⁻¹ in middle latitudes ($f \approx 10^{-4}$ s⁻¹). This is of the order of 10^5 times the molecular kinematic viscosity.

As the wind turns in response to the trigonometric terms of

Eq. (18.21) the amplitudes of these terms decrease exponentially. At $z = 1$ km, $e^{-az} \approx 1/23$. Thus the wind speed and direction converge rather rapidly on the geostrophic values. This fact removes an implied contradiction between the assumptions and one of the boundary con-

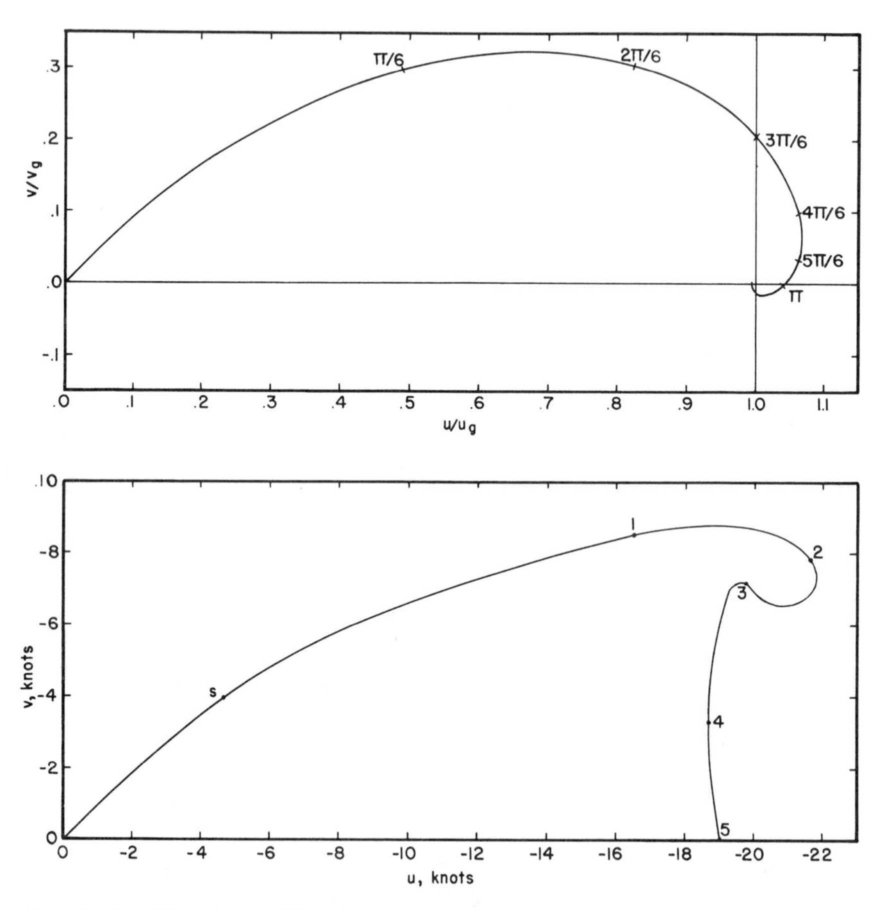

FIG. 18.4. Theoretical Ekman spiral (top) on which points are marked with values of the parameter az. Observed wind hodograph at Tallahassee, Florida, 0500 GMT, 15 June 1958 (bottom) on which points are marked with values of height in thousands of feet.

ditions. We have assumed constant density, pressure gradient, and eddy viscosity throughout the layer. But the second boundary condition applies at $z = \infty$ which implies that these quantities must be constant through the entire atmosphere, a clearly unrealistic requirement. We now see, however, that for all practical purposes the wind becomes geostrophic in 1 km rather than at an infinite height. Thus ρ, $\partial p/\partial x$, and K need be constant only in the lowest kilometer, and this is a far less stringent requirement.

The nature of the solution may be seen graphically by means of a polar-coordinate plot of wind direction tangentially against wind speed radially for successive heights. This is called a *hodograph* and may be described also as a Cartesian-coordinate plot of v against u for successive elevations. Such a hodograph is shown at the top of Fig. 18.4. This result is called the *Ekman spiral*, after W. F. Ekman, who first obtained an analogous result for the surface layers of the ocean.

The bottom of Fig. 18.4 gives the actual wind variation with elevation on a selected day. The resemblance of this observed wind structure in the lowest 3000 ft to the theoretical structure derived here is obvious. It must be pointed out, however, that such clear-cut resemblances cannot be expected on all occasions because one or more of the assumptions made in deriving the Ekman spiral may not be satisfied. We shall mention only two examples. First, the Austausch coefficient is often variable with elevation. This should not prevent the wind from turning clockwise with height in the friction layer but may cause it to turn at a different rate than the theoretical one. Second, if strong geostrophic cold or warm advection is present in addition to friction then the pressure gradient will change rapidly with height instead of remaining constant. In the case of cold air advection, which turns the geostrophic wind counterclockwise with height, the clockwise turning due to viscosity may be completely reversed.

18.8 Diffusion of Other Properties

The preceding sections of this chapter have dealt with viscous and turbulent diffusion of momentum. The same ideas can be applied to the diffusion of water vapor, heat, atmospheric contaminants, and the like; it will be the purpose of this section to elaborate the theory of such diffusion and present a few applications.

Consider some identifiable atmospheric property, $\mathcal{J}$, which will be measured in terms of a quantity per gram of air. For example, if water vapor is being considered, $\mathcal{J}$ may be grams of vapor per gram of air (specific humidity) and if smoke is being considered $\mathcal{J}$ might be particles of smoke per gram of air. We can calculate from Fig. 18.5 the rate at which $\mathcal{J}$ is transported in any one of the coordinate directions, for example upward. The upward transfer per unit time of $\mathcal{J}$ through the bottom face is the mass transport of air multiplied by the value of $\mathcal{J}$ at that face: $\mathcal{J}\rho w dxdy$. Similarly the upward transport through the top face is $\mathcal{J}\rho w dxdy + \partial/\partial z\ (\mathcal{J}\rho w)\ dxdydz$. The difference between the rate of gain through the bottom and the loss through the top is

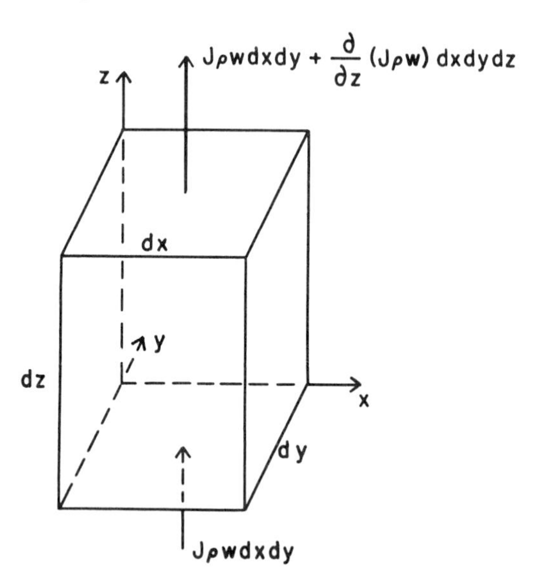

FIG. 18.5. Parallelopiped for calculation of the convergence of the transport of a property $\mathcal{J}$

$-\ \partial/\partial z\ (\mathcal{J}\rho w)\ dxdydz$, and dividing by the volume of the parellopiped gives the net rate of increase of the quantity per unit volume: $-\ \partial/\partial z$ $(\mathcal{J}\rho w)$. Analogous calculations can be performed for the other two coordinate directions and when the results are added we get an expression for the local rate of increase of the property in question per unit volume:

$$\frac{\partial}{\partial t}(\rho\mathcal{J}) = -\frac{\partial}{\partial x}(\mathcal{J}\rho u) - \frac{\partial}{\partial y}(\mathcal{J}\rho v) - \frac{\partial}{\partial z}(\mathcal{J}\rho w) \tag{18.22}$$

We next resolve the instantaneous values into mean and eddy values: $\mathcal{J} = \bar{\mathcal{J}} + \mathcal{J}'$, $w = \bar{w} + w'$ and ignore turbulent fluctuations of density which are relatively small. If we insert these results into Eq. (18.22) and average with respect to time, remembering that $\bar{\mathcal{J}}' = \bar{w}' = 0$, we get

$$\begin{aligned}\frac{\partial}{\partial t}(\rho\bar{\mathcal{J}}) = &-\left[\frac{\partial}{\partial x}(\rho\bar{\mathcal{J}}\,\bar{u}) + \frac{\partial}{\partial y}(\rho\bar{\mathcal{J}}\,\bar{v}) + \frac{\partial}{\partial z}(\rho\bar{\mathcal{J}}\,\bar{w})\right] \\ &-\left[\frac{\partial}{\partial x}(\rho\overline{\mathcal{J}'u'}) + \frac{\partial}{\partial y}(\rho\overline{\mathcal{J}'v'}) + \frac{\partial}{\partial z}(\rho\overline{\mathcal{J}'w'})\right]\end{aligned} \tag{18.23}$$

If we expand the left side and the first brackets of the right side of Eq. (18.23) and utilize the mean mass-continuity equation

$$\frac{\partial \bar{\rho}}{\partial t} = - \left[\frac{\partial}{\partial x} (\rho \bar{u}) + \frac{\partial}{\partial y} (\rho \bar{v}) + \frac{\partial}{\partial z} (\rho \bar{w}) \right]$$

we obtain

$$\begin{aligned} \frac{\partial \bar{\mathcal{J}}}{\partial t} &= - \left[\bar{u} \frac{\partial \bar{\mathcal{J}}}{\partial x} + \bar{v} \frac{\partial \bar{\mathcal{J}}}{\partial y} + \bar{w} \frac{\partial \bar{\mathcal{J}}}{\partial z} \right] \\ &\quad - \frac{1}{\rho} \left[\frac{\partial}{\partial x} (\rho \overline{\mathcal{J}'u'}) + \frac{\partial}{\partial y} (\rho \overline{\mathcal{J}'v'}) + \frac{\partial}{\partial z} (\rho \overline{\mathcal{J}'w'}) \right] \end{aligned} \tag{18.24}$$

This form asserts that the local rate of change of the mean value of $\mathcal{J}$ is produced by two effects. First is mean advection of $\bar{\mathcal{J}}$, and second is convergence of the eddy transport of $\mathcal{J}$ by the turbulent wind. In the derivation of Eq. (18.24) we have implicitly assumed that no production or destruction of the property in question occurs in the volume considered. If such sources or sinks do exist (for example, production of contaminants by a smoke stack or consumption of a gas like SO_2 by photochemical processes) then the strength of these sources and sinks will have to be added to the right side of Eq. (18.24) with appropriate signs.

We shall now concentrate on the eddy terms in Eq. (18.24) and introduce the Prandtl mixing length, as we have already done for diffusion of momentum. Since in Prandtl's theory the property transported by a turbulent eddy is assumed to remain unchanged in value until it has moved a distance l', the quantities most suitable for application of this theory are those which are conserved, such as water vapor (in the absence of condensation) and potential temperature (under adiabatic conditions). By the same argument which gave rise to Eq. (18.13) and those following we find that

$$- \rho \overline{\mathcal{J}'w'} = \rho \overline{w'l'} \frac{\partial \bar{\mathcal{J}}}{\partial z} = A_z \frac{\partial \bar{\mathcal{J}}}{\partial z}$$

where $A_z = - \rho \overline{w'l'}$ is the eddy exchange coefficient for vertical diffusion of the property in question. One can derive similarly expressions for diffusion in the two horizontal directions with the result that Eq. (18.24) can be written:

$$\begin{aligned} \frac{\partial \bar{\mathcal{J}}}{\partial t} = - \left[\bar{u} \frac{\partial \bar{\mathcal{J}}}{\partial x} + \bar{v} \frac{\partial \bar{\mathcal{J}}}{\partial y} + \bar{w} \frac{\partial \bar{\mathcal{J}}}{\partial z} \right] + \frac{1}{\rho} \left[\frac{\partial}{\partial x} \left(A_x \frac{\partial \bar{\mathcal{J}}}{\partial x} \right) + \frac{\partial}{\partial y} \left(A_y \frac{\partial \bar{\mathcal{J}}}{\partial y} \right) \right. \\ \left. + \frac{\partial}{\partial z} \left(A_z \frac{\partial \bar{\mathcal{J}}}{\partial z} \right) \right] \end{aligned} \tag{18.25}$$

where the last terms on the right in brackets represent turbulent flux-convergence of $\mathcal{J}$ into the volume.

One may expect that different values of the exchange coefficients will be needed for each property being diffused. Observations indicate this is so, but because the variability among values of A for various properties is comparable to the error in measuring A, one frequently assumes the same exchange coefficients for momentum, water vapor, and heat.

We shall now consider an application of these results to the vertical turbulent diffusion of water vapor in the absence of mean advection. Our purpose will be to obtain a formula from which evaporation from an ocean or lake can be calculated using standard observations above and at the water surface. The rate of upward eddy transport of specific humidity q is

$$T_q = -A_z \frac{\partial \bar{q}}{\partial z}$$

Since A_z is positive, an upward transport will occur when $\partial \bar{q}/\partial z$ is negative—that is, when the specific humidity decreases upward. This is an example of the general rule that turbulent mixing of conservative properties produces a transport from high to low values of the property. Because we are concerned with the lowest 50 m or so of the atmosphere we shall make an assumption parallel to that made for diffusion of momentum, namely the transport is constant in this lowest layer. Since the evaporation E is the upward transport at or near $z = 0$, we must calculate this height-independent transport. We shall assume that the eddy viscosity (exchange coefficient for momentum) is equal to the eddy diffusivity A_z (exchange coefficient for vapor). Then from Eq. (18.14),

$$A_z = \mu_{ex} = \rho\, l_x^2 \left| \frac{\partial \bar{u}}{\partial z} \right|$$

But from Eq. (18.15), $\partial u/dz = u_*/l_x$; therefore $A_z = \rho l_x u_*$. Since in this lowest layer we have previously assumed $l_x = kz$, we find that $A_z = \rho k u_* z$, and

$$E = T_q = -\rho k u_* z \frac{\partial \bar{q}}{\partial z}$$

If we write this in finite-difference form between mean deck level (6 m) and z_0, the roughness parameter, we get

$$E = -\frac{\rho k u_*}{\ln (6/z_0)} (\bar{q}_6 - \bar{q}_{z_0})$$

The roughness of the sea surface is a function of the wind speed, and Rossby has found that for $u \geq 6.5$ m s^{-1} the sea surface is aerodynamically rough with $z_0 = 0.6$ cm. Furthermore, from the logarithmic wind law $u_* = k\bar{u}_6/\ln(6/z_0)$. Therefore,

$$E = -\frac{\rho k^2}{(\ln 1000)^2}\,\bar{u}_6\,(\bar{q}_6 - \bar{q}_{z_0})$$

But $q = .622\ e/p$, and if we take $p = 1013$ mb,

$$E = -\left[\frac{.622\ \rho\ k^2}{1013\ (\ln 1000)^2}\right]\bar{u}_6\,(\bar{e}_6 - \bar{e}_{z_0})$$

Finally, if we take a standard value of density at sea level, $\rho = 1.3 \times 10^{-3}$ g cm^{-3}, $k = 0.38$, and express E in cm of liquid water per hour, u_6 in m s^{-1}, and e in mb, we get

$$E = -\,8.7 \times 10^{-4}\,\bar{u}_6\,(\bar{e}_6 - \bar{e}_{z_0}) \qquad (18.26)$$

The numerical coefficient will be about 1/3 as large for winds below 6.5 m s^{-1} because the sea surface will then be aerodynamically smooth.

Equation (18.26) gives the evaporation rate in terms of quantities which can be measured easily from a ship. The wind speed is obtainable from an anemometer at or near deck level, e_s is obtainable from a psychrometer, and e_{z_0} is the vapor pressure so close to the surface that the air may be considered saturated at the temperature of the water. Thus e_{z_0} is a function only of the water temperature, which is easily and routinely measured.

Jacobs has determined the mean distribution of evaporation from the sea by analysis of the heat budget. When the climatological means of $\bar{u}_6$, $\bar{e}_6$, and $\bar{e}_{z_0}$ are substituted in Eq. (18.26) together with the independently calculated values of E, the value of the constant of proportionality becomes 6×10^{-4}, which is somewhat lower than the theoretical value. This discrepancy is partially explained by the fact that some of the time the wind is below 6.5 m s^{-1} even though the average is generally above this value. Thus the climatological value of the constant represents a combination of smooth and rough seas. Various investigators have reported observationally determined values of the constant in Eq. (18.26) which differ appreciably. Some of the factors contributing to these differences are variations in the roughness of the water surface, variations in the common elevation of the observing instruments, and different elevations for the anemometer and psychrometer. The important fact is that observations bear out the theoretical conclusion that evaporation increases with increasing wind speed and with decreasing moisture content of the air at deck level.

As our second application we shall consider the turbulent diffusion of kinetic energy. We shall obtain from this choice a criterion for the growth, maintenance, or decay of turbulence, ignoring mean advection of energy and concerning ourselves only with mean motion in the x-direction and vertical diffusion of its energy. In addition, since energy can be produced by work or transformed to other forms, we shall have to consider possible sources and sinks, S. Equation (18.25) then is to be written

$$\frac{\partial \bar{\mathcal{J}}}{\partial t} = -\frac{1}{\rho}\frac{\partial}{\partial z}(\rho\, \overline{\mathcal{J}'w'}) + S \tag{18.27}$$

Then $\bar{\mathcal{J}} \equiv \overline{u^2}/2$, and since $u = \bar{u} + u'$, $\bar{\mathcal{J}} = \frac{1}{2}(\bar{u}^2 + \overline{u'^2})$. We must find also a suitable expression for $\mathcal{J}' \equiv \frac{1}{2}\,(u^2)'$. Since $u^2 = \overline{u^2} + (u^2)'$, $\mathcal{J}' = \frac{1}{2}\,(u^2 - \overline{u^2})$. But from $u = \bar{u} + u'$ we have $u^2 = \bar{u}^2 + 2\bar{u}u' + u'^2$ and $\overline{u^2} = \bar{u}^2 + \overline{u'^2}$. Therefore, $\mathcal{J}' = \frac{1}{2}\,(2\bar{u}u' + u'^2 - \overline{u'^2})$ and Eq. (18.27) becomes

$$\frac{\partial}{\partial t}\left(\frac{\bar{u}^2 + \overline{u'^2}}{2}\right) = -\frac{1}{\rho}\frac{\partial}{\partial z}\left[\rho\, \overline{\left(\bar{u}\, u' + \frac{u'^2}{2}\right) w'}\right] + S \tag{18.28}$$

We now consider the case in which the mean flow is steady, so $\bar{u}^2$ is independent of time. Further, consider the two terms in brackets on the right: $\rho\, \bar{u}\, \overline{u'w'}$ and $\rho\overline{(u'^2/2)w'}$. The latter term is the eddy flux of eddy-kinetic energy. There are several circumstances in which this term may be negligibly small. First, if the eddy velocities are appreciably smaller than the mean velocity, the second term is small compared to the first term. Second, even if the eddy speeds are comparable to the mean motion, the second term contains the mean of the product of a positive quantity (u'^2) and a quantity with randomly alternating sign (w'). Therefore its value will tend to be small compared to the first term which contains the mean product of two quantities which tend to be opposite in sign. Third, in the case when no difference in magnitude of the two terms arises from these causes, the eddy flux of eddy energy may have no important vertical gradient while the first term ordinarily will have a vertical gradient because of the presence of $\bar{u}$. On these three grounds we shall neglect the vertical derivative of the eddy flux of eddy energy. With these two simplifying assumptions Eq. (18.28) becomes

$$\frac{\partial}{\partial t}\left(\frac{\overline{u'^2}}{2}\right) = -\frac{1}{\rho}\frac{\partial}{\partial z}(\rho\, \bar{u}\, \overline{u'w'}) + S$$

Since $\tau_{zx} = -\rho\, \overline{u'w'}$, this may be written as

$$\frac{\partial}{\partial t}\left(\frac{\overline{u'^2}}{2}\right) = \frac{\tau_{zx}}{\rho}\frac{\partial \bar{u}}{\partial z} + \frac{\bar{u}}{\rho}\frac{\partial \tau_{zx}}{\partial z} + S \tag{18.29}$$

This is an equation for the local rate of change of the eddy-kinetic energy per unit mass.

We now consider possible sources and sinks of kinetic energy.

One of these is the rate at which work is done by the turbulent forces, S_1. From the mean equation of motion in the x-direction the turbulent force per unit mass owing to vertical diffusion is $(1/\rho)\,\partial\tau_{zx}/\partial z$ and the rate at which the turbulent force contributes energy to the mean motion is $(\bar{u}/\rho)\,\partial\tau_{zx}/\partial z$. Since this is done at the expense of the eddy energy it is a sink, and

$$S_1 = -\frac{\bar{u}}{\rho}\frac{\partial\tau_{zx}}{\partial z}$$

The second contribution comes from the combined effect of pressure forces and gravity. We shall neglect all effects of horizontal pressure gradients and consider only the nonhydrostatic difference between the vertical pressure-gradient force per unit mass and gravity. In Chapter 7 we found that when a parcel of air is displaced from position 1 (at $z + l'$) to position 2 (at z) the unbalanced force per unit mass is $g\,(T_1 - T_2)/T_2$, where T_1 is the temperature of the displaced parcel after it reaches position 2. But for dry adiabatic processes this is equal to

$$g\,\frac{\bar{\theta}_{z+l'} - \bar{\theta}_z}{\bar{\theta}_z}$$

where θ is potential temperature. But

$$\bar{\theta}_{z+l'} = \bar{\theta}_z + l'\frac{\partial\bar{\theta}}{\partial z}$$

if l' is small. Therefore $\theta' = \bar{\theta}_{z+l'} - \bar{\theta}_z = l'\,\partial\bar{\theta}/\partial z$ and the unbalanced acceleration is $(gl'/\bar{\theta})\,\partial\bar{\theta}/\partial z$. The rate at which this vertical force does work is $w'l'\,(g/\bar{\theta})\,\partial\bar{\theta}/\partial z$ and the mean rate is $\overline{w'l'}\,(g/\bar{\theta})\,\partial\bar{\theta}/\partial z$. When $\partial\bar{\theta}/\partial z$ is positive (stability), the eddies will lose energy to the gravity-pressure field. Thus $S_2 = -\,\overline{w'l'}\,(g/\bar{\theta})\,\partial\bar{\theta}/\partial z$.

All other forces will be neglected, since they are not ordinarily important in turbulent phenomena. The chief of these is the Coriolis force which is not significant in the small-scale turbulence we are concerned with.

With these results Eq. (18.29) becomes

$$\frac{\partial}{\partial t}\left(\frac{\overline{u'^2}}{2}\right) = \frac{\tau_{zx}}{\rho}\frac{\partial\bar{u}}{\partial z} + \frac{\bar{u}}{\rho}\frac{\partial\tau_{zx}}{\partial z} - \frac{\bar{u}}{\rho}\frac{\partial\tau_{zx}}{\partial z} - \overline{w'l'}\,\frac{g}{\bar{\theta}}\frac{\partial\bar{\theta}}{\partial z}$$

in which the second and third terms on the right cancel. Now $\tau_{zx} = A_M\,\partial\bar{u}/\partial z$ and $\overline{w'l'} = A_H/\rho$, where A_M and A_H are the eddy-diffusion constants for momentum and heat respectively. Thus

$$\frac{\partial}{\partial t}\left(\frac{\overline{u'^2}}{2}\right) = \frac{A_M}{\rho}\left(\frac{\partial\bar{u}}{\partial z}\right)^2 - \frac{A_H}{\rho}\frac{g}{\bar{\bar{\theta}}}\frac{\partial\bar{\theta}}{\partial z}$$

Here the left side is the local rate of change of the eddy-kinetic energy per unit mass. If this rate of change is positive the turbulence increases in magnitude; if it is zero the turbulence is constant; and if it is negative the turbulence decreases in magnitude. Thus if we define the non-dimensional *Richardson number*[6],

$$Ri = \frac{\dfrac{g}{\bar{\theta}} \dfrac{\partial \bar{\theta}}{\partial z}}{\left(\dfrac{\partial \bar{u}}{\partial z}\right)^2}$$

we find that

$$Ri < \frac{A_M}{A_H} \qquad \text{for turbulence increasing with time,}$$

$$Ri = \frac{A_M}{A_H} \qquad \text{for turbulence constant with time,}$$

$$Ri > \frac{A_M}{A_H} \qquad \text{for turbulence decreasing with time.}$$

One can understand physically how this comes about because the denominator of Richardson's number contains the shear of the mean wind, and the bigger this shear the greater the tendency for eddies to form; further, the numerator of *Ri* contains the thermal stability, and the more stable the lapse rate the more turbulence is suppressed.

The problem remains of determining the appropriate value of the ratio A_M/A_H. If the diffusion constants for heat and momentum are taken to be equal the ratio is unity. This assumption is often made. However, attempts to verify the Richardson criteria experimentally suggest that the value of A_M/A_H may lie between 1 and ½ or be even lower. The variations observed are not unexpected because, as has been pointed out earlier, eddy-diffusion coefficients are dependent upon the wind field and are not unique constants.

One of the phenomena which can be understood qualitatively by means of Richardson's criteria is the vertical wind structure which commonly occurs on a cold, clear winter night. Under such conditions the layer of air near the ground is extremely stable. That is, the numerator of *Ri* is quite large. At the same time a pronounced increase of wind speed with height is observed, since the surface wind decreases to nearly zero while the wind aloft remains strong. Nevertheless, turbulence does not develop because the great thermal stability permits *Ri* to remain above the critical value.

---

[6] Note that the parameter *Ri* used by Deacon on page 279 is proportional to the Richardson number.

PROBLEMS

1. Suppose one wished to make a crude model of the general circulation of the atmosphere consisting of a flat circular pan filled with water, rotating in a horizontal plane about its center, and heated at the outer edges while being cooled at the axis of rotation. If the pan rotates at the rate of 10 revolutions per minute and has a diameter of 30 cm, what must be the ratio of the relative water speeds developed in the model to corresponding wind speeds in order that one kind of dynamic similitude shall exist?

2. In Fig. 18.3, on Aug 31, the values of $\bar{u}$ for heights of 2 m and 16 m were 8.34 m s^{-1} and 11.28 m s^{-1} respectively. Calculate τ_{zx} and z_0 in this case.

3. Re-derive the Ekman spiral in the case where the geostrophic wind is not constant but $u_g = u_{g_0} + Cz$, where C is the vertical shear of the geostrophic wind. Compute the numerical values of u/u_{g_0} and v/u_{g_0} for az ranging from 0° to 210° by increments of 30° in the three cases:

$$C = \frac{au_{g_0}}{4}, \qquad C = 0, \qquad C = -\frac{au_{g_0}}{4}.$$

Plot the hodograph of these three results.

4. Suppose a forest fire has released a large number of smoke particles of uniform size into the air. These try to settle downward at a constant rate but also are diffused upward by turbulence. Assume all advective effects are negligible, the concentration of smoke at every level remains constant for several days, and K, the kinematic coefficient of eddy viscosity, is constant. Given that the smoke concentration is zero at great heights and has the value $\mathcal{J}_0$ at the ground, show that at other heights

$$\mathcal{J} = \mathcal{J}_0\, e^{-\frac{w}{K} z}$$

where w is the terminal velocity of fall of the particles. If $K = 10^5$ cm^2 s^{-1} and $\mathcal{J} = \mathcal{J}_0/2$ at $z = 1$ km, calculate the terminal velocity of the smoke particles.

CHAPTER 19

ENERGY AND STABILITY RELATIONSHIPS

19.1 The Energy Equation

As has been pointed out earlier, the ultimate source of virtually all of the energy of the atmosphere is solar energy. When radiation is absorbed at the earth's surface or in the atmosphere, it appears as internal energy. One of the basic problems of atmospheric science is to determine how this internal energy is converted into other atmospheric forms (potential, kinetic, latent), and to formulate the laws governing the relationships among these various forms.

One important relationship of this sort can be obtained from the equations of motion. If we multiply the x, y, and z tangent-plane Eq. (11.10) by u, v, and w respectively, we get

$$\frac{d}{dt}\left(\frac{u^2}{2}\right) = fuv - uw\,(2\,\Omega\cos\phi) - u\alpha\frac{\partial p}{\partial x} + uF_x$$

$$\frac{d}{dt}\left(\frac{v^2}{2}\right) = -fuv - v\alpha\frac{\partial p}{\partial y} + vF_y$$

$$\frac{d}{dt}\left(\frac{w^2}{2}\right) = uw(2\Omega\cos\phi) - w\alpha\frac{\partial p}{\partial z} - wg + wF_z$$

where F_x, F_y, F_z represent the general viscous and frictional forces per unit mass. When we add all three equations we get

$$\frac{d}{dt}\left(\frac{u^2+v^2+w^2}{2}\right) = -\alpha\left(u\frac{\partial p}{\partial x} + v\frac{\partial p}{\partial y} + w\frac{\partial p}{\partial z}\right) - wg + (uF_x + vF_y + wF_z)$$

Let $u^2 + v^2 + w^2 = c^2$ where c is the magnitude of the three-dimensional velocity vector. Furthermore $w = dz/dt$, so that

$$\frac{d}{dt}\left(\frac{c^2}{2} + gz\right) + \alpha\left(u\frac{\partial p}{\partial x} + v\frac{\partial p}{\partial y} + w\frac{\partial p}{\partial z}\right) - (uF_x + vF_y + wF_z) = 0 \qquad (19.1)$$

The two terms in the first parentheses are the specific kinetic energy and the specific gravitational potential energy of an atmospheric parcel. The remaining two sets of terms in parentheses represent, respectively, the work done on a unit mass by the pressure field as a parcel crosses isobaric surfaces (nongeostrophic flow) and the energy dissipated by the frictional forces. The latter term will ordinarily be negative because this effect converts kinetic energy into heat. Note that the Coriolis terms cancel each other—a consequence of the property of the Coriolis force whereby it can change the direction of a parcel but not its speed or energy.

Equation (19.1) deals only with mechanical forms of energy since it is derived from the equations of motion. In order to include thermal energy we must use the First Law of Thermodynamics, Eq. (3.11), in the form

$$\frac{dh}{dt} = c_v \frac{dT}{dt} + p \frac{d\alpha}{dt} \tag{19.2}$$

where dh represents an increment in specific heat added to a parcel in the time dt. Since this last may be converted to mechanical energy as well as thermal forms we must add the left side of Eq. (19.1) to the right side of Eq. (19.2). We shall at the same time utilize the fact that

$$u \frac{\partial p}{\partial x} + v \frac{\partial p}{\partial y} + w \frac{\partial p}{\partial z} = \frac{dp}{dt} - \frac{\partial p}{\partial t}$$

We then obtain

$$\frac{dh}{dt} = \frac{d}{dt}\left(\frac{c^2}{2} + gz\right) + \alpha \frac{dp}{dt} - \alpha \frac{\partial p}{\partial t} - (uF_x + vF_y + wF_z) + c_v \frac{dT}{dt} + p \frac{d\alpha}{dt}$$

But

$$\alpha \frac{dp}{dt} + p \frac{d\alpha}{dt} = \frac{d(\alpha p)}{dt}$$

therefore

$$\frac{dh}{dt} = \frac{d}{dt}\left(\frac{c^2}{2} + \alpha p + gz + c_v T\right) - \alpha \frac{\partial p}{\partial t} - (uF_x + vF_y + wF_z) \tag{19.3}$$

If the motion is adiabatic, if a steady state exists, and if viscous effects may be neglected, Eq. (19.3) may be integrated with respect to time with the following result:

$$\frac{c^2}{2} + \alpha p + gz + c_v T = \text{constant} \tag{19.4}$$

Since we are integrating an individual derivative in the steady state, the combination on the left of Eq. (19.4) is constant along a streamline. The constant may vary from one streamline to another.

Either Eq. (19.3) or Eq. (19.4) is called the *atmospheric energy equation.* The four terms on the left are various forms of energy, and Eq. (19.4) simply says that the sum of these energies is constant for a parcel under the conditions assumed. This is a special case of Bernoulli's equation for an incompressible fluid:

$$\frac{c^2}{2} + \alpha p + gz = \text{constant}$$

As we follow an air parcel along a streamline (in the steady state) the pressure may have to change as the elevation of the parcel changes. For example, in flow of air over a hill the air rises and generally the speed increases, resulting in a decrease in pressure. Barometric observations from such a location then may not be representative of the static pressure for the area. A similar phenomenon occurs in flow over a building. The resulting pressure drop may amount to a few mb.

19.2 Internal and Potential Energies

The internal energy, per unit horizontal area, of a layer of air dz thick is $dE_I = \rho c_v T dz$. The internal energy, per unit horizontal area, of an atmospheric column extending from sea level to height h is

$$E_I = c_v \int_0^h \rho T dz \qquad (19.5)$$

From the hydrostatic equation $\rho dz = - dp/g$, therefore

$$E_I = -\frac{c_v}{g} \int_{p_0}^{p_h} T dp \qquad (19.6)$$

Either of these expressions may be used to compute E_I.

The potential energy per unit horizontal area, of a layer of atmosphere dz thick is $dE_P = \rho\, g\, z\, dz$, where z is the mean elevation of the layer above sea level. The corresponding quantity for a layer from sea level to h is

$$E_P = g \int_0^h \rho z dz \qquad (19.7)$$

Once again, use of the hydrostatic equation permits us to replace height by pressure as the variable of integration:

$$E_P = - \int_{p_0}^{p_h} z dp \qquad (19.8)$$

A simple relationship exists between E_I and E_P. Integration of Eq. (19.8) by parts yields

$$E_P = - p_h h + \int_0^h p dz$$

But $p = \rho RT$, therefore

$$E_P = -p_h h + R \int_0^h \rho T dz$$

With the use of Eq. (19.5), this becomes

$$E_P = -p_h h + \frac{R}{c_v} E_I \tag{19.9}$$

Thus the potential energy of a column of height h is proportional to its internal energy. If the column extends to great heights where p_h is essentially zero, then

$$E_P = \frac{R}{c_v} E_I$$

The physical meaning of these relationships is easily seen. If the temperature of a column of air rises, so that E_I increases, then the column must expand vertically. This raises its center of mass and increases the value of E_P also.

19.3 Frictional Dissipation of Kinetic Energy

It is of considerable interest to estimate the rate at which turbulence and friction against the earth's surface dissipate the kinetic energy of the atmosphere. To make such an estimate, consider a case in which the isobars are oriented east-west at all elevations with vertically constant pressure-gradient force per unit mass. We shall assume horizontal motion and a balance of forces at all levels. Thus, below the gradient wind level H the Ekman spiral solution applies and above H geostrophic balance exists. Since $w = 0$, $\partial p/\partial x = 0$, $-\alpha\, \partial p/\partial y = fu_g$, and no accelerations are permitted, the energy equation (19.1) reduces to

$$0 = -fvu_g - (uF_x + vF_y) \tag{19.10}$$

Below H the term in parentheses represents the rate of frictional dissipation of energy and the term $-fvu_g$ represents the work done by the pressure field on air crossing the isobars. Thus the pressure field adds energy at exactly the rate necessary to cancel the frictional loss. Above H, $v = 0$ and friction is negligible so Eq. (19.10) is satisfied because the two terms on the right both vanish. Thus all the frictional loss occurs below H in this model. The total rate of change of kinetic energy per unit horizontal area owing to friction is

$$\frac{dE_K}{dt} = \int_0^H \rho\,(uF_x + vF_y)\, dz = -f \int_0^H \rho u_g v dz$$

Within this layer ρ is essentially constant, $v = u_g e^{-az} \sin az$ from Eq. (18.21), and u_g is constant by assumption. Therefore

$$\frac{dE_K}{dt} = -\rho f u_g^2 \int_0^H e^{-az} \sin az \, dz$$

Since the gradient level is the height at which the wind first becomes parallel to the isobars, $H = \pi/a$ and

$$\frac{dE_K}{dt} = -\frac{\rho f u_g^2 H}{2\pi}(1 + e^{-\pi})$$

The average gradient level is at $H \approx 1$ km, and near the surface $\rho \approx 1.3 \times 10^{-3}$ g cm^{-3}. Therefore

$$\frac{dE_K}{dt} = -21.6 f u_g^2 \tag{19.11}$$

where dE_K/dt will be in ergs cm^{-2} s^{-1} if f and u_g are expressed in c-g-s units.

This dissipation rate must be compared with the total kinetic energy of the entire atmospheric column in order to be meaningful. In the friction layer the kinetic energy is

$$E_{K1} = \rho \int_0^H \frac{u^2 + v^2}{2} dz$$

but here $u = u_g(1 - e^{-az} \cos az)$ and $v = u_g e^{-az} \sin az$. Therefore

$$E_{K1} = \frac{\rho u_g^2}{2} \int_0^H (1 - 2e^{-az} \cos az + e^{-2az}) \, dz$$

or

$$E_{K1} = \frac{\rho H u_g^2}{2}\left[1 - \frac{(1 + e^{-\pi})^2}{2\pi}\right] = 53.8 \, u_g^2$$

where E_{K1} will be in ergs cm^{-2} if u_g is in c-g-s units.

Above the gradient level $v = 0$ and the kinetic energy is

$$E_{K2} = \int_H^\infty \rho \frac{u_g^2}{2} dz = \frac{u_g^2}{2g} \int_0^{p_H} dp$$

Thus

$$E_{K2} = \frac{p_H}{2g} u_g^2 = 458 \, u_g^2$$

since the pressure at 1 km is 899 mb in the U.S. Standard Atmosphere. Here E_{K2} will be in ergs cm^{-2} if u_g is in c-g-s units. Finally, the total kinetic energy is

$$E_K = E_{K1} + E_{K2} = 512 \, u_g^2$$

Consequently, the fractional rate of dissipation is

$$\frac{1}{E_K}\frac{dE_K}{dt} = -\frac{21.6 f}{512} = -4.22 \times 10^{-6} \text{ s}^{-1}$$

in middle latitudes where $f \approx 10^{-4}$ s^{-1}.

If the rate of loss were completely uncompensated, the entire atmospheric column would lose energy at the rate of 36 per cent per day. This would reduce the energy to

$$\frac{1}{e} \text{ of its initial value in } 2.7 \text{ days,}$$

$$\frac{1}{10} \text{ of its initial value in } 6.4 \text{ days,}$$

$$\frac{1}{100} \text{ of its initial value in } 12.8 \text{ days.}$$

The above values are very rough, tending to be underestimates of the dissipation times because we have not included any turbulent loss of energy above the friction layer. They also tend to be overestimates because u_g normally increases with height and we have not incorporated this into our crude model. If we assume these errors nearly balance, it is reasonable to conclude that turbulent dissipation alone can reduce the atmospheric kinetic energy to a small fraction of its original value in about one week.

It is of interest to compare the dissipation rate to the rate of absorption of solar energy by the earth-atmosphere system, since ultimately the sun supplies the energy to run the atmospheric circulation. The mean rate of absorption is $(1 - A)S/4$, where A is the albedo (0.35), S is the solar constant (2.00 ly min^{-1}), and the factor 1/4 comes from the fact that a beam of sunlight of cross-sectional area πa^2 is intercepted by the earth but this energy is spread over the surface area $4\pi a^2$. This mean absorption rate is 2.2×10^5 ergs cm^{-2} s^{-1}. The dissipation rate given by Eq. (19.11) is about 2.2×10^3 ergs cm^{-2} s^{-1} if $u_g =$ 10 m s^{-1}. Therefore the rate of dissipation of kinetic energy is roughly 1 per cent of the rate of absorption of sunlight. Thus the conversion of no more than a few per cent of the available solar energy will suffice to compensate for turbulent destruction of organized kinetic energy. If we view the atmosphere as an engine for the conversion of radiant and thermal energy into kinetic energy with a compensating degradation of kinetic energy by friction, we conclude that the engine operates at a low level of efficiency.

19.4 The Conversion of Potential and Internal Energies to Kinetic Energy

We shall consider a classical model first studied by Margules.[1] Consider two dry air masses of different temperatures arranged side

[1] M. Margules, Zur Sturmtheorie, *Met. Zeit.*, 23, 1906, pp. 481-497.

by side, as in Fig. 19.1a, and initially at rest. Suppose they are allowed to rearrange themselves within the same volume (we assume fixed top and sides to the volume) until a minimum of potential energy is reached. This will occur when the air mass of cooler potential temperature lies completely under the air mass of warmer potential temperature, as in Fig. 19.1b. Since no energy will be allowed to enter or leave the volume,

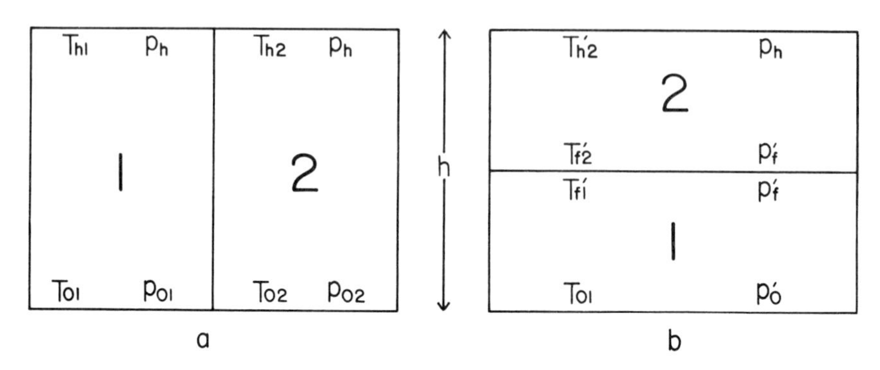

FIG. 19.1. (a) Two air masses of different potential temperature arranged side by side and (b) the same two air masses after a dry adiabatic rearrangement in which the colder air mass has been allowed to sink completely under the warmer one

and since we shall assume all energy to be in the form of kinetic, potential, and internal energies with no turbulent or frictional losses, $E_K + E_P + E_I =$ constant for the system. Since initially the air is at rest there is no kinetic energy at the start. Furthermore, from Eq. (19.9) $E_P = -p_h h + (R/c_v)\, E_I$. Therefore,

$$-p_h h + E_I\left(\frac{R}{c_v} + 1\right) = E'_K - p_h h + E'_I\left(\frac{R}{c_v} + 1\right)$$

where a prime indicates the value after rearrangement of the air masses. Consequently

$$E'_K = \frac{c_p}{c_v}\,(E_I - E'_I) \tag{19.12}$$

In order to calculate the internal energy before and after rearrangement we need to know the temperature distribution in each layer. For simplicity, we shall assume the lapse rates are dry adiabatic. Thus, at any level after rearrangement Poisson's equation (Eq. 3.16) requires that

$$T'_1 = T_1\left(\frac{p'_1}{p_1}\right)^{\kappa} \qquad \text{and} \qquad T'_2 = T_2\left(\frac{p'_2}{p_2}\right)^{\kappa} \tag{19.13}$$

Because the two layers conserve their mass but spread over twice their former horizontal area, the final pressure at any level in mass 2 is given by

$$p_2' - p_h = \tfrac{1}{2}(p_2 - p_h)$$

and the final pressure at any level in mass 1 is given by

$$p_1' - p_h = \frac{p_1 - p_h}{2} + \frac{p_{02} - p_h}{2}$$

Therefore,

$$\begin{aligned} p_2' &= p_2 - \frac{p_2 - p_h}{2} \\ p_1' &= p_1 + \frac{p_{02} - p_1}{2} \end{aligned} \tag{19.14}$$

Substitution of Eq. (19.14) into Eq. (19.13) yields

$$T_1' = T_1 \left[1 + \frac{p_{02} - p_1}{2p_1}\right]^{\kappa}$$

$$T_2' = T_2 \left[1 - \frac{p_2 - p_h}{2p_2}\right]^{\kappa}$$

Provided the layers are not very deep, the second terms in brackets will be small compared to unity, so we may expand each of these expressions by means of the binomial theorem and retain only the first two terms:

$$\begin{aligned} T_1' &\approx T_1 \left[1 + \kappa \frac{p_{02} - p_1}{2p_1}\right] \\ T_2' &\approx T_2 \left[1 - \kappa \frac{p_2 - p_h}{2p_2}\right] \end{aligned} \tag{19.15}$$

From Eq. (19.6), the total internal energies before and after rearrangement of the air masses are

$$\begin{aligned} E_I &= \frac{Ac_v}{g} \left[\int_{p_h}^{p_{01}} T_1\, dp_1 + \int_{p_h}^{p_{02}} T_2\, dp_2\right] \\ E_I' &= \frac{2Ac_v}{g} \left[\int_{p_f'}^{p_0'} T_1'\, dp_1' + \int_{p_h}^{p_f'} T_2'\, dp_2'\right] \end{aligned} \tag{19.16}$$

where it is assumed that initially each air mass occupies the same horizontal area, A. We wish to express E_I' in terms of unprimed temperatures and pressures. Elimination of T_1' and T_2' is easily accomplished by means of Eq. (19.15), and from Eq. (19.14) we find that $dp_1' = dp_1/2$ and $dp_2' = dp_2/2$. Thus

$$E_I' = \frac{Ac_v}{g} \left\{\int_{p_h}^{p_{01}} T_1 \left[1 + \kappa \frac{p_{02} - p_1}{2p_1}\right] dp_1 + \int_{p_h}^{p_{02}} T_2 \left[1 - \kappa \frac{p_{02} - p_h}{2p_2}\right] dp_2\right\}$$

where the limits of integration extend over the entire range of the variables p_1 and p_2 just as they do for the variables p_1' and p_2' in the second of Eq. (19.16). Thus from Eq. (19.12) the kinetic energy becomes

$$E_K' = \frac{RA}{2g}\left[-\int_{p_h}^{p_{01}} T_1 \frac{p_{02}-p_1}{p_1}\,dp_1 + \int_{p_h}^{p_{02}} T_2 \frac{p_2-p_h}{p_2}\,dp_2\right] \quad (19.17)$$

Margules has shown that Eq. (19.17) is accurate to the second order even through second order terms were omitted from Eq. (19.15).

In order to evaluate these integrals, we note from the hydrostatic equation that

$$\int_{p_h}^{p_{01}} T_1 \frac{dp_1}{p_1} = \int_{p_h}^{p_{02}} T_2 \frac{dp_2}{p_2} = \frac{g}{R}\,h$$

Furthermore we may introduce mean temperatures of the two layers before rearrangement as follows:

$$\int_{p_h}^{p_{01}} T_1\,dp_1 = \bar{T}_1(p_{01}-p_h) \qquad \text{and} \qquad \int_{p_h}^{p_{02}} T_2\,dp_2 = \bar{T}_2(p_{02}-p_h)$$

Consequently Eq. (19.17) becomes

$$E_K' = \frac{RA}{2g}\left[\bar{T}_1(p_{01}-p_h) + \bar{T}_2(p_{02}-p_h) - (p_{02}-p_h)\frac{g}{R}\,h\right] \quad (19.18)$$

This may be simplified by using the integrated hydrostatic equation in the form

$$p_{01} = p_h\, e^{\frac{gh}{R\bar{T}_1}} \qquad \text{and} \qquad p_{02} = p_h\, e^{\frac{gh}{R\bar{T}_2}}$$

where T_1 and T_2 are reciprocal mean temperatures which differ insignificantly from the mean temperatures of Eq. (19.18). Since we may expand an exponential in a power series:

$$e^u = 1 + u + \frac{1}{2!}\,u^2 + \ldots.$$

and since we are considering layers which are not too thick, we find

$$\begin{aligned} p_{01} &\approx p_h\left[1 + \frac{gh}{R\bar{T}_1} + \frac{1}{2}\left(\frac{gh}{R\bar{T}_1}\right)^2\right] \\ p_{02} &\approx p_h\left[1 + \frac{gh}{R\bar{T}_2} + \frac{1}{2}\left(\frac{gh}{R\bar{T}_2}\right)^2\right] \end{aligned} \quad (19.19)$$

where the approximations include second order terms because Eq. (19.17) is correct to the second order. With these results Eq. (19.18) becomes

$$E_K' = \frac{RAp_h}{2g}\left[\frac{1}{2}\frac{g^2h^2}{R^2}\left(\frac{1}{\bar{T}_1}+\frac{1}{\bar{T}_2}\right) - \frac{g^2h^2}{R^2\bar{T}_2} - \frac{1}{2}\frac{g^3h^3}{R^3\bar{T}^2}\right]$$

The last term inside the brackets is third order in magnitude and may be neglected. Thus we get

$$E_K' = \frac{Ap_h gh^2}{4R}\left[\frac{\bar{T}_2-\bar{T}_1}{\bar{T}_1\bar{T}_2}\right] \quad (19.20)$$

Now the kinetic energy may be written as $E'_K = \frac{1}{2} M\bar{c}^2$, where M is the total mass of air in the system and $\bar{c}$ is a mean wind speed which can be looked upon as a measure of the energy of the resulting flow. Furthermore,

$$M = \frac{A}{g} [(p_{02} - p_h) + (p_{01} - p_h)]$$

or, with the use of Eq. (19.19),

$$M = \frac{Ap_h}{g} \left[\frac{gh}{R\bar{T}_2} + \frac{1}{2}\left(\frac{gh}{R\bar{T}_2}\right)^2 + \frac{gh}{R\bar{T}_1} + \frac{1}{2}\left(\frac{gh}{R\bar{T}_1}\right)^2 \right]$$

The squared terms are of higher order and may be neglected. Thus

$$M = \frac{Ahp_h}{R} \frac{\bar{T}_1 + \bar{T}_2}{\bar{T}_1 \bar{T}_2}$$

and Eq. (19.20) becomes

$$\bar{c}^2 = \frac{gh}{2} \frac{\bar{T}_2 - \bar{T}_1}{\bar{T}_1 + \bar{T}_2}$$

Since the lapse rates are the same in the two air masses $\bar{T}_2 - \bar{T}_1 = \Delta T$, the temperature difference between the two masses at any level. Furthermore $\bar{T}_1 + \bar{T}_2 \approx 2\bar{T}$, where $\bar{T}$ is the average temperature of the two air masses. Thus we find that

$$c = \frac{1}{2}\sqrt{gh \frac{\Delta T}{\bar{T}}} \qquad (19.21)$$

Equation (19.21) gives the average wind speed developed in the two air masses when a maximum amount of potential and internal energy is converted to kinetic energy and there are no viscous losses. Table 19.1, adapted from Margules, gives the results for $\bar{c}$ for typical values of h, $\bar{T}$, and ΔT.

TABLE 19.1

AVERAGE WIND SPEEDS DEVELOPED BY CONVERSION OF POTENTIAL AND INTERNAL ENERGIES

h, (m)	$\bar{T}$, (°K)	ΔT, (°C)	$\bar{c}$, (m s^{-1})	h, (m)	$\bar{T}$, (°K)	ΔT, (°C)	$\bar{c}$, (m s^{-1})
3000	246	5	12.2	6000	216	5	18.4
3000	248	10	17.3	6000	218	10	26.4

These speeds are not completely unreasonable, although they are somewhat high compared with what we normally observe at low levels. This condition is easily understood, since frictional dissipation of kinetic energy was ignored and because a maximum conversion of potential and internal energies takes an appreciable number of days to accomplish. During this time friction can be expected to take an appreciable toll from the kinetic energy being generated. Margules' theory undoubtedly provides a good description, in principle, of one important mechanism for the generation of atmospheric motion, namely the subsidence of a cold air mass accompanied by the lifting of an adjacent warm air mass. This is precisely what we see on weather maps during the normal life history of an extra-tropical frontal storm. However, the atmosphere generates kinetic energy in other important ways.

19.5 The Mechanical Generation of Kinetic Energy

In order to gain further insight into the physics of the production of atmospheric kinetic energy, we shall study a simple model first proposed by Starr.[2] In this model thermal effects are not included directly and only mechanical production of kinetic energy is involved. Consider a mass of air defined by vertical sides extending from the level surface of the earth to so great a height as to include essentially all the atmosphere. The equations of horizontal motion are

$$\frac{du}{dt} = fv - \alpha \frac{\partial p}{\partial x} + F_x$$

$$\frac{dv}{dt} = -fu - \alpha \frac{\partial p}{\partial y} + F_y$$

If we multiply the first of these by u and the second by v and then add the two equations, we obtain an equation for the kinetic energy of horizontal motion:

$$\rho \frac{\partial}{\partial t}\left(\frac{c^2}{2}\right) + \rho u \frac{\partial}{\partial x}\left(\frac{c^2}{2}\right) + \rho v \frac{\partial}{\partial y}\left(\frac{c^2}{2}\right) + \rho w \frac{\partial}{\partial z}\left(\frac{c^2}{2}\right) = -\left(u \frac{\partial p}{\partial x} + v \frac{\partial p}{\partial y}\right) + \rho u F_x + \rho v F_y$$

where $c^2 = u^2 + v^2$ is proportional to the kinetic energy of horizontal motion. Since the equation of continuity is

$$\frac{\partial \rho}{\partial t} + \frac{\partial \rho u}{\partial x} + \frac{\partial \rho v}{\partial y} + \frac{\partial \rho w}{\partial z} = 0$$

[2] Victor P. STARR, "Applications of Energy Principles to the General Circulation." In *Compendium of Meteorology*, T. F. Malone, ed., Amer. Meteor. Soc., Boston, 1951, pp. 568-574.

the energy equation may be written as

$$\frac{\partial K}{\partial t} + \frac{\partial uK}{\partial x} + \frac{\partial vK}{\partial y} + \frac{\partial wK}{\partial z} = -\left(\frac{\partial pu}{\partial x} + \frac{\partial pv}{\partial y}\right) + p\left(\frac{\partial u}{\partial x} + \frac{\partial v}{\partial y}\right) - \mathfrak{d} \quad (19.21)$$

where K is the horizontal kinetic energy per unit volume, $\rho c^2/2$, and $-\mathfrak{d}$ is the rate at which K is decreasing owing to turbulence and viscosity, $\mathfrak{d} = -(\rho u F_x + \rho v F_y)$.

We now integrate Eq. (19.21) over the entire volume so as to obtain an expression for the local rate of change of the total kinetic energy within the prescribed volume extending from the surface to the "top" of the atmosphere. The result is

$$\begin{aligned}\frac{\partial}{\partial t}\iiint K dxdydz = &-\iiint\left[\frac{\partial uK}{\partial x} + \frac{\partial vK}{\partial y} + \frac{\partial wK}{\partial z}\right] dxdydz \\ &-\iiint\left[\frac{\partial pu}{\partial x} + \frac{\partial pv}{\partial y}\right] dxdydz \\ &+\iiint p\left[\frac{\partial u}{\partial x} + \frac{\partial v}{\partial y}\right] dxdydz - \mathfrak{D} \qquad (19.22)\end{aligned}$$

where $-\mathfrak{D} = -\iiint \mathfrak{d}\, dxdydz$ is the dissipation rate for the entire volume. The first two integrals on the right of Eq. (19.22) may be transformed by means of Gauss's divergence theorem which asserts that if A_x, A_y, A_z are the three Cartesian components of any vector, then

$$\iiint\left[\frac{\partial A_x}{\partial x} + \frac{\partial A_y}{\partial y} + \frac{\partial A_z}{\partial z}\right] dxdydz = \iint A_n\, d\sigma$$

where A_n is the outward-directed component of the vector perpendicular to any element, $d\sigma$, of the surface bounding the volume over which the integration is performed. That is, Gauss's theorem says that the integral over a certain volume of the divergence of a vector is equal to the integral, over the surface bounding the volume, of the outward-directed normal component of the vector. The truth of this theorem can be seen intuitively if **A** is the velocity vector, **V**, for then the left side represents the overall fluid divergence in a given volume, and such net divergence will produce a certain rate of export of fluid through the boundary surface of the volume. But the right side is an expression for just this rate of export. Gauss's theorem is proved in some degree of generality in Appendix II.

In the first integral on the right of Eq. (19.22) we have the divergence of the vector $K\mathbf{V}$. Therefore

$$\iiint\left[\frac{\partial Ku}{\partial x} + \frac{\partial Kv}{\partial y} + \frac{\partial Kw}{\partial z}\right] dxdydz = \iint KV_n\, d\sigma$$

In the second integral on the right of Eq. (19.22) we have the divergence of the two-dimensional vector whose components are pu and pv. Therefore, applying a two-dimensional form of the divergence theorem

$$\int\left(\iint\left[\frac{\partial pu}{\partial x}+\frac{\partial pv}{\partial y}\right]dxdy\right)dz=-\iint p\,(vdx-udy)\,dz$$

Thus Eq. (19.22) becomes

$$\frac{\partial}{\partial t}\iiint Kdxdydz=-\iint KV_n\,d\sigma+\iint p\,(vdx-udy)\,dz$$
$$+\iiint p\left(\frac{\partial u}{\partial x}+\frac{\partial v}{\partial y}\right)dxdydz-\mathfrak{D} \tag{19.23}$$

The first term on the right represents advection of new fluid carrying its kinetic energy across the boundary of the volume and transport out of the volume of fluid carrying some of the kinetic energy once contained within it. Consequently, this term represents a *redistribution* in space of existing kinetic energy. The second term on the right represents the effect of work done by pressure forces along the boundary as air moves across it. Although this a form of production of kinetic energy, it is a purely boundary effect. The third term on the right represents the primary source of production of kinetic energy within the volume. The last term represents the dissipation by frictional and turbulent forces.

Equation (19.23) can be simplified considerably if we consider a mechanically closed system. If, for example, the volume under consideration is enclosed by rigid vertical walls, then no energy can be advected across them and no work can be done at the solid boundary by pressure forces. Equation (19.23) then becomes

$$\frac{\partial}{\partial t}\iiint Kdxdydz=\iiint p\left(\frac{\partial u}{\partial x}+\frac{\partial v}{\partial y}\right)dxdydz-\mathfrak{D}$$

This simple form also applies if the volume considered is the entire atmosphere, for then there are no lateral boundaries. Consequently, we see that the term

$$\iiint p\left(\frac{\partial u}{\partial x}+\frac{\partial v}{\partial y}\right)dxdydz$$

is indeed the primary source of generation of kinetic energy which, in the long run for the entire atmosphere, is balanced by the dissipation term. Since the term $-\mathfrak{D}$ must be negative, the generation term must be positive, if a long-term balance is to be maintained. The generation term may be looked upon as the vertical sum of the contributions from all the horizontal layers and written as

$$\int \left[\iint p \left(\frac{\partial u}{\partial x} + \frac{\partial v}{\partial y} \right) dxdy \right] dz$$

Now the surface integral of the horizontal divergence,

$$\iint \left(\frac{\partial u}{\partial x} + \frac{\partial v}{\partial y} \right) dxdy$$

must vanish in each layer if there is no flow across the lateral boundaries, as is true for rigid walls, or if there are no lateral boundaries, as is true for the entire atmosphere. Thus the generation term will vanish unless, on the whole, p tends to be large where the horizontal divergence is positive and small where the horizontal divergence is negative (convergence). Thus it is clear that areas of divergence are the places where kinetic energy is generated while areas of convergence are places where such energy is lost. The operation of this mechanism can be seen clearly in low-level weather maps. There high-pressure cells are marked by divergence and low-pressure centers by convergence, thus kinetic energy can be generated.

It is of considerable interest to point out that these results indicate one cannot have areas of generation of kinetic energy (divergent areas) without also having areas of destruction of this energy (convergent areas), completely independent of frictional effects. For in a mechanically closed system the net divergence must be zero since the normal velocities on the boundaries must be zero. Thus if there is a divergent area in the system it must be counterbalanced by an equivalent convergent area. In such convergences kinetic energy vanishes by purely hydrodynamic means. The net production rate for energy is the smaller difference between a large generation rate and an inevitable, large, nonfrictional destruction rate.

The implication that divergent highs are the source of kinetic energy may seem contrary to general experience, since high-pressure cells are not ordinarily the centers of major activity. Starr has pointed out, however, that if one considers a volume which encloses only a divergent high-pressure cell, the energy produced within it is exported rapidly by the work done through pressure forces at the boundary and through advection by the outward-directed winds at the boundary. The large amount of kinetic energy often found in low-pressure cells is the result of work done by pressure forces on the convergent winds at the boundaries of the low, and of convergent transport of energy into the low.

19.6 Inertial Stability

In Chapter 7 we considered the hydrostatic stability of parcels displaced vertically. Stability was found to prevail if the algebraic sum of the vertical pressure-gradient and gravity forces produced a net restoring force on a displaced parcel, and instability if these stresses led to a net force in the direction of this displacement. It is profitable to consider an extension of this technique to the horizontal case, in which we can conceive of an environment in which the horizontal forces are in balance (for example, geostrophic equilibrium) and then investigate whether a restoring or a displacing force appears when a parcel is moved horizontally from its equilibrium position.

In order to get a firm grasp on the essential physics of this problem, consider a cylindrical vessel partly filled with water and rotating about a vertical axis. If the rotation rate is large, the centrifugal force will greatly exceed gravity and the water will pile up around the sides of the vessel, as shown in Fig. 19.2. We may then neglect gravity compared to centrifugal force and consider the horizontal balance of forces to exist between the outward-directed centrifugal force and the inward-directed pressure-gradient force. The water surface is then nearly vertical and the motions to be considered will be horizontal. A small disturbance applied to the water surface will cause surface waves to form and lateral fluid displacements to take place. The problem is to determine whether these displacements, hence the waves, will be stable or not.

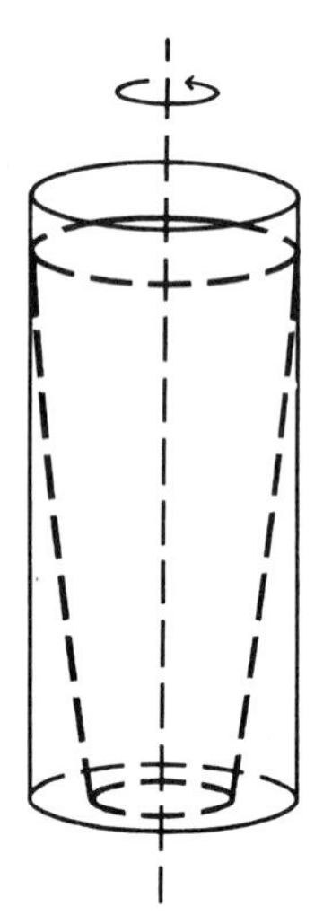

FIG. 19.2. A mass of fluid (heavy lines) in a rapidly rotating vessel

Let R be the distance of any chosen fluid parcel from the axis of rotation and let ω be the angular velocity of rotation at R. The angular momentum per unit mass of such a parcel will be conserved during a radial displacement, provided no tangential forces act:

$$\omega R^2 = \text{constant}$$

If the parcel is displaced outward a distance δR (not necessarily small) the angular velocity becomes $\omega + \Delta\omega$ and, because of conservation of angular momentum,

$$\omega R^2 = (\omega + \Delta\omega)(R + \delta R)^2$$

That is,

$$\omega + \Delta\omega = \frac{\omega R^2}{(R + \delta R)^2}$$

The centrifugal force per unit mass on the displaced parcel is then

$$(\omega + \Delta\omega)^2 (R + \delta R) = \frac{\omega^2 R^4}{(R + \delta R)^3}$$

The environmental fluid at $R + \delta R$ has angular velocity $\omega + \delta\omega$, where $\delta\omega$ represents an increment in the ω of the environment. Thus the centrifugal force per unit mass acting on the nondisplaced fluid at $R + \delta R$ is $(\omega + \delta\omega)^2 (R + \delta R)$. This centrifugal force is exactly balanced by the horizontal pressure gradient. If the parcel arrives at its new position with a smaller centrifugal force than it finds in its new environment, it will be subject to a net restoring force by the pressure gradient. That is, the displacement will be stable if

$$(\omega + \delta\omega)^2 (R + \delta R) > \frac{\omega^2 R^4}{(R + \delta R)^3}$$

By taking square roots this becomes

$$(\omega + \delta\omega)(R + \delta R)^2 > \omega R^2 \tag{19.24}$$

This means that such a rotating fluid will be stable with respect to parcel displacements when the angular momentum increases outward. Clearly the fluid will be unstable if ωR^2 decreases outward and neutral if ωR^2 is constant along a radius.

We have arrived at the criteria, in this simple case, for what is called *inertial stability*. The name is appropriate because this stability or instability depends upon the centrifugal acceleration which, in turn, arises from the inertial terms of the equations of motion.

We now turn to consideration of a simple meteorological situation.[3] Suppose we have a zonal geostrophic current which is steady and which, at any given level, varies only in the y-direction: $u_g(y)$. If we neglect friction the horizontal equations of motion may be written

$$\frac{du}{dt} = fv \qquad \text{and} \qquad \frac{dv}{dt} = f(u_g - u)$$

since there is no east-west pressure gradient. Here u and v are the actual horizontal velocity components, which may be nongeostrophic. We wish to determine whether a parcel of air displaced latitudinally will be forced back to or will accelerate away from its original position. This is the same as determining if the south-north kinetic energy of the parcel will decrease or increase with time. Therefore, we wish to investigate the sign of $d/dt(v^2/2)$. We assume the atmosphere initially to be in

[3] A more complete survey of meteorological inertial stability may be found in C. L. GODSKE, T. BERGERON, J. BJERKNES, R. C. BUNDGAARD, *Dynamic Meteorology and Weather Forecasting*, Amer. Meteor. Soc. and Carnegie Inst. of Wash., Boston, 1957, pp. 294-311.

geostrophic balance. If a parcel is then displaced north or south, its new zonal velocity will be

$$u = u_{g0} + \frac{du}{dt}\,dt$$

where u_{g0} is the geostrophic wind at the initial time and position, and dt is the time during which the lateral displacement occurs. From the first equation of motion above, this may be written

$$u = u_{g0} + fvdt \tag{19.25}$$

The geostrophic wind at the new location of the displaced parcel is approximately

$$u_g = u_{g0} + \frac{\partial u_g}{\partial y}\,dy$$

where dy is the distance displaced, and higher order terms in the series expansion of u_g have been neglected. Since $dy = vdt$, this may be written as

$$u_g = u_{g0} + \frac{\partial u_g}{\partial y}\,vdt \tag{19.26}$$

Subtraction of Eq. (19.25) from Eq. (19.26) gives

$$u_g - u = -\left(f - \frac{\partial u_g}{\partial y}\right) vdt$$

But we may eliminate $u_g - u$ between this result and the second equation of motion above, with the result:

$$\frac{dv}{dt} = -\left(f - \frac{\partial u_g}{\partial y}\right) fvdt$$

If we multiply through with v we get

$$\frac{d}{dt}\left(\frac{v^2}{2}\right) = -\left(f - \frac{\partial u_g}{\partial y}\right) fv^2\,dt$$

which is the desired result. Since f is positive in the Northern Hemisphere, v^2 is necessarily positive, and dt is positive, this means the stability criteria are

$$\begin{aligned} f - \frac{\partial u_g}{\partial y} &> 0 : \text{inertial stability,} \\ f - \frac{\partial u_g}{\partial y} &= 0 : \text{inertial neutrality,} \\ f - \frac{\partial u_g}{\partial y} &< 0 : \text{inertial instability.} \end{aligned} \tag{19.27}$$

In the first case, the meridional energy will decrease with time; in the second, it will remain constant with time; and in the third, it will increase with time. Since $-\partial u_g/\partial y$ is the relative geostrophic vorticity, this is equivalent to saying that inertial stability prevails when the absolute geostrophic vorticity of a zonal current is positive, and inertial instability prevails when the absolute vorticity is negative. Although these criteria have been established for a northward displacement, they are easily shown to apply also to a southward displacement. Furthermore, the criteria are valid in the Southern Hemisphere if all the inequality signs are reversed.

We find normally that the relative vorticity is smaller in magnitude than the Coriolis parameter. Thus inertial stability predominates. The outstanding place where instability may occur is on the equatorward side of a strong jet stream at high levels. There the anticyclonic shear ($\partial u_g/\partial y > 0$) may be large enough to exceed f and so produce a negative absolute vorticity. If this condition exists over a large enough area, the resulting instability will lead to lateral mixing which will tend to produce a uniform field of absolute vorticity, that is, inertial neutrality will be established. This is analogous to the case of hydrostatic parcel instability ($\partial\theta/\partial z < 0$) which leads to vertical mixing which tends toward a uniform potential temperature in the vertical (hydrostatic neutrality).

The physical basis of these inertial criteria may be clarified further by demonstrating the equivalence between the meteorological stability criterion, Eq. (19.27) and the stability criterion, Eq. (19.24), which we obtained for a simple rotating fluid in the laboratory. Inequality (19.24) for stability may be expanded as follows:

$$\omega R^2 + 2\omega R\,\delta R + \omega(\delta R)^2 + R^2\,\delta\omega + 2R\,\delta\omega\,\delta R + \delta\omega(\delta R)^2 > \omega R^2$$

If we cancel ωR^2 on both sides of the inequality and divide by $R\,\delta R$, we get

$$2\omega + R\frac{\delta\omega}{\delta R} + \frac{\omega}{R}\,\delta R + 2\,\delta\omega + \frac{\delta\omega\,\delta R}{R} > 0$$

If we now let δR (and therefore $\delta\omega$) approach zero, the last three terms become vanishingly small and

$$2\omega + R\frac{\partial\omega}{\partial R} > 0$$

in the limit. Now the linear speed is $c = \omega R$, so $R\,\partial\omega/\partial R = \partial c/\partial R - \omega$ and the criterion for stability becomes $\omega + \partial c/\partial R > 0$. But ω is the vorticity owing to curvature and $\partial c/\partial R$ is the vorticity arising from shear. Thus (19.24) becomes identical with the first of expressions (19.27) because positive total vorticity means stability. Thus the same result obtains whether the viewpoint is that of centrifugal acceleration or kinetic energy of disturbances.

PROBLEMS

1. Assume that air flows over a broad building 10 m high in an incompressible fashion. It is observed that the speed at ground level is 5 m s^{-1} while at the roof top it is 9 m s^{-1}. If $\rho = 1.3 \times 10^{-3}$ g cm^{-3}, what is the pressure difference (in mb) between the ground and roof level? How much of this is purely hydrostatic and how much is a dynamic reduction in pressure?

2. Suppose that over a rectangular area L cm on a side, the pressure distribution is given by

$$p = p_0 + p_1 \sin \frac{2\pi}{L} x \sin \frac{2\pi}{L} y$$

where p_0 and p_1 are positive constants. This means a pattern of alternating highs and lows in a checkerboard pattern with two highs and two lows in the given area. Suppose further that the horizontal divergence is given by

$$\frac{\partial u}{\partial x} + \frac{\partial v}{\partial y} = D_0 \sin \frac{2\pi}{L} x \sin \frac{2\pi}{L} y$$

where D_0 is a positive constant. This means divergence in the highs and convergence in the lows. Show that the mean kinetic energy per unit volume in a vertical layer 1 cm thick increases at the rate of

$$\frac{\partial \bar{E}_K}{\partial t} = \frac{p_1 D_0}{4}$$

If $D_0 = 10^{-5}$ s^{-1} and the central pressures of the highs and lows are 1025 and 1005 mb respectively, calculate the numerical value of

$$\frac{\partial \bar{E}_K}{\partial t}$$

CHAPTER 20

NUMERICAL WEATHER PREDICTION

20.1 Introduction

One of the ultimate goals of theoretical meteorology is the development of means of computing the future state of the atmosphere from the basic theoretical equations which govern that state. This goal has been pursued by many of the greatest hydrodynamicists and meteorologists of the past, including von Helmhotz, V. Bjerknes, and Richardson. However, the first indications of real success have come only since 1948.

About 1910 Richardson began to explore the possibility of utilizing the complete set of hydrodynamic equations for this purpose. These equations may be written, for a nonviscous adiabatic atmosphere, in the form

$$\begin{aligned}
\frac{\partial u}{\partial t} &= -\left(u\frac{\partial u}{\partial x} + v\frac{\partial u}{\partial y} + w\frac{\partial u}{\partial z}\right) - \frac{1}{\rho}\frac{\partial p}{\partial x} + 2\Omega\,(v\sin\phi - w\cos\phi) \\
\frac{\partial v}{\partial t} &= -\left(u\frac{\partial v}{\partial x} + v\frac{\partial v}{\partial y} + w\frac{\partial v}{\partial z}\right) - \frac{1}{\rho}\frac{\partial p}{\partial y} - 2\Omega u\sin\phi \\
\frac{\partial w}{\partial t} &= -\left(u\frac{\partial w}{\partial x} + v\frac{\partial w}{\partial y} + w\frac{\partial w}{\partial z}\right) - \frac{1}{\rho}\frac{\partial p}{\partial z} - 2\Omega u\cos\phi - g \qquad (20.1) \\
\frac{\partial \rho}{\partial t} &= -\left(u\frac{\partial \rho}{\partial x} + v\frac{\partial \rho}{\partial y} + w\frac{\partial \rho}{\partial z}\right) - \rho\left(\frac{\partial u}{\partial x} + \frac{\partial v}{\partial y} + \frac{\partial w}{\partial z}\right) \\
\frac{\partial p}{\partial t} &= -\left(u\frac{\partial p}{\partial x} + v\frac{\partial p}{\partial y} + w\frac{\partial p}{\partial z}\right) - \gamma p\left(\frac{\partial u}{\partial x} + \frac{\partial v}{\partial y} + \frac{\partial w}{\partial z}\right)
\end{aligned}$$

where the first three are the equations of motion, and the other two are the equation of continuity and the law of conservation of energy for an adiabatic process. It will be noted that this set is complete; there are five independent equations in five dependent variables: u, v, w, p, ρ. These equations have been written with all time derivatives on the left and all space derivatives on the right. Richardson[1] realized that the possibility of writing these equations in this manner meant that the instantaneous local rates of change of each of the dependent variables could be computed from a knowledge of the distribution in space of these variables at a given moment. Since it is theoretically possible to measure the spatial distribution of the variables at any time, it should

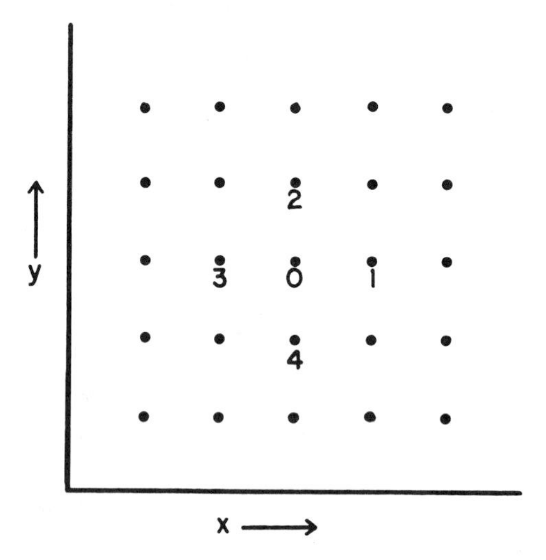

FIG. 20.1. Example of a finite-difference grid

be possible to determine their rates of change from Eq. (20.1). This constitutes a prediction of the values of these variables at an infinitesimally small increment of time later.

Richardson realized further that these equations could be solved by the *method of finite differences*, consisting of four steps. First, in the region under consideration, a grid of points is laid out. These points are usually equally spaced in one or more of the coordinate directions. At each grid point, at a given moment the values of the several variables are determined, either by direct measurement or by interpolation from analyzed fields.

Second, whenever a first derivative of a variable is needed, it is approximated at every grid point by taking the finite difference in the variable between surrounding grid points and dividing by the distance

[1] L. F. RICHARDSON, *Weather Prediction by Numerical Process*, Cambridge Univ. Press, London, 1922.

separating these two points. Thus in the grid of Fig. 20.1, an approximation to the value of $\partial p/\partial y$ at the point 0 is $(p_2 - p_4)/2d$, where d is the distance separating grid points. In fact we may write

$$\frac{\partial p}{\partial y} = \lim_{d \to 0} \frac{p_2 - p_4}{2d}$$

This equation states that $\partial p/\partial y$ becomes equal to this ratio of differences only when the distance separating points approaches zero. Nevertheless, the derivative will be approximated sufficiently well even if d remains finite, so long as d is small compared to the distance which separates maxima and minima of the variable. With these restrictions

$$\frac{\partial p}{\partial y} \approx \frac{p_2 - p_4}{2d}$$

Similar expressions can be written for all other derivatives. Thus $\partial p/\partial t \approx (p_1 - p_0)/(t_1 - t_0)$, where subscripts 0 and 1 respectively refer to the initial moment and some future moment a finite time later.

Third, all derivatives in the set of governing equations are replaced by such ratios of finite differences. The resultant finite-difference equations are then solved for the values of the variables at the future moment. The great strength of this procedure is that it reduces a set of differential equations, which may not be solvable directly, to a set of algebraic equations, whose solution can always be obtained. This solution may be at the expense of large amounts of numerical computation, but involved computation has become a far smaller problem with the advent of high-speed electronic computing machines.

Finally, one takes advantage of the fact that all this computation results in the values of the variables at each grid point at a future time. Since this is the kind of information that was needed to begin the process, it is apparent that one may now repeat the procedure so as to extend the results farther in time.[2] As a matter of fact, in theory the process can be repeated as many times as desired until a forecast for a predetermined, large, future time is reached.

Richardson used essentially the equations and procedure outlined above to make the first real attempt at numerical weather prediction. After a great deal of laborious hand computation he obtained results that were grossly disappointing. He predicted changes in the pressure, for example, which were from ten to a hundred times too large. As a result of this initial failure and the extensive labor required for such forecasts little further work was done for nearly a quarter of a century.

[2] It is also necessary to know the values of the variables on the boundary of the region being considered at the future time. This difficulty and ways of handling it will be discussed in section 20.6.

20.2 The Reasons for Richardson's Failure

It is clear that any scheme of numerical forecasting depends for its success upon the accuracy and extent of the observational data which specify the initial conditions. We now know that in Richardson's day the observations of the upper air were completely inadequate to reveal significant phenomena affecting the forecast. On this ground alone his experiment was doomed to failure. In addition, modern investigators feel that the set of equations solved by Richardson have certain grave practical drawbacks, despite their theoretical adequacy. The main consideration here is the matter of compensation of terms in the equations. For example, the third of Eq. (20.1), which deals with vertical forces and accelerations, is dominated by two large terms. Gravity and the vertical pressure-gradient force are each larger than the other terms by a factor of 10^4 for most meteorologically significant phenomena. These two terms must then be of opposite sign and must compensate each other almost completely in order to satisfy the equation. The small difference between these two is of the magnitude of the remaining terms. But this means that if we wish to determine $\partial w/\partial t$ to an accuracy of 10 per cent we must measure gravity and the vertical pressure-gradient force to an accuracy of one part in a hundred thousand. Since we cannot measure pressure throughout much of the atmosphere with a precision in excess of one part in ten thousand ($\pm$ 0.1 mb at best) we fall far short of the necessary accuracy.

This particular example can be taken care of simply by neglecting vertical accelerations and related factors, in which case the equation reduces to an expression of hydrostatic balance. Indeed, this is what Richardson did. However, the effects of compensation appear in the horizontal equations of motion as well, although less severely. Here the problem arises because of the near gradient balance which exists in middle and high latitudes. For example, in curved flow the three terms on the right of the second of Eq. (20.1) nearly cancel among themselves. Thus each must be measured with considerable precision in order that $\partial v/\partial t$ may be determined with adequate accuracy. Richardson did not take this into account and it is one important reason for the nature of his results.

The final major obstacle to Richardson's success was a purely mathematical one. It would seem to be true intuitively that if the finite increments of distance and of time are allowed to become smaller the solution of the finite-difference equations will approach the correct solution to the differential equations more closely. It may be shown, however, that not every combination of small space and time increments will serve. The meteorological equations are ones which permit a wide variety of wave motions ranging from fast sound waves to slow Rossby waves. In the solution of such equations we now know that the adopted grid interval, d, must be related to the chosen time interval, Δt, by the relation: $d \geq c\Delta t$, where c is the speed of the fastest waves permitted by the equations. That is, the grid distance must not be less than the distance a sound wave can travel in the finite time interval chosen. If this condition (discovered by Courant, Friedrichs, and Lewy[3]) is not satisfied, small errors of observation or of numerical round-off will be amplified during the iterative solution of the finite-difference equations until they mask the physical solution sought—a situation known as *computational instability*. To see this effect in a meteorological case, let us choose a grid distance of 100 miles, approximately the distance between observing stations. Since sound moves at the rate of about 700 miles per hour we must choose to forecast in steps of less than ten minutes. This result is economically unacceptable since it means that a 24-hr forecast would require nearly 150 iterations of the prediction process. The amount of numerical work would be very large. This result is unacceptable from a physical point of view also. No experienced meteorologist feels that the large-scale weather patterns change enough in ten minutes to make necessary a repetition of the forecast in so short a time. Experience dictates a choice for Δt closer to two hr. To avoid computational instability the grid distance d should then be no smaller than 1400 miles. But every meteorologist knows that this is too coarse a grid to represent many meteorological systems of importance. The trouble lies in the generality of our equations which contain such meteorologically nonsignificant phenomena as sound waves. Richardson was unaware of these computational considerations.

In retrospect, it is now believed that Richardson's experiment failed for three reasons:

1. Observations of inadequate accuracy, representativeness, and coverage, especially aloft.

2. Compensation among terms in the equations of motion. These always include the pressure terms.

3. Computational instability arising from his choice of grid and time intervals.

---

[3] Über die partiellen Differentialgleichungen der mathematische Physik, *Math. Ann.*, 100, 1928, pp. 32-74; or see P. D. THOMPSON, Notes on the Theory of Large-scale Disturbances in Atmospheric Flow with Applications to Numerical Weather Prediction, *Geoph. Res. Papers*, Geophysics Research Directorate, Cambridge, Mass., No. 16, 1952.

20.3 The Basis of Modern Numerical Weather Prediction

In recent years developments in meteorology have shown how each of the difficulties outlined above may be overcome, at least partially. Today we have available a network of upper-air sounding stations which is hemispheric in coverage and which far exceeds anything available in the 1920's. There is reason to believe that the data coverage is still inadequate, especially over the oceans, but the way is open to remedy such deficiencies should it prove necessary.

Means have been found, also, to remedy the problem of compensation of terms in the equations of motion. This compensation involves the pressure-gradient terms. When a vertical-vorticity theorem is formed from the equations of motion, the pressure appears only in the solenoid term, which expresses the torque due to the noncoincidence of isobaric and isosteric surfaces. In the case of barotropy this term is zero, and even in baroclinic situations the contribution of the solenoid term to the rate of change of vorticity is ordinarily small compared to the contributions of the other terms. Thus, to this degree of approximation, the vorticity theorem is free of pressure-gradient terms and their compensation of Coriolis terms is avoided. It appears therefore that the vorticity theorem for near barotropic flow is more suitable for numerical computation than are the equations of motion from which it is derived.

The most severe problem is that of computational instability. The current approach to this matter is to simplify the equations in such a way as to eliminate or "filter out" the rapidly moving sound waves and long gravitational waves. Such waves are presumably of little meteorological significance and no great loss will be suffered if they are eliminated, provided the remaining meteorological waves are not also affected by the simplifications. The remaining waves will be sufficiently slow that the Courant-Friedrichs-Lewy condition can be met with reasonable choices of d and Δt.

Charney[4] has found that systematic but restrained use of the hydrostatic and geostrophic approximations will accomplish the desired end. Without going into details, it is possible to see that this procedure is reasonable, since the waves to be excluded are neither hydrostatic nor geostrophic. Under appropriate conditions[5] the vorticity equation reduces to the simple form, $f + \zeta =$ constant, at some level in the atmosphere.

[4] J. Charney, On the Scale of Atmospheric Motions, *Geofys. Publ.*, 17, No. 2, 1948.

[5] J. Charney and A. Eliassen, A Numerical Method for Predicting the Perturbations of the Middle Latitude Westerlies, *Tellus*, 1, 1949, pp. 38-54.

This is simply the equation for the conservation of absolute vertical vorticity which was discussed in Chapter 16.

It should be realized that, as a governing equation, the law of conservation of absolute vorticity is a considerable oversimplification. All baroclinic effects are neglected, divergence and vertical motions are omitted, and the geostrophic assumption does much more than filter out sound and long gravity waves. The equation will serve, however, as an example of procedure in numerical solution.

20.4 Numerical Solution of the Law of Conservation of Vorticity

If we make the geostrophic assumption on a constant pressure surface and ignore small variations in f, the relative vertical vorticity becomes

$$\zeta = \frac{g}{f}\left(\frac{\partial^2 z}{\partial x^2} + \frac{\partial^2 z}{\partial y^2}\right) = \frac{g}{f}\,\nabla^2 z$$

where z is the height of the constant pressure surface and $\nabla^2 z$ is a symbol for the LaPlacian of z. When this is substituted in the expanded expression for the conservation of vorticity we get

$$\frac{g}{f}\frac{\partial \nabla^2 z}{\partial t} + u\,\frac{\partial}{\partial x}\left(\frac{g}{f}\,\nabla^2 z + f\right) + v\,\frac{\partial}{\partial y}\left(\frac{g}{f}\,\nabla^2 z + f\right) = 0$$

or

$$\frac{\partial(\nabla^2 z)}{\partial t} = -\frac{f}{g}\left(u\,\frac{\partial}{\partial x} + v\,\frac{\partial}{\partial y}\right)\left(\frac{g}{f}\,\nabla^2 z + f\right) \qquad (20.2)$$

Equation (20.2) says that $\nabla^2 z$, which is proportional to the geostrophic relative vorticity, can change only because of horizontal advection of absolute vorticity. Note that this equation has the desirable property discussed earlier, that the time derivative of a variable ($\nabla^2 z$) is expressed in terms of instantaneous space derivatives of known variables. For in the geostrophic case, u and v can be expressed as height derivatives, and f is a known function of latitude.

The first step in solving Eq. (20.2) numerically in a given case is to lay out on an analyzed chart of isobaric contours a grid of equally spaced points as in Fig. 20.1. The 500-mb surface is the one most commonly chosen. From the height values at each point one calculates the LaPlacian of the height from the appropriate finite-difference formula. For example, the approximate value of the LaPlacian of height at point 0 in Fig. 20.1 is

$$(\nabla^2 z)_0 \approx \frac{z_1 + z_2 + z_3 + z_4 - 4z_0}{d^2} \qquad (20.3)$$

Second, one converts $\nabla^2 z$ to absolute vorticity by multiplying by g/f and then adding f. The resultant field is analyzed by drawing isolines of the absolute vorticity.

Third, the lines of constant absolute vorticity are advected with the geostrophic speed along the contours for the distance a particle of air would travel in a selected time interval (say $\Delta t = 3$ hr). This yields a predicted absolute vorticity field. The difference between the predicted and initial absolute vorticity fields is the change in relative vorticity ($\Delta\zeta$) which is forecasted for each grid point (since f does not change with time at a fixed point.)

Fourth, the values of $\Delta\zeta$ are converted to changes in $\nabla^2 z$ by multiplication by f/g at each point.

Fifth, by one of several possible processes of numerical integration one goes from a field of changes in $\nabla^2 z$ to a field of changes in z. One method for doing this will be described below.

Finally, by adding the height-change field so obtained to the original z-field, one obtains a new contour pattern which is the field predicted by the law of conservation of vorticity. The whole process may be repeated indefinitely so as to obtain a prediction for any future time, in theory.

20.5 Integration by the Method of Relaxation

The problem which was deferred above is that of obtaining the values of a function $G(x, y)$ at each grid point given the values of $\nabla^2 G$ at each point. It is necessary also to have the values of G along the boundary of the region under consideration. These boundary values are not known because they apply at the future time for which we are trying to predict G. For the sake of discussion we shall assume the boundary values to be given, and we shall show later how they may be approximated sufficiently well.

Since $\nabla^2 G$ is a known function, say $F(x, y)$, we may write the approximate finite-difference form:

$$\frac{G_1 + G_2 + G_3 + G_4 - 4G_0}{d^2} = F(x, y)$$

where the values of G are not known. We then guess at these values, and if we guess correctly

$$\frac{G_1 + G_2 + G_3 + G_4}{4} - G_0 - \frac{d^2 F}{4} = 0$$

If, as will almost always happen, our guesses of the field of G are not completely correct, these three terms will yield not zero but some residual error, $E(x, y)$:

$$\frac{G_1 + G_2 + G_3 + G_4}{4} - G_0 - \frac{d^2F}{4} = E_0$$

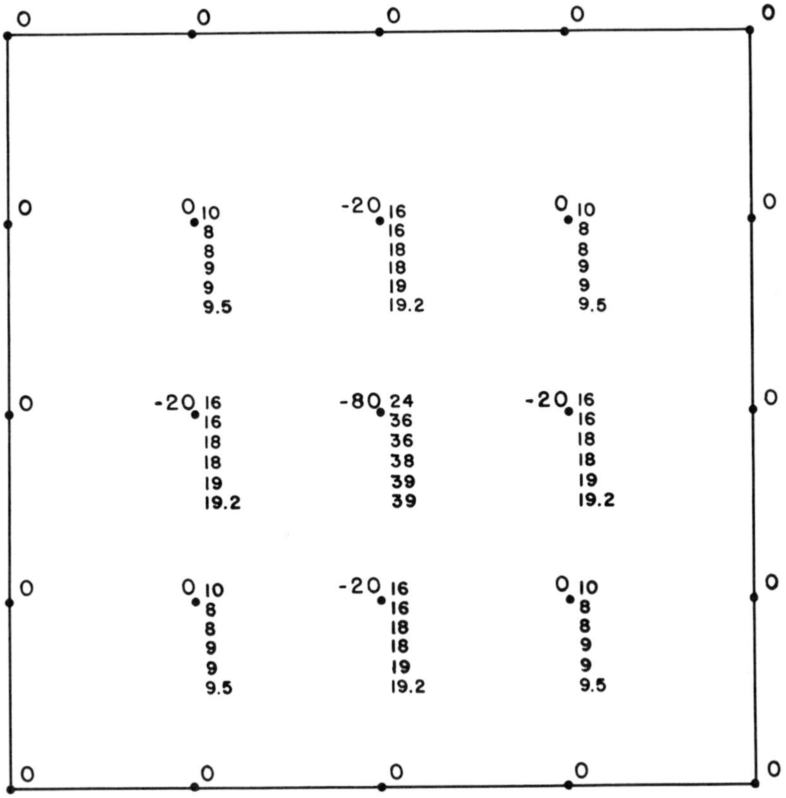

0	0	0	0	0
0	0 · 10 8 8 9 9 9.5	-20 · 16 16 18 18 19 19.2	0 · 10 8 8 9 9 9.5	0
0	-20 · 16 16 18 18 19 19.2	-80 · 24 36 36 38 39 39	-20 · 16 16 18 18 19 19.2	0
0	0 · 10 8 8 9 9 9.5	-20 · 16 16 18 18 19 19.2	0 · 10 8 8 9 9 9.5	0
0	0	0	0	0

FIG. 20.2. Example of relaxation. Along the boundary the height changes are assumed to be zero. At each interior point $\nabla^2(\Delta z)$ has been entered to the upper left of the point, and successive height-change estimates have been entered to the right of the point with the later estimates below.

Our aim is to make successive estimates of the G's which will reduce the residual error to zero or a negligibly small amount. It has been shown[6] that if we add E_0 to G_0 at each point to get new estimates of the G's, recompute the residuals, and then repeat the process several times, the results will converge on the correct values of the field of G

[6] R. V. SOUTHWELL, *Relaxation Methods in Theoretical Physics*, Clarendon Press, Oxford, 1946, 248 pp.

regardless of the initial guesses made. Of course, the more judicious one's initial guesses the fewer will be the number of repetitions needed. The whole process is one of "relaxing" the artificialities of the guessed solution and is therefore known as the *method of relaxation.*

An example of relaxation to solve the equation $\nabla^2(\Delta z) = F(x, y)$ is given in Fig. 20.2. The numbers are in arbitrary units but for the sake of realism may be looked upon as height changes in tens of feet. The distance d is taken to be one unit to simplify the arithmetic. Since the given values of $\nabla^2(\Delta z)$ become more negative towards the center of the region (a change towards anticyclonic vorticity) an initial field is guess in which the height change increases toward the center. The residual error is calculated for each point and then added to the guessed value. For example, the first relaxation computation for the central point is:

$$E_1 = \frac{16 + 16 + 16 + 16}{4} - 24 - \frac{(-80)}{4} = 12$$

Therefore the second estimate of Δz at this point is $24 + 12 = 36$. This first relaxation is carried out at all points in the grid to obtain second estimates everywhere. The second relaxation computation at the central point is

$$E_2 = \frac{16 + 16 + 16 + 16}{4} - 36 - \frac{(-80)}{4} = 0$$

That is, the third estimate of Δz is unchanged from the second estimate.

In practice, one decides in advance on a residual error which is considered acceptable. When a stage is reached where the estimates at every point in the grid change by less than the acceptable residual, the computation is stopped. In Fig. 20.2 one unit of Δz was considered an acceptable residual; this was reached after five relaxations.

The final result in Fig. 20.2 is one in which height change increases towards the center as expected and as required by the boundary conditions and the central maximum of anticyclonic vorticity.

20.6 Establishment of the Future Boundary Values

To round out this simple example of numerical weather prediction presented here, we have only to show how one obtains adequate future values of height along the boundary of the forecast region. This is a problem only if one is dealing with a restricted portion of the earth, for if the forecast area is the whole earth there are no lateral boundaries.

Since we do not now gather adequate data for the whole earth (and since the geostrophic approximation is totally inadequate in the tropics) the problem remains.

If one simply guesses at future conditions on the boundary a certain error will result. This error will be carried from the boundary into the forecast region a certain distance with each time increment over which a forecast is made. That is, the error will be propagated inward with a certain finite speed and, on physical grounds, this speed cannot much exceed the speed of the fastest waves permitted by the governing equations. Since in the example discussed here this speed is low (comparable to the wind speed) it may be expected that much of the original forecast area will remain unaffected by the errors along the boundary. Thus the procedure is to operate over a region sufficiently larger in all directions than the area with which one is concerned; then one can make any reasonable assumption about the heights along this distant boundary—for example, these heights may be assumed to be invariant with time. Then the forecast results are rejected for a strip inside the boundary wide enough to encompass all the boundary error.

20.7 Synopsis of the Procedure

In summary, the steps in this method of numerical prediction using the barotropic, nondivergent vorticity theorem are:

(1) Analyze the height field of an isobaric surface and interpolate the height at each point in an equally spaced rectangular grid covering somewhat more than the forecast region.

(2) Calculate the LaPlacian of the height from Eq. (20.3).

(3) Obtain the absolute vertical vorticity by multiplying $\nabla^2 z$ with g/f and adding f to the result at each grid point.

(4) Advect the absolute vorticity values with the initial geostrophic wind for whatever small Δt has been chosen as the basic time increment.

(5) Subtract the old vorticity field from this new one to get $\Delta(f + \zeta)$ (same as $\Delta\zeta$).

(6) Multiply by f/g to convert to changes in $\nabla^2 z$ (same as $\nabla^2[\Delta z]$).

(7) Integrate $\nabla^2[\Delta z]$ by relaxation or other methods assuming $\Delta z = 0$ on the boundary. This gives Δz everywhere.

(8) Add Δz to the initial z field to get a new z field at time Δt later.

(9) Repeat the whole process until a forecast for whatever desired future time is reached. Discard the results just inside the boundary.

20.8 Conclusion

The discussion above does not include all the refinements and subtleties which can be found in the original papers in the literature of numerical forecasting. For example, the use of the nondivergent barotropic vorticity theorem can be shown to be far more reasonable physically than would appear from the presentation above. The conditions under which the equations $f + \zeta = K$ can be applied to an "equivalent-barotropic" model of the atmosphere have been discussed by Charney and Eliassen.[7]

They show that if one integrates throughout the depth of the atmosphere, under appropriate conditions, the resultant simple vorticity equation is one which will apply at one level in the atmosphere. This level is close to 500 mb, so that all attempts to predict numerically with this model have used 500 mb data.

The computational schemes which have actually been used to solve the equation $f + \zeta = K$ differ in details from that presented here in order to facilitate certain of the operations. A graphical procedure which is convenient has been described by Fjørtoft.[8]

The results of prediction for 24 hr with this simple model are about equal in accuracy to forecasts made by experienced meteorologists using the usual subjective techniques. Both kinds of forecasts tend to fail when pronounced development of systems occurs. It is not surprising that this numerical technique should fail during cyclogenesis, because advection of vorticity is little more than a reasonable extrapolation procedure. No increase in the extreme values of vorticity can be predicted by such physical extrapolation. But when development occurs, the centers of vorticity grow and extreme values appear that were not to be found before. Such situations must deviate from the conditions of the equivalent-barotropic model.

One important way this simple model deviates from the real atmosphere is through the combined effects of the geostrophic and barotropic assumptions. Under these restrictions the wind cannot vary with elevation, and the effect of differences in flow at various levels is omitted. Various multilayer models have been proposed in which the wind is constant within each layer but changes from layer to layer.[9] These

[7] CHARNEY and ELIASSEN, loc. cit.

[8] R. FJØRTOFT, On a Numerical Method of Integrating the Barotropic Vorticity Equation, *Tellus*, 4, 1952, pp. 179-194.

[9] J. G. CHARNEY and N. A. PHILLIPS, Numerical Integration of the Quasi-Geostrophic Equations for Barotropic and Simple Baroclinic Flows, *J. Meteor.*, 10, 1953, pp. 71-99.

models are capable of predicting development, since in them vorticity is not conserved.

The present capabilities of high-speed computing machines are so great that the many thousands of calculations which go into a 24-hr forecast can be completed in an hour or less. Far more time is consumed in plotting and analyzing the data, and preparing it for the machines. Means have been devised by which the observational data may be fed directly into the computing machines so that the machines may perform a rapid objective analysis.[10] There is great difficulty in this because of the complexity of the processes by which a human analyst decides whether a datum is incorrect or unrepresentative and so should be ignored. It is not easy to instruct a machine to do this job.

The great advantages that numerical weather prediction enjoys over ordinary subjective forecasting are as follows: (1) The unswerving consistency of a machine, which can do no other than follow the instructions it has received. A properly functioning machine never forecasts in an illogical or inconsistent fashion, provided it has been supplied with logical and consistent instructions. (2) The tireless nature of a machine, which is not subject to human frailties. (3) The far more vast amount of data that can be absorbed and dealt with by a machine than by a human. (4) The possibilities that exist for improving the models upon which the machine computations are based. We can see opportunities for removing many of the assumptions and crudities of our various numerical models, but the outlook for improving some kinds of subjective forecasts are far more murky.

Machine computation of forecasts has certain disadvantages. They are: (1) The high cost of the necessary machines. This limits their use to a very few centers. (2) Machine errors. There are so many units of a machine involved in each operation that, even though each unit has a very high dependability, the chance of an error in the final result is not negligible. Fortunately these errors are usually random and simple repetition of the calculation suffices to detect them. (3) The present limitation of the machines to a forecast of the field of motion. The important problem of predicting hydrometeors remains a subjective one. Some aid along these lines is given by the machines which can generate, using certain dynamic models, a forecast of the large-scale field of vertical motion. This is intimately connected to clouds and precipitation. (4) The inapplicability of current numerical models to significant small-scale phenomena such as thunderstorms, tornadoes, and the like.

---

[10] See, for example, B. GILCHRIST and G. P. CRESSMAN, An Experiment in Objective Analysis, *Tellus*, 6, 1954, pp. 309-318.

PROBLEM

Suppose in the example given in Fig. 20.2, the prescribed boundary conditions are in error, and the value at the middle point of the lefthand boundary should be —10 instead of 0. All other boundary values remain unchanged. Form an estimate of the importance of the boundary prescription by repeating the relaxations for this new boundary condition. The relaxation should be stopped when no point in the grid changes by more than one unit. Compare the results with those of Fig. 20.2.

CHAPTER 21

THE GENERAL CIRCULATION

21.1 Scale of Atmospheric Motions

The motions that take place in the atmosphere exhibit a continuous gradation in space and time. The smallest and most rapid motions are those involved in molecular diffusion and sound waves; next are those occuring in the usual macroscopic turbulence. Still larger are dust devils, billow clouds, thunderstorms, and tornadoes. Further up the scale are the smaller frontal disturbances, tropical disturbances, and hurricanes, which are generally smaller than the major perturbations of the principal frontal zones, the classical frontal waves. The largest and slowest of all are the Rossby waves of the mid-troposphere and the associated semipermanent centers of action (the Aleutian low, the Icelandic low, and the subtropical anticyclones). The characteristic sizes of these modes of motion vary from a fraction of a centimeter to several thousand kilometers.

Each of the various scales of motion has a varying degree of influence upon all the others. When we average the behavior of the entire atmosphere over a long period of time (or simultaneously over large distances), we in effect suppress the faster and smaller phenomena, revealing only the slowest and largest modes of motion. The study of *the general circulation* is, accordingly, the description and explanation of these long-term mean states. Clearly the general circulation, because of the mutual interaction of all scales of motion, cannot be studied completely without reference to the smaller scales. Nevertheless much knowledge can be gained by initially neglecting all but the largest phenomena.

We shall begin by giving a brief description of the general circulation from known observations; then proceed to recognize the fundamental physical questions raised by the known facts; and finally present the elaborations and theories, however incomplete at present, which are available to explain these facts.

21.2 Longitudinally Averaged Flow

We know that the daily radiation received at the top of the atmosphere is essentially independent of longitude. Since the latitudinal distribution of this radiation "drives" the atmosphere, it is reasonable first to examine the general circulation averaged over a long period of years and over all longitudes. This average gives the mean flow as a function of latitude and height alone, and smooths out the longitudinal variations of (1) the fraction of incident radiation reaching the surface, (2) the radiation re-emitted to space, and (3) the character of the earth's surface—primarily the distribution of mountains and of sea and land.

Figure 21.1 gives the mean zonal flow in summer and winter for both Northern and Southern Hemispheres. In both seasons and hemispheres there are surface belts of easterly winds in low latitudes and near the poles. In middle latitudes there are surface westerlies. These facts will be basic to our later discussion of the angular-momentum budget of the earth-atmosphere system.

Aloft, the most striking feature is a belt of very strong west winds at about 20 cb (200 mb) in both seasons and hemispheres, the well-known primary *jet stream.* In the Northern Hemisphere these wind features are stronger in winter than summer, when the jet stream is also closer to the equator. Similar statements can be made about seasonal variations in the Southern Hemisphere, except that not much change seems to occur in the strength of the tropical easterly trade winds.

These cross sections give no information about mean meridional motions, because the time-mean north-south velocities tend to alternate sign with longitude. Accordingly, summation over longitude causes almost complete cancellation. The residue is usually so small that it cannot be determined with adequate accuracy. Mean vertical velocities likewise cannot be represented, since from the continuity equation they must connect the small meridional circulations and provide a long-term steady state with respect to mass. Since the mean meridional motions are small and undetermined (except in the trade winds) the mean vertical velocities almost everywhere must also be small and unspecified.

The nature of these mean data raises a number of fundamental questions. Each hemisphere can be characterized clearly as having three cells: shallow polar easterlies, deep middle-latitude westerlies, and deep tropical easterlies. Why does the general circulation break

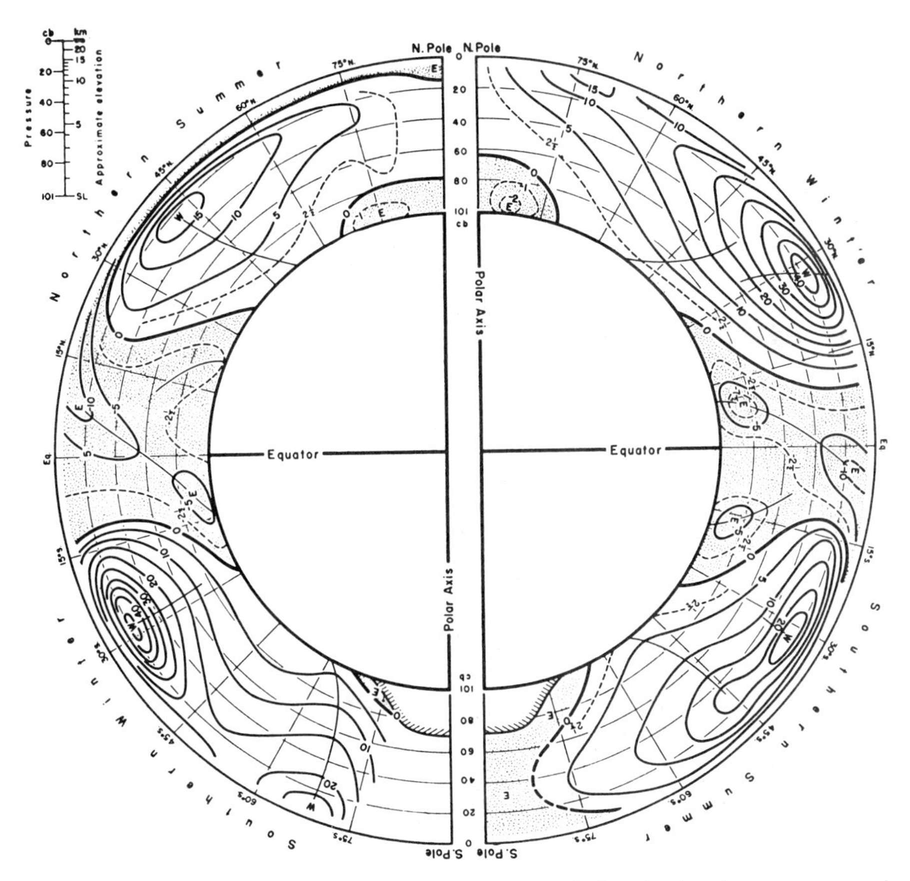

FIG. 21.1. Mean zonal wind averaged over all longitudes in summer and winter. Isotachs are in meters per second. Regions of east wind are shaded. Because of the linear scale of pressure the stratosphere is greatly compressed and only the troposphere can be examined in detail.

into these three cells? What mechanisms drive each of them? What is the role of the poorly known circulation in vertical meridional planes in maintaining the kinetic energy and angular momentum of these motions? What is the role of smaller scale motions (which we have removed by averaging) in this maintenance?

21.3 Longitudinally Varying Flow

If we do not remove longitudinal variations we must examine a number of time-mean maps, each of which will show the distribution of the motion in latitude and longitude for a given elevation. Figure 21.2

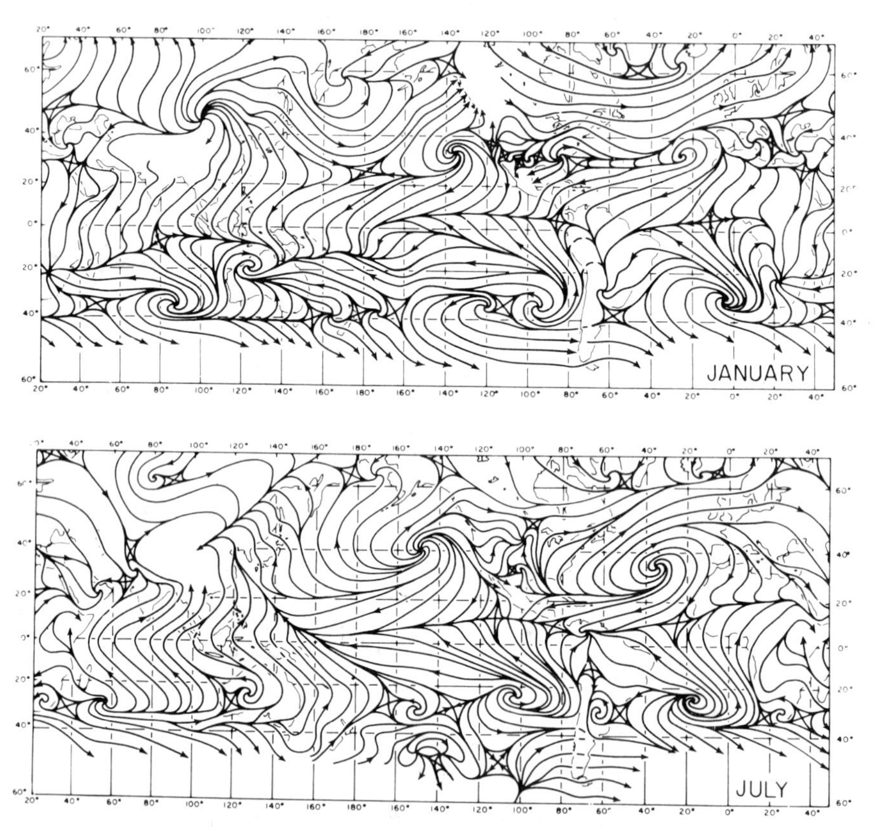

FIG. 21.2. Mean surface streamlines in January and July

gives the streamlines of the mean surface wind in January and July.[1] In both seasons these maps are characterized by large divergent anticyclones over the oceans (warm subtropical highs) and marked equatorial convergence zones. In Northern Hemisphere winter we find a strong divergent anticyclone over Siberia (cold high) and convergent cyclones

---

[1] Y. MINTZ and G. DEAN, The Observed Field of Motion of the Atmosphere, *Geoph. Res. Papers*, No. 17, Geophysics Research Directorate, Cambridge, Mass., 1952.

in the Pacific and Atlantic Oceans near 60° N (Aleutian and Icelandic lows). These cyclones are less impressive on the surface streamline maps than on corresponding maps of mean sea-level pressure because the surface wind is markedly altered by friction. One should also note the monsoonal change with season over Asia. Here the divergent anticyclone of January gives way to convergence of a partly cyclonic nature in July—a direct and simple response to the low continental temperatures in winter and the high continental temperatures in summer.

Figure 21.3 gives the mean contour distribution for January and July at 500 mb.[2] From this one can judge only the distribution of the geostrophic wind, but evidence exists that for long-term averages the actual and geostrophic winds are very nearly the same. This level, which is also close to the middle of the vertical mass distribution of the atmosphere, will be used here to represent the mid-troposphere. The outstanding feature of both maps is the broad continuous belt of circumpolar westerlies in middle latitudes, which is stronger and wider in January than in July, being associated with a low-height center west of Greenland and south of the pole in both seasons. In January there is another low center west of Kamchatka which has its counterpart as an open trough east of Kamchatka in July. These are the upper-level reflections of the surface Icelandic and Aleutian lows. We also find a belt of "high" heights on the southern side of the westerlies which is broken into individual cells corresponding to the surface subtropical anticyclones.

Further examination of these mean values reveals that the westerly current is not purely zonal but exhibits meridional undulations of appreciable amplitude. If one scans along latitude 45° N, one can find four complete sinusoidal waves around the hemisphere in January, and six in July. These are clearly the reflections in the mean of the largest scale of atmospheric motions, the Rossby waves.

Figures 21.2 and 21.3 give rise to further fundamental questions. What is the role of the semipermanent centers (polar cyclones and subtropical anticyclones)? Why are they located at the indicated longitudes? Are the fixed positions of the mean Rossby waves due to the distributions of land and sea (and therefore to a longitudinal variation in heat input), or to the distribution of mountains? Unfortunately it is not possible at present to give complete answers to these questions.

[2] R. A. Bryson, J. F. Lahey, W. L. Somervell, Jr., and E. W. Wahl, Normal 500 mb charts for the Northern Hemisphere, *Sci. Report No. 8, USAF Contract AF* 19(604)992, U. of Wisc. Dept. of Meteor., 1957.

21.4 Constraints on Theories of the General Circulation

Examination of the global distribution of the meteorological elements averaged over a decade or so reveals that the atmospheric circulation is subject to certain important long-term balances. The first of these is the heat balance discussed in Section 7.10, where it was shown that an excess of incoming over outgoing radiation is found on the equatorward side of middle latitudes and the reverse on their poleward side. Nevertheless the equatorial atmosphere does not continually rise in temperature, nor does the polar atmosphere continually cool over the years, because the atmospheric circulation carries the equatorial excess of heat energy poleward and makes up the polar

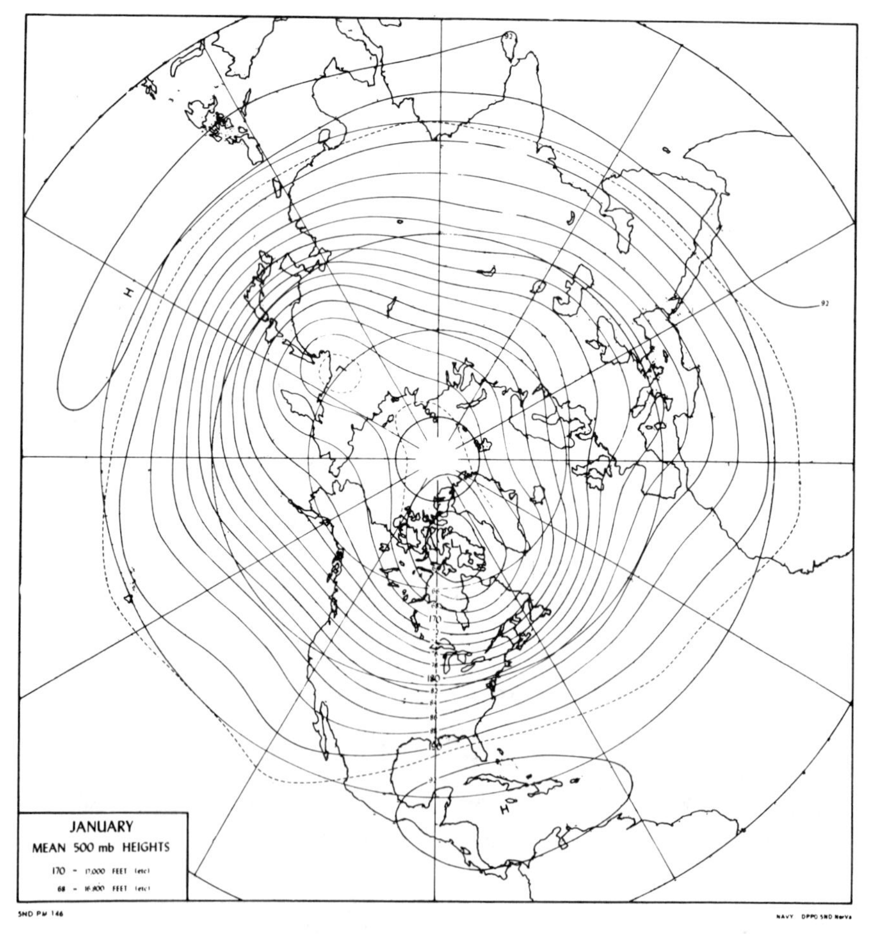

FIG. 21.3A. Mean contours of the 500 mb surface for the Northern Hemisphere in January

deficit. Thus any general circulation theory must provide for an adequate set of mechanisms for this latitudinal transport of heat.

Secondly, we note that the planet Earth, in its rotation on its axis, is subject to no significant external torques about that axis. It follows from the law of conservation of angular momentum that no changes with time can occur in the angular momentum of the earth-atmosphere system. The earth, however, is in mechanical contact with the atmosphere at their interface and it is possible for a transfer of angular momentum from one to the other to occur through frictional interaction. A significant net flow of such momentum occuring in either direction over a period of years would result in a change in the angular velocity of the earth. But earth's rotation rate, when measured

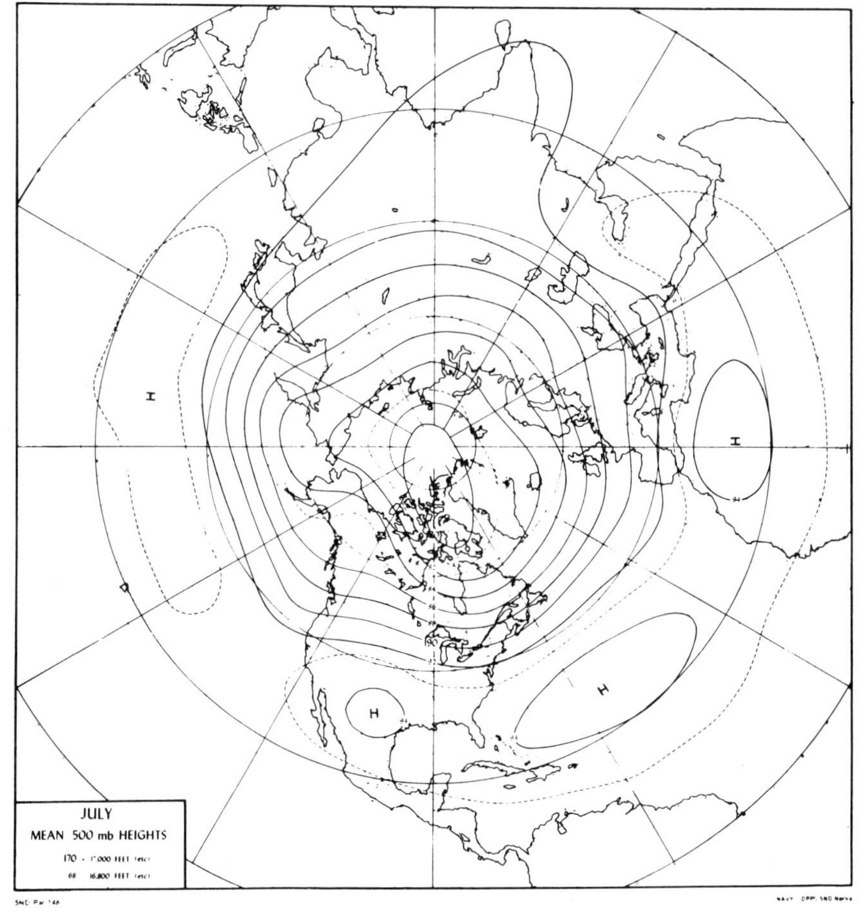

FIG. 21.3B. Mean contours of the 500 mb surface for the Northern Hemisphere in July

with great precision, discloses no important time variations. Hence, in the long run, the angular momentum of the atmosphere itself is constant.

Since there are belts of surface easterlies in the tropics and the arctics it follows that friction transfers angular momentum from the earth to the atmosphere in those latitudes. In order for the total atmospheric momentum to remain constant there must be westerlies in middle latitudes which will transfer angular momentum back to the earth at essentially the same rate that it is being extracted elsewhere. But since the mean surface easterlies and westerlies are not dying out with time, any good theory must explain how atmospheric angular momentum is transported towards the belt of surface westerlies from the belts of easterlies.

A third requirement of such a theory is that it must provide for a long-term steady distribution of atmospheric mass. Since we observe that there are no appreciable secular variations in pressure, it follows from the tendency equation that there can be no net mean convergence of mass.

In principle, a fourth independent condition could be imposed on any theoretical explanation of the circulation from the requirement that the long-run distribution of water vapor must be constant with time. We could simply take the latitudinal mean evaporation from the surface (rate of addition of vapor to the atmosphere) and subtract the latitudinal mean rainfall (rate of loss of vapor by the atmosphere) to obtain the net rate of gain or loss at each latitude. We could then require that our theories explain the lateral transport of water vapor needed to keep the concentration steady everywhere. Unfortunately the available estimates of mean evaporation and rainfall are quite inadequate and no reliable result can presently be obtained this way.

It is important to note that these general conditions are largely independent of each other because the latitudinal and vertical distributions of enthalpy, angular momentum, mass, and water vapor are mutually independent. Thus a transfer mechanism adequate to explain the distribution of one may be inadequate for another. The fact is that no single simple mechanism can suffice.

A complete theory of the general circulation should begin with the dynamical equations governing the motion, the equation of continuity for atmospheric mass, the First Law of Thermodynamics, and the equation of continuity for water vapor. To these must be added the externally supplied distribution of incoming radiation and the internally imposed distribution of vertical radiative transfer and loss to space.

Next three types of effects of the lower boundary must be

taken into account: (1) *mechanical* effects such as friction and perturbations by mountain ranges; (2) *radiative* effects of varying albedo and emissivity; (3) *thermal* effects of the different heat storage capacities of land and sea.

Finally, the four general restraints discussed above must be imposed. It is quite possible that more than one solution to the governing equations exists which will satisfy the four restraints—that is, as certain parameters (i.e. Coriolis parameter, temperature gradient) take on various values, there may be different modes of circulation which will satisfy the four physical constraints. As we shall see, experimental evidence suggests that this is so.

A general theory to be successful must satisfy all the conditions listed above and will automatically explain the salient features of the general circulation discussed earlier in this chapter. Such a general theory will obviously be extremely complicated; it is hardly surprising that we have not yet achieved such a complete understanding of the general circulation.

21.5 A Meridional Circulation Model

The oldest basic theoretical explanation, initiated in its essence by Hadley in 1735, has been developed by numerous investigators since. This approach neglects initially the longitudinal perturbations of the flow and emphasizes the latitudinal variations. Since the equatorial regions are heated by the sun more strongly than are the poles, it is natural to think of zonal rings of air rising in the tropics and sinking near the poles. In order that mass continuity be satisfied, there must be a poleward-directed current aloft and an equatorward-directed current near the ground, resulting in a direct solenoidal circulation. If this effect occurs at all longitudes it is possible to deduce simply the effect of the earth's rotation. For there can be no net west-east pressure gradient acting on a zonal ring and, except for friction, no torques can act on such rings. Thus the absolute angular momentum per unit mass is conserved:

$$\left(\Omega + \frac{u}{a \cos \phi}\right) a^2 \cos^2 \phi = \text{const.}$$

or

$$u = \frac{1}{a \cos \phi} [\text{const.} - \Omega\, a^2 \cos^2 \phi]$$

where u is the relative zonal wind speed which may be generated. It follows that u will increase in rings of air moving poleward and u will decrease in rings of air moving equatorward. Thus initially the upper poleward-directed current will develop a strong westerly component and the lower equatorward-directed current will develop a strong easterly component, as shown in Fig. 21.4*A*.

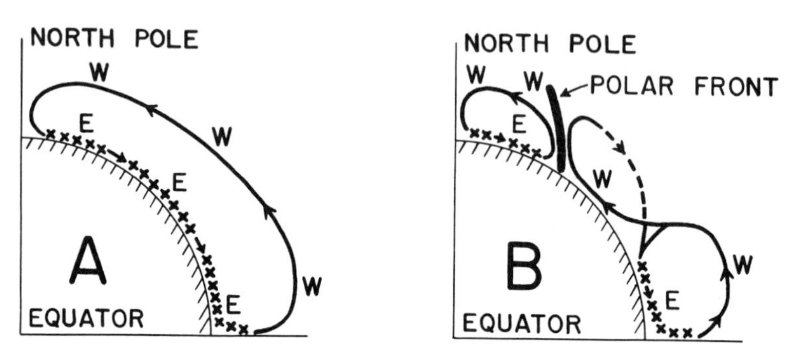

FIG. 21.4. Evolution of three cells in the general circulation according to a Hadley-meridional circulation theory

It is clear that this preliminary picture cannot be the final one. The easterlies implied at all latitudes near the surface would constantly extract angular momentum from the earth and slow its rotation rate, in contradiction to observation. Furthermore, this does not take into account the radiative loss of heat from the upper layers of the atmosphere. In particular the northward-moving equatorial air cools at a rate of 1° to 2° C per day as shown in Fig. 10.4 and 10.5. As a result this upper current will cool enough by the time it reaches latitude 30° to sink, warm quasi-adiabatically, and spread out horizontally at the ground. Part of this subsiding air may be expected to return equatorward and part to spread poleward. The equatorward branch of this subsided air can be expected to lose rapidly its initial westerly momentum through friction and to develop an easterly component through the tendency toward conservation of angular momentum. Thus a separate direct cell is established in the tropics by means of which the northeast trade winds of the Northern Hemisphere and the southeast trade winds of the Southern Hemisphere may be explained.

The west winds of the poleward branch of this subsiding current should increase in strength as the air moves towards higher latitudes. On the other hand, friction with the earth's surface must remove angular momentum and a balanced state with moderate surface westerlies can be established. We see, therefore, that radiative processes require the simple one-cell circulation to break down; the initial surface easterlies of middle latitudes must give way to westerlies, and this radiatively imposed change is qualitatively just what is needed to satisfy the mechanical requirement that the net flux of angular momentum between earth and atmosphere be zero.

Near the poles there is a net radiative loss of heat by the atmosphere. Just as in the one-cell case, this cold air must subside and spread equatorward with east winds. When this cold, low-level, easterly current meets the warmer westerly current of middle latitudes the polar front is established with the necessary cyclonic shear. Thus radiative considerations force a breakdown of the single cell of Fig. 21.4*a* into the three cells of Fig. 21.4*b*. The equatorial and polar cells are direct solenoidal circulations and are, therefore, energy producing; the middle latitude cell is an indirect solenoidal circulation and so is energy consuming.

At this point in the argument the theory is in qualitative agreement with observations. The polar and tropical easterlies are explained, the surface westerlies of middle latitudes are accounted for, and the polar front is a logical necessity. However, the theory fails to account for the fact that the westerlies of middle latitudes increase with height. Indeed, the equatorward-bound air aloft in the middle cell should develop an easterly component if angular momentum is to be conserved. To get around this difficulty, Rossby[3] suggested that large-scale turbulence in this region, arising from eddies lying in horizontal planes, might link the upper westerlies of the equatorial and polar cells to the upper troposphere of middle latitudes. Then, by turbulent interaction, the middle-latitude air will be impelled eastward, as observed. The essential point, for our purposes, is that such eddies are no longer zonal rings of air. East-west pressure gradients must act and the concept of conservation of absolute angular momentum must be abandoned. Indeed, a closer examination of the observed circulations elsewhere strongly suggests that such complete zonal rings are rare and conclusions based upon conservation of angular monumentum must be considered dubious. An acceptable theory of the general circulation must take into account longitudinal variations of the flow.

[3] C.-G. Rossby, The Scientific Basis of Modern Meteorology. In *Climate and Man*, U. S. Dept. of Agriculture, Washington, D. C., 1941, pp. 599-655.

21.6 An Experimental Approach

Since we have seen that a model of the general circulation based upon meridional circulations involving zonal rings of air lacks complete realism, we must begin anew, this time requiring some hints from nature as to the basis for our approach. In the present section we shall consider the suggestions that arise from certain laboratory experiments with small models of the atmosphere. In the section that follows, we shall see what leads are offered by hemispheric meteorological observations.

Fultz[4] and his colleagues have studied the behavior of a simple laboratory model of the atmosphere having the following physical characteristics. A flat circular pan of about 30 cm diameter containing water 4 cm deep as a working fluid, is rotated about its center in a horizontal plane. The pan is heated at the rim (which corresponds to the equator) and cooled at the axis of rotation (which corresponds to the pole). The circulations set up in the water are studied through a viewing device which optically stops the rotation and produces the same effect as if the observer were rotating with the pan. Such "dishpan" experiments obviously do not have complete geometric and dynamic similitude with the atmosphere. The spherical shape of the earth is replaced by the flat cylindrical shape of the pan bottom. Thus, for example, the vertical component of the angular velocity vector no longer varies as the sine of the latitude but is constant with radial distance from the center of the pan. Nevertheless, enough of the essential dynamic factors are retained with sufficient verisimilitude that the system can produce circulations remarkably similar to those observed in the atmosphere.

The type of general circulation produced in a rotating "dishpan" depends primarily upon two factors, the fluid speed in relation to the rotation rate and the size of the pan (the Rossby number) and the rates of equatorial heating and polar cooling. For fixed heating and cooling rates, four distinct regimes have been discerned as the rotation rate is increased.

1. *Low rotation rate* (large Rossby number). Here the circulation is primarily that of the Hadley type corresponding to the first stage of the meridional model described in the previous section of this chapter. The water at the upper surface moves poleward and develops a strong westerly component. The water at the bottom of the pan moves outward

---

[4] D. Fultz, A survey of Certain Thermally and Mechanically Driven Fluid Systems of Meteorological Interest, *Proc. First Sympos. on Geophysical Models*, Baltimore, Johns Hopkins Univ., 1953.

and has a pronounced easterly component. This motion is rather steady under the correct experimental conditions and is nearly symmetric about the center of the pan.

2. *Medium rotation rate* (moderately large Rossby number). As the rotation rate is increased the meridional flow breaks down and is replaced by a system in which jet streams are found near the upper surface and a very steady pattern of Rossby waves exists in these westerlies. The number of waves varies from 3 to 4 and then to 5, as the rotation rate is increased further. The rotation rate for these conditions is sufficiently high that the motions are quasi-geostrophic.

3. *Large rotation rates* (moderately small Rossby number). The flow is similar to the preceding case except that the jets are shorter in length and not continuous, the flow is markedly unsteady, and 5 to 6 Rossby waves are found. Low-level cyclones appear with evidence of sharp discontinuities suggestive of fronts. Cold anticyclones and many other entities analogous to observed atmospheric features make their appearance. In particular, the troughs in the wave patterns are tilted from northeast to southwest instead of lying along a meridian. The dynamical significance of this will be discussed later.

4. *Very high rotation rates* (small Rossby number). The higher the rotation rate the smaller are the physical circulation entities which appear in a "dishpan" experiment. At very high rotation rates these entities are so small they may be considered convective elements, especially if there is adequate heating from below. The convective cells have dimensions about one-tenth the size of the circulation features in the previous two cases. The cells are quite unsteady and ephemeral. They are connected by short-lived narrow jets winding in and out among the cells. There is a distinct resemblance to the classical convection picture of Benard cells. This quasi-Benard state occurs at such large values of the Coriolis parameter that it is not observed in the atmosphere.

The two regimes described above which most resemble atmospheric phenomena are the first and third. The nearly steady meridional flow at low rotation rates is comparable to the observed circulation in the tropics where the steady trades exist at low levels. This is quite reasonable, since at low latitudes the Coriolis parameter is small, corresponding to a slow rate of rotation about the vertical. The quasi-geostrophic unsteady regime which goes with large rotation rates in the "dishpan" closely resembles the atmospheric situation in middle and high latitudes. This resemblance is physically understandable, since the Coriolis parameter is large in middle and high latitudes, corresponding to a sizable rate of rotation about the vertical. The outstanding characteristics of the mode of circulation with high rotation rate are the absence of recognizable meridional circulations and the clear presence of sizable eddies lying in essentially horizontal planes. These characteristics strongly suggest, as did the arguments presented in criticism of the Hadley meridional model, that at least two mechanisms must be considered in evolving a complete theory of the general circulation. We should now ascertain whether the ageostrophic meridional mode of flow in low latitudes and the quasi-geostrophic horizontal mode of flow in middle and high latitudes are adequate mechanisms for the necessary transport of heat, angular momentum, mass, and water vapor.

21.7 The Angular-Momentum Balance

We shall discuss here only the theory and data concerning the transport and balance of angular momentum. The absolute angular momentum per unit mass of air is

$$m = (\Omega\, a \cos\phi + u)\, a \cos\phi$$

or

$$m = \Omega\, a^2 \cos^2\phi + u\, a \cos\phi \tag{21.1}$$

The first term on the right side of Eq. (21.1) is called the *Ω-momentum* and represents the angular momentum owing to the rotation of the earth. The second term on the right is called the *relative momentum* and represents the angular momentum owing to the zonal motion of air relative to the earth.

It follows from Newton's second law of motion that changes in angular momentum can come about only through the action of torques (tangential forces possessing a moment about the axis of rotation). The only such torques possible are those due to east-west pressure gradients and friction. The pressure gradient torque per unit mass is $-\alpha\, \partial p/\partial x\; a \cos\phi$, and the frictional torque per unit mass is $F_x a \cos\phi$. Therefore

$$\frac{dm}{dt} = \left(-\alpha \frac{\partial p}{\partial x} + F_x\right) a \cos\phi$$

This is really a modified form of the equation of motion in the x-direction. When we multiply both sides by ρ we get

$$\rho \frac{dm}{dt} = \left(-\frac{\partial p}{\partial x} + \rho F_x\right) a \cos\phi \tag{21.2}$$

The left side of Eq. (21.2) may be written as

$$\rho \frac{dm}{dt} = \rho \frac{\partial m}{\partial t} + \rho u \frac{\partial m}{\partial x} + \rho v \frac{\partial m}{\partial y} + \rho w \frac{\partial m}{\partial z}$$

or

$$\rho \frac{dm}{dt} = \frac{\partial(\rho m)}{\partial t} + \frac{\partial(\rho u m)}{\partial x} + \frac{\partial(\rho v m)}{\partial y} + \frac{\partial(\rho w m)}{\partial z} - m\left[\frac{\partial \rho}{\partial t} + \frac{\partial(\rho u)}{\partial x} + \frac{\partial(\rho v)}{\partial y} + \frac{\partial(\rho w)}{\partial z}\right]$$

But from the equation of continuity the term in brackets is zero, so that Eq. (21.2) may be written:

$$\frac{\partial(\rho m)}{\partial t} = -\left[\frac{\partial(\rho u m)}{\partial x} + \frac{\partial(\rho v m)}{\partial y} + \frac{\partial(\rho w m)}{\partial z}\right] - \frac{\partial p}{\partial x} a \cos\phi + \rho F_x a \cos\phi \quad (21.3)$$

which is an equation for the local rate of change of the absolute angular momentum per unit volume (ρm). If we integrate Eq. (21.3) over V, the entire volume of atmosphere poleward of a certain latitude ϕ_1, we get

$$\frac{\partial}{\partial t}\int \rho m\, dV = -\int \rho m\, V_n\, d\sigma - \int a \cos\phi \frac{\partial p}{\partial x} dV + \int \rho F_x a \cos\phi\, dV \quad (21.4)$$

where V_n is the outward-directed component of velocity perpendicular to the surface bounding the volume V, $d\sigma$ is an element of area of the surface of that volume, and the volume integral of the term in brackets in Eq. (21.3) has been converted to a surface integral by means of Gauss's divergence theorem. (See Appendix II.)

The volume V is bounded by the earth's surface, the top of the atmosphere, and a vertical surface at latitude ϕ_1. The contribution to the first integral on the right of Eq. (21.4) must be zero along the earth's surface because the kinematic boundary condition requires the normal component of velocity to be zero at a solid boundary. The contribution at the top of the atmosphere is zero because there $\rho = 0$. Thus the only contribution can come from the component of velocity perpendicular to the vertical surface at ϕ_1. If V_n there is positive it is directed out of the volume or equatorward. This will transport angular momentum per unit volume (ρm) out of the volume causing a decrease in the amount of this quantity remaining, as indicated by the minus sign. Thus this is a *meridional-transport term.*

The second term on the right of Eq. (21.4) may be written

$$-\int a \cos\phi \frac{\partial p}{\partial x} dV = -\iiint a \cos\phi \frac{\partial p}{\partial x} dx\, dy\, dz$$

If we perform the integration with respect to x, holding y and z constant, we get

$$-\int a\cos\phi\frac{\partial p}{\partial x}\,dV = -\iint (a\cos\phi\,\Delta p)\,dy\,dz$$

If the surface of the earth were completely horizontal, Δp would be zero because the value of $\partial p/\partial x$ integrated around a latitude circle must be zero. If mountains are present however, the pressure on the west sides of the mountains need not be the same as the pressure on their east sides. Then Δp need not be zero. This term represents the torque arising from asymmetrical pressure distributions about mountain barriers, and may be referred to as the *mountain-pressure torque.*

In dealing with the third term of Eq. (21.4) we recall that the frictional force exerted on a unit mass of air may be represented essentially as $F_x = \alpha\,\partial\tau/\partial z$, where τ is the east-west eddy stress. Therefore this term is

$$\int \rho F_x\,a\cos\phi\,dV = \iiint a\cos\phi\frac{\partial\tau}{\partial z}\,dx\,dy\,dz$$

If we integrate with respect to z, holding x and y constant, we get

$$\int \rho F_x\,a\cos\phi\,dV = \iint (\tau\,a\cos\phi)\,dx\,dy$$

Here the surface integral is to be evaluated only at the lower boundary of the atmosphere, since τ is zero at the top of the atmosphere. Since the stress acting on the air is opposite to the direction of the wind, this term will be negative for positive zonal winds (westerlies) and positive for negative zonal winds (easterlies). This is the *frictional-torque term.*

In the middle-latitude belt of westerlies we observe that the mountains extract angular momentum from the atmosphere via the mountain-pressure torque, and that the earth's surface extracts momentum from the atmosphere through the frictional-torque term. If the westerlies are not to die out in time, this momentum must be restored by the only term left, the meridional-transport term. We shall now expand the transport term to see how it achieves this balance. From our previous argument about this term we know that the only possible transport into such a polar cap is through the vertical bounding surface of the edge of the polar cap at latitude ϕ_1. Thus the meridional-transport term may be written as

$$-\int \rho\,m\,V_n\,d\sigma = \iint \rho\,m\,v\,dx\,dz$$

where v is the northward-directed meridional speed which is opposite in sign to the outward-directed normal velocity in the Northern Hemisphere. This equation may also be written

$$\int_0^\infty\int_0^{2\pi} \rho\,m\,v\,a\cos\phi\,d\lambda\,dz = a^2\cos^2\phi\int_0^\infty\int_0^{2\pi}(u+\Omega\,a\cos\phi)\,\rho v\,d\lambda\,dz$$

where $d\lambda$ is an increment in longitude. We now make the substitution $\rho dz = -dp/g$ from the hydrostatic equation and extract the average value of the variables around a latitude circle. The transport term becomes

$$\frac{2\pi a^2 \cos^2\phi}{g} \int_0^{p_0} (\overline{uv} + \overline{\Omega v \cos\phi})\, dp =$$
$$\frac{2\pi a^2 \cos^2\phi}{g} \int_0^{p_0} (\bar{u}\bar{v} + \overline{u'v'} + \Omega\, a\bar{v} \cos\phi)\, dp \qquad (21.5)$$

where p_0 is the pressure at $z = 0$, and $\overline{uv} = \bar{u}\,\bar{v} + \overline{u'v'}$ since a departure (indicated by a prime) has the value zero when averaged.

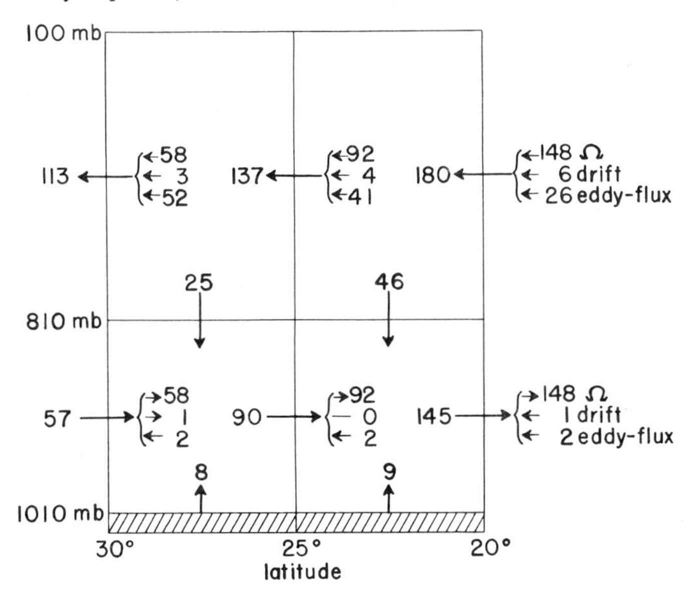

FIG. 21.5. Estimates of the mean rates of latitudinal and vertical transfer of absolute angular momentum in the tropics (after Riehl and Yeh). Numbers are in units of 10^{25} g cm^2 s^{-2}.

The terms in the integrand of the right side of Eq. (21.5) are $\bar{u}\,\bar{v}$, called the *drift term*; $\overline{u'v'}$, called the *eddy-flux term*; and $\Omega\bar{v}\cos\phi$, called the *Ω-transport term*. The drift term is so named because $\bar{v}$ represents the slow, nongeostrophic meridional drift. If such a mean meridional flow exists in a layer, it will transport the angular momentum of the mean zonal current, thus producing a flux of angular momentum into a polar cap. The eddy-flux term is so named because it represents the transport into the polar cap of the zonal eddy momentum by meridional eddies. This term will vanish unless there is a correlation between u' and v'. Finally, the Ω-transport term is so named because it represents the meridional flux of angular momentum of the earth's rotation by the mean meridional velocity, $\bar{v}$.

Riehl and Yeh[5] have made estimates of the mean transport of angular momentum by these three terms in January in the subtropics. Their results are shown in Fig. 21.5. Since the mean meridional motion, $\bar{v}$, can presently be determined only in the lowest layers, for lack of adequate upper air data, the observed surface meridional transport was assumed to decrease linearly with height and to become zero 200 mb above the surface. In order to insure continuity of mass, a compensating return flow was assumed from 810 mb to 100 mb near the tropopause. Thus Fig. 21.5 is divided into a shallow lower layer and a deep upper layer. For each layer and for three latitudes the values of the estimated transport by the three mechanisms and the total transport are given. In addition, the estimated frictional transport to the easterlies from the surface are indicated, and the vertical flux through the 810-mb level necessary to produce balance in the lower layer are entered. These vertical fluxes are nearly those necessary to produce a balance in the upper layer also; the small departures from balance in the upper layer are within the errors of the estimates of horizontal transport.

These data clearly show in the lower layer a net equatorward transport of angular momentum which is due primarily to the Ω-transport term. This is the effect of the steady northeast trade winds. In the upper layer there is a net poleward transport, much of which is due to Ω-transport by the return current but to which there is a significant contribution by the eddy-flux term. We draw the conclusion that mean meridional circulation is highly important in the subtropics but that large-scale horizontal mixing processes are not negligible in the mid-troposphere even in these low latitudes.

At higher latitudes we have reason to believe that most of the transport is via the eddy term. If we make the geostrophic assumption in Eq. (21.5) the quantity

$$\bar{v} = \frac{1}{2\pi}\int_0^{2\pi} v\, d\lambda \quad \text{becomes} \quad \bar{v} = \frac{g}{2\pi fa\cos\phi}\int_0^{2\pi}\frac{\partial z}{\partial\lambda}\, d\lambda = 0$$

Thus the geostrophic assumption automatically causes the drift term and the Ω-transport term to vanish and leaves the eddy-flux term the only contributor to the transport of angular momentum. The geostrophic assumption then represents a convenient way of isolating the eddy-flux term in order to determine its adequacy in producing the needed transport.

[5] H. Riehl and T. C. Yeh, The Intensity of the Net Meridional Circulation, *Quart. J. Royal Meteor. Soc.*, 76, 1950, p. 82.

If geostrophic eddy flux is the primary process at work, a simple conclusion can be drawn about the synoptic patterns which will produce a poleward flux of angular momentum. When $\overline{u'v'}$ is positive, the flux will be poleward. This means that strong zonal flow must occur where strong poleward flow exists. Such a correlation of u' and v' is found if the trough and ridge axes in an upper-level pattern of isobars or contours are not meridional, but slope from northeast to southwest in the Northern Hemisphere as shown in Fig. 21.6. Under these conditions the strong

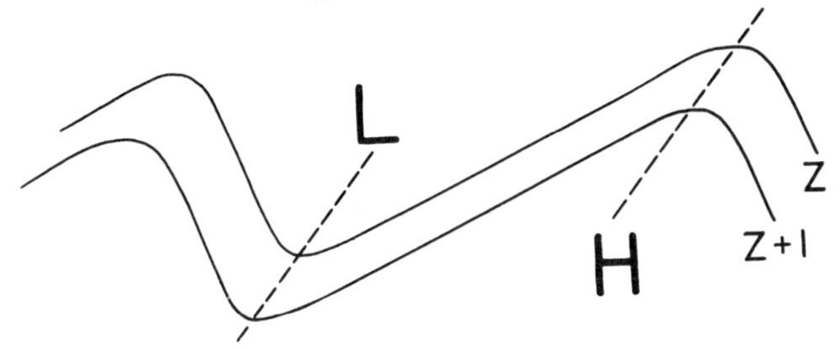

FIG. 21.6. Schematic representation of troughs and ridges in the westerlies sloping from northeast to southwest

pressure gradients are found east of the trough, the weak pressure gradients west of the troughs. This means that large positive values of u are found where v is positive, while small positive values of u occur where v is negative, so that the sum of $u'v'$ around a latitude circle will be positive. Synoptic experience indicates that the type of pattern shown in Fig. 21.6 is the predominant one in middle latitudes. Thus the disturbances of the middle-latitude westerlies must transport some angular momentum poleward by means of the eddy-flux term. Is this transport sufficient to balance the angular momentum budget?

Various investigators have sought to measure the north-south flux of angular momentum averaged over a month or more. We present here only the results of Mintz.[6] His values of the mean geostrophic poleward flux of angular momentum over the Northern Hemisphere are shown in Fig. 21.7. The maximum flux is found near 200 mb at about latitude 32° N. Clearly there is convergence of this flux at 200 mb north of latitude 32° and divergence south of this latitude. Thus the expected accumulation of angular momentum does occur in the belt of westerlies.

The accumulated momentum must be transported downward and transferred to the earth's surface by friction. Mintz calculated the total rate of accumulation and the rate of transfer by friction and found them

[6] Y. MINTZ, The Geostrophic Poleward Flux of Angular Momentum in the Month of January 1949, *Tellus*, 3, 1951, pp. 195-200.

to agree very roughly. Because of the crudeness of both calculations no more exact agreement could be expected. We thus have an observational verification of the suggestion that eddy flux of momentum may be adequate to explain the momentum balance in middle latitudes.

South of the location of the maximum there must be divergence of the flux and depletion of angular momentum. During the month studied, however, the maximum westerlies were found south of 32° N. This jet must have been maintained by another mechanism than geostrophic

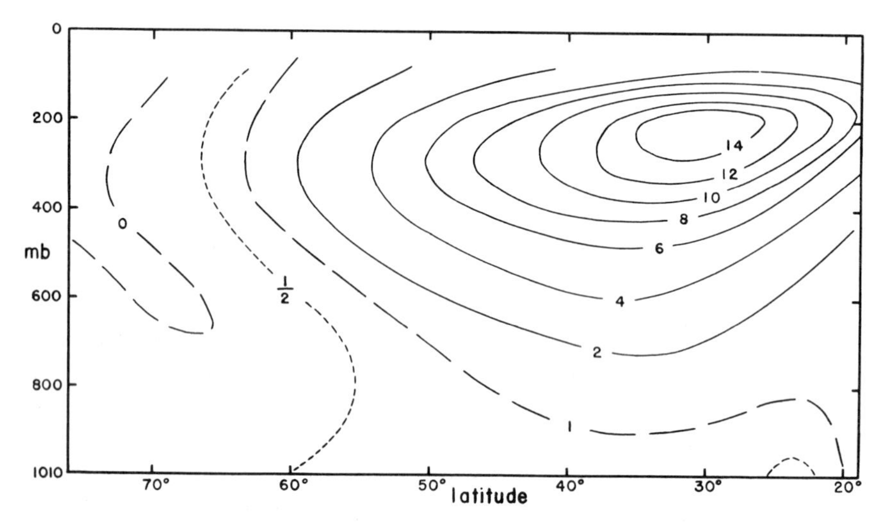

FIG. 21.7. Mean geostrophic poleward flux of absolute angular momentum in January, 1949, averaged over all longitudes (after Mintz). Numbers are in units of 10^{24} g cm^2 s^{-2} per 10 mb layer.

eddy flux of angular momentum. Presumably the mean meridional flow increases toward the equator and becomes significant in transporting angular momentum, as indicated by the estimates of Riehl and Yeh.

The evidence currently available suggests that the horizontal transport of angular momentum needed to achieve a momentum balance is accomplished by at least two different mechanisms: quasi-geostrophic eddy flux in middle latitudes and nongeostrophic meridional flow in low latitudes. There exists a border zone in which these two processes seem to be of comparable importance, and in which the jet stream is often located—at least in the mean.

21.8 A Numerical Experiment

An important advance in the effort to explain the general circulation by means of the fundamental physical laws governing the atmosphere was recently made by Phillips.[7] He set out to determine whether the techniques developed in the field of numerical prediction could be applied to the problem of the general circulation. The model he adopted is probably the simplest one capable of developing many of the observed features of the atmospheric circulation. It has two layers, one extending from 1000 mb to 500 mb and the other from 500 mb to the top of the atmosphere. Each layer was assumed to have its own individual vertically invariant flow pattern characterized by the motion at 750 mb and 250 mb respectively. The dynamical equation governing each layer is the vorticity equation in the form

$$\frac{d}{dt}(f + \zeta) = -f\left(\frac{\partial u}{\partial x} + \frac{\partial v}{\partial y}\right) + A_v\left(\frac{\partial^2 \zeta}{\partial x^2} + \frac{\partial^2 \zeta}{\partial y^2}\right)$$

Here ζ has been neglected compared to f in the first term on the right, and A_v is a lateral kinematic eddy viscosity coefficient. The last term represents an attempt to introduce the effects of turbulence as a smoothing and dissipating phenomenon. The horizontal divergence was then replaced by the vertical motion from the continuity equation. Everywhere else it was assumed that the geostrophic approximation could be applied. In addition, the model incorporated the First Law of Thermodynamics with a nonadiabatic heat supply which is a linear function of north-south distance. This was arranged to supply heat at low latitudes and extract it at the same rate at high latitudes, so that no net heating or cooling occurred. Nevertheless, since heat was added at high temperature and extracted at low temperature a certain fraction of the thermal energy became available for the development of organized flow, as indicated by the Second Law of Thermodynamics.

The area dealt with extended 10,000 km north-south (about the distance from equator to pole) and almost 5000 km east-west. However, the electronic computer was instructed to regard the solution as periodic in x, so the results repeat themselves indefinitely in the x-direction every 5000 km. The geometry chosen was Cartesian rather than spherical.

The nonadiabatic function adopted amounted to about 0.23° C per day of heating at the equator and cooling at the pole. From existing linear theory this differential heating rate would have to be maintained

[7] N. A. Phillips, The General Circulation of the Atmosphere: A Numerical Experiment, *Quart. J. Royal Meteor. Soc.*, 82, 1956, pp. 124-164.

for about four months to yield a latitudinal temperature gradient which would produce unstable waves of the observed variety. Thus the computation was started with an atmosphere of uniform temperature at rest. As the temperature gradient built up, motions developed. In order to save computing time, no variations with x were permitted for the first 130 days. At the end of this period, waves would become unstable because of the baroclinity present, so random perturbations were imposed and variation with x permitted. Thereafter the stable random waves died out gradually and the unstable random waves grew in amplitude until they dominated the flow.

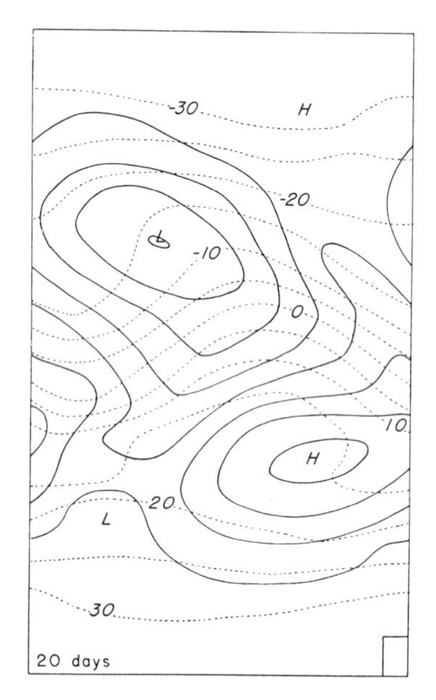

Fig. 21.8. Computed distribution of 1000-mb contour height at 200-foot intervals (solid lines) and 500-mb temperature at 5° C intervals (dashed lines) 20 days after variations with x were permitted. The height and temperature fields are very similar to those around an occluded frontal system.

At the end of the preliminary 130-day period a zonal flow of appreciable magnitude had developed together with slow meridional motions poleward aloft and equatorward below. At this stage, however, the westerlies did not exhibit the narrow maximum characteristic of a jet stream. Within ten days after eddies were permitted, a definite jet stream appeared which became quite marked in another ten days. Furthermore, the flow patterns developed were remarkably like those associated with real middle-latitude disturbances. An example is shown in Fig. 21.8. The cyclones produced sloped from northeast to southwest, as required to transport angular momentum northward, and went through a life cycle remarkably like the occlusion process.

In order to study the general circulation of the model, Phillips formed time-means of various important characteristics of the flow. From these it was found that with the introduction of eddies the single direct meridional circulation had broken down into three cells—an indirect cell in middle latitudes flanked by two weak direct cells—in agreement with observation.

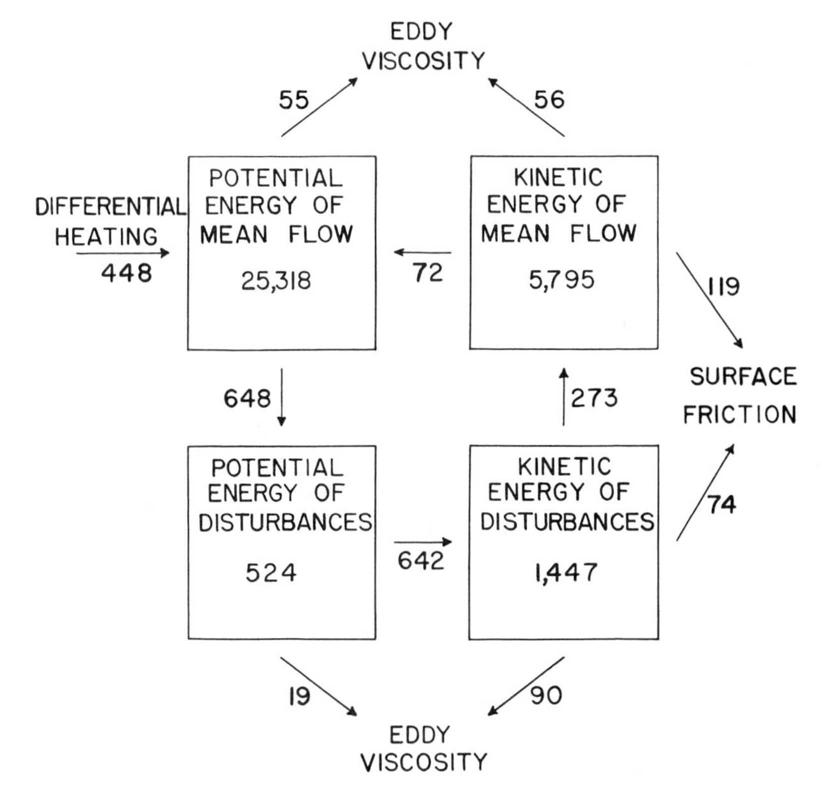

FIG. 21.9. The mean energy transformation diagram from Phillips' numerical experiment. The energies are given in arbitrary units.

Phillips made a very interesting analysis of the energy transformations in this model which is shown in Fig. 21.9. The energy of this model may appear in potential or kinetic form. In addition, one may distinguish between energy residing in the mean motion and energy of the transient eddies. Thus Fig. 21.9 shows the average amount of energy found in four different categories. At the upper left of the diagram there is given the rate at which energy is supplied to the potential energy of the mean flow by differential heating. This amounts to 448 units per day and is the only source of energy in this model. The arrows show the direction of transformation of energy from one form to another and losses owing to eddy viscosity and surface friction.

The first thing to note in Fig. 21.9 is that the calculated losses are nearly equal to the gain from differential heating. Thus even in the short period used, the model approaches a state of energy balance. Second, the energy transformations all proceed in a counterclockwise direction on the diagram. Thus the potential energy of the mean flow is transformed to potential energy of the disturbances at a fairly rapid rate (648 units per day). This is essentially the process of injection of polar air masses into southerly latitudes and northward flow of tropical air aloft which is characteristic of the life history of frontal storms. The potential energy of disturbances is transformed at almost the same rate into kinetic energy of disturbances. This transformation is largely the classical conversion process, described by Margules, in which cold air sinks and warm air rises with an attendant increase in kinetic energy.

The disturbance kinetic energy is found to pass over into kinetic energy of mean motion at a moderate rate. This indicates, for example, that the energy of a mean jet stream is maintained by intermittent impulses supplied by growing disturbances, which synoptic experience confirms.

Next, there is a slow transformation of mean kinetic energy into mean potential energy—which is superficially surprising, since the direction of the energy flow is opposite to that required by the direct circulations of the Margules theory. However, the transformation is due to the indirect meridional circulation of middle latitudes which acts to raise the cold air and lower the warm air thus increasing potential energy at the expense of kinetic energy. Observational evidence indicates that this same direction of energy transformation occurs in the real atmosphere, although two processes go on simultaneously in middle latitudes. At the same time that the disturbances are converting potential into kinetic energy, the mean motion is consuming kinetic energy to create potential energy. Thus the indirect cell of the general circulation is a characteristic of the mean state only. The generation of kinetic energy by the disturbances greatly exceeds the consumption of kinetic energy by the mean motion.

The losses of energy are worthy of some attention. Clearly both forms of kinetic energy will be diminished by surface friction. Moreover, eddy viscosity in the free atmosphere will reduce the two kinetic energies by degrading the organized motion into heat. But, the eddy process will also diminish both forms of potential energy because turbulent mixing reduces the two lateral temperature gradients. Since these temperature gradients on isobaric surfaces are a direct indication of baroclinity, and therefore of potential energy, it follows that eddy viscosity of the form assumed must drain away potential energy.

Phillips' experiment represents a significant improvement in our theoretical understanding of the general circulation. It fortifies our earlier conclusion that lateral eddy transport is a significant factor in the general circulation and helps to clarify the role of mean meridional circulations. No doubt many of the now unavoidable crudities and artificialities of his model will be removed in time, and we may expect this approach to yield further improvements in our understanding of the complex mechanisms of the general circulation.

APPENDIXES AND INDEX

NUMERICAL CONSTANTS AND CONVERSIONS

---

equatorial radius of earth = 6378.39 km

polar radius of earth = 6356.91 km

radius of a sphere having the same volume as earth = 6371.22 km

angular velocity of rotation of earth, $\Omega = 7.292 \times 10^{-5}\ s^{-1}$

acceleration of gravity at sea level and 45° latitude, $g_{45} = 980.616\ cm\ s^{-2}$

solar constant = 2.00 ly min^{-1}

universal gas constant, $R^* = 8.3144 \times 10^7\ erg\ mol^{-1}\ ^\circ K^{-1}$

gas constant for dry air, $R^*/\bar{m} = 2.8704 \times 10^6\ erg\ g^{-1}\ ^\circ K^{-1}$

mean molecular weight of dry air, $\bar{m} = 28.966\ g\ mol^{-1}$

molecular weight of water, $m_w = 18.016\ g\ mol^{-1}$

specific heat capacity of dry air:

at constant pressure, $c_p = 7R/2 = 0.240\ cal\ g^{-1}\ ^\circ K^{-1}$

at constant volume, $c_v = 5R/2 = 0.171\ cal\ g^{-1}\ ^\circ K^{-1}$

ratio of the specific heats of dry air, $c_p/c_v = 1.400$

Stefan-Boltzmann constant, $\sigma = 8.128 \times 10^{-11}\ ly\ ^\circ K^{-4}\ min^{-1}$

1° latitude (at 45° N or S) = 111.1 km = 59.96 nautical miles = 69.05 statute miles

1 m s^{-1} = 1.94 knots = 2.24 miles hr^{-1} = 3.28 ft s^{-1} = 3.60 km hr^{-1}

Note: 1 m $s^{-1} \approx$ 2 knots.

1 atmosphere = 1013.25 mb = 760.00 mm of mercury = 29.921 inches of mercury = 1.01325×10^6 dynes cm^{-2}

1 calorie = 4.186×10^7 ergs

melting point of ice = 0.00° C = 32.00° F = 273.16° K

DERIVATION OF GAUSS'S DIVERGENCE THEOREM

Gauss's divergence theorem asserts that for a vector **A** with Cartesian components A_x, A_y, A_z

$$\iiint\left[\frac{\partial A_x}{\partial x}+\frac{\partial A_y}{\partial y}+\frac{\partial A_z}{\partial z}\right] dx\, dy\, dz = \iint A_n\, d\sigma$$

where the left-hand integral extends over some given volume of space, the right-hand integral extends over the surface bounding that volume, and A_n is the outward-directed normal component of **A** at an element of surface area $d\sigma$.

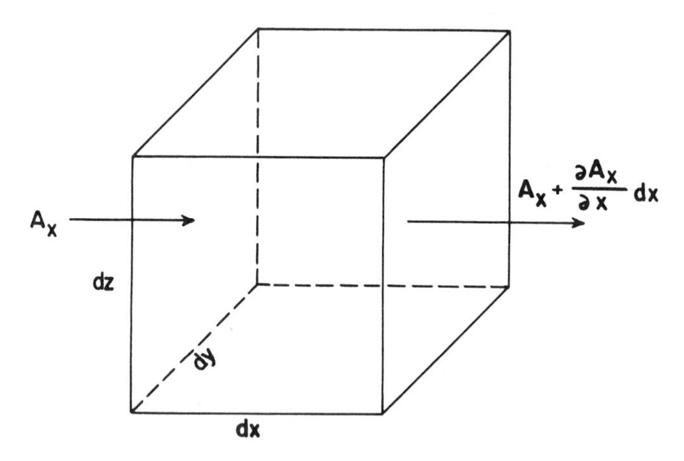

To prove this theorem, consider an infinitesimal increment of volume as in the accompanying figure. Let the normal component of **A** for the left-hand face be A_x. Then the normal component for the right-hand face is $A_x + \partial A_x/\partial x\, dx$. If we multiply by $dy\, dz$, the area of these faces, and add the area-weighted *outward-directed* normals for the two faces, we get $\partial A_x/\partial x \, dx \, dy \, dz$. Similar results may be obtained for faces normal to the y and z axes. If we sum over all faces of the increment of volume we find that

$$\left[\frac{\partial A_x}{\partial x} + \frac{\partial A_y}{\partial y} + \frac{\partial A_z}{\partial z}\right] dx \, dy \, dz = A_n \, d\sigma$$

where A_n represents the outward-directed normal for any of the faces and $d\sigma$ is the area of any of the faces.

We now consider a finite volume of which the above increment is a small part. If we sum over the entire volume the contributions of the right side of the last equation will vanish for all interior elements of area, because adjacent elements of volume will have a common face, but A_n will be directed inward for one element and outward for the other. Thus only the contributions from the surface area of the finite volume appear and

$$\iiint \left[\frac{\partial A_x}{\partial x} + \frac{\partial A_y}{\partial y} + \frac{\partial A_z}{\partial z}\right] dx \, dy \, dz = \iint A_n \, d\sigma$$

which is Gauss's divergence theorem.

INDEX

LIST OF SYMBOLS

A	area; albedo; Angstrom(s); eddy diffusion coefficient
a	radius of earth; acceleration; constant; characteristic length; parameter in the theory of Ekman spiral
a_λ	fractional absorption
b	specific enthalpy; constant in Wien's radiation law
C	heat capacity; circulation
C_x, C_y, C_z	velocity components of a moving point
c	specific heat capacity; speed of light; horizontal wind speed
c_1, c_2	constants in Planck's law
c^*	any of the three Cartesian components of velocity
c_g	geostrophic wind speed
c_p, c_v	specific heat capacities at constant pressure and constant volume
c_w	specific heat capacity of liquid water
D	horizontal divergence; depth of an air column
$\mathfrak{D}$	differential at a point moving with arbitrary velocity; rate of dissipation of horizontal kinetic energy for a given volume
$\mathfrak{d}$	rate of dissipation of horizontal kinetic energy per unit volume
E	energy of a quantum; rate of evaporation
E_I	internal energy per unit horizontal area
E_K	kinetic energy per unit horizontal area
E_P	potential energy per unit horizontal area
E_λ	monochromatic intensity of emission of radiation
e	natural base of logarithms
e	water substance pressure; vapor pressure
e_s	saturation vapor pressure
F	a function; force; area enclosed by a projection on the equatorial plane; Froude number
F_x, F_y, F_z	viscous and frictional forces per unit mass
f	Coriolis parameter
G	force of Newtonian gravitation per unit mass
g	acceleration of gravity
H	heat; height of the homogeneous atmosphere; gradient wind level
h	heat per unit mass; Planck's constant
I_λ	monochromatic intensity of radiation
i	square root of minus one
$\mathcal{J}$	any atmospheric property per unit mass
K	constant
k	constant; von Karman's constant
k_λ	absorption coefficient
L	latent heat of a phase change; wave length
$[L]$	the dimension of length
l_λ	smoothed absorption coefficient
l_x, l_y, l_z	root - mean - square mixing lengths
l'	distance moved by a parcel in the mixing-length theory
M	mass of air
$[M]$	the dimension of mass
m	molecular weight; mass of a body; non-dimensional exponent; absolute angular momentum
$\bar{m}$	mean molecular weight of a mixture
n	a natural coordinate
p	pressure
Q	heat energy
q	specific humidity; any atmospheric property
R	specific gas constant; radial distance from an axis of rotation
R^*	universal gas constant
R_e	Reynolds number
R_i	Richardson number
R_o	Rossby number

r	relative humidity; radial distance in polar coordinates
S	the solar constant; source or sink of an atmospheric property; a displacement
s	a natural coordinate
T	Kelvin temperature; period of an oscillation
T^*	absolute virtual temperature
$[T]$	the dimension of time
T_d	dew-point temperature
T_e	equivalent temperature
T_q	rate of upward eddy transport of specific humidity
T_w	wet-bulb temperature
t	time, centigrade temperature
U	internal energy; constant westerly basic current; representative wind speed
u	specific internal energy; optical depth; eastward component of velocity
$\bar{u}$	mean eastward speed
u'	perturbation value of eastward speed; instantaneous departure from mean eastward speed
u_*	friction velocity
u_g	geostrophic eastward speed
u_0	eastward component of pure translation
V	volume, velocity
v	northward component of velocity
v'	perturbation value of northward speed
v_0	northward component of pure translation
v_r	radial speed
v_θ	tangential speed
W	work
w	specific work, mixing ratio; vertical velocity
w_l	specific liquid water content
w_s	saturation mixing ratio
x	Cartesian coordinate; eastward coordinate in the tangent plane system; eastward distance along a level surface
y	Cartesian coordinate; northward coordinate in the tangent plane system; northward distance along a level surface
z	Cartesian coordinate; vertical distance in the tangent plane system; elevation above mean sea level
z_0	roughness parameter
ר	(Hebrew reysch) a representative density
∇^2	LaPlacian operator
α	specific volume; angle between wind and isobars
β	rate of change of the Coriolis parameter with distance northward
Γ_d	dry adiabatic lapse rate
Γ_s	saturation pseudo-adiabatic lapse rate
γ	lapse rate of temperature; ratio of the specific heat capacities at constant pressure and volume
γ_n	lapse rate for neutral equilibrium
ϵ	ratio of the molecular weights of water and dry air
ζ	a Cartesian coordinate; relative vorticity about the vertical
η	a Cartesian coordinate
θ	angle; potential temperature; zenith angle; a polar coordinate
$[\theta]$	the dimension of temperature
θ_e	equivalent potential temperature
θ_w	wet-bulb potential temperature
κ	the ratio of the specific gas constant to the specific heat capacity for dry air
λ	wave length
μ	micron(s); dynamic coefficient of viscosity
μ_e	dynamic coefficient of eddy viscosity or Austausch coefficient
ν	frequency; kinematic coefficient of viscosity
ρ	density
ρ_v	absolute humidity
σ	the Stefan-Boltzmann constant; area
τ	period; optical thickness; stress
τ_I	transmission function
Φ	normal probability integral
ϕ	specific entropy; latitude
χ	a Cartesian coordinate
Ψ	geopotential measured in geopotential meters
ψ	geopotential; stream function
Ω	angular velocity of rotation of the earth; characteristic frequency
ω	angular velocity; individual derivative of pressure with respect to time